Hydraulic Engineering: Emerging Trends and Technologies

Hydraulic Engineering: Emerging Trends and Technologies

Edited by Will Shelton

Clanrye International
www.clanryeinternational.com

Clanrye International,
750 Third Avenue, 9th Floor,
New York, NY 10017, USA

ISBN: 978-1-63240-926-3

Cataloging-in-Publication Data

Hydraulic engineering : emerging trends and technologies / edited by Will Shelton.
p. cm.
Includes bibliographical references and index.
ISBN 978-1-63240-926-3
1. Hydraulic engineering. 2. Hydraulics. 3. Engineering. 4. Fluid mechanics. I. Shelton, Will.
TC153 .H93 2020
627--dc23

For information on all Clanrye International publications visit our website at www.clanryeinternational.com

CLANRYE INTERNATIONAL

Printed in China.

Contents

Preface

In my initial years as a student, I used to run to the library at every possible instance to grab a book and learn something new. Books were my primary source of knowledge and I would not have come such a long way without all that I learnt from them. Thus, when I was approached to edit this book; I became understandably nostalgic. It was an absolute honor to be considered worthy of guiding the current generation as well as those to come. I put all my knowledge and hard work into making this book most beneficial for its readers.

Hydraulic engineering is a sub-discipline of civil engineering that is concerned with the flow and conveyance of fluids. This field is particularly relevant in the design of water and sewage systems, which are important aspects of urban planning. It applies the principles of fluid mechanics to design strategies for the efficient storage, collection, measurement, regulation, transport and use of water. The architectural planning and design of spillways and outlet paths for dams, canals, culverts, irrigation structures and cooling water facilities are under the purview of hydraulic engineering. The use of computer-aided design and computational fluid dynamics, as well as GPS mapping and laser-based surveying tools have brought tremendous advancements in hydraulic engineering. This book elucidates the concepts and innovative models around prospective developments with respect to hydraulic engineering. The topics included in this book on hydraulic engineering are of utmost significance and bound to provide incredible insights to readers. It attempts to assist those with a goal of delving into this field.

I wish to thank my publisher for supporting me at every step. I would also like to thank all the authors who have contributed their researches in this book. I hope this book will be a valuable contribution to the progress of the field.

Editor

1

TRACE METAL CONCENTRATION AND SPECIATION IN STORM WATER RUNOFF ON IMPERVIOUS SURFACES

Enrico MANCINELLI[a], Edita BALTRĖNAITĖ[b], Pranas BALTRĖNAS[b], Dainius PALIULIS[b], Giorgio PASSERINI[c], Åsgeir R. ALMÅS[d]

[a]*Faculty of Civil Engineering, Università Politecnica delle Marche, via Brecce Bianche 12, 60131 Ancona, Italy*
[b]*Department of Environmental Protection, Vilnius Gediminas Technical University, Saulėtekio al. 11, 10223 Vilnius, Lithuania*
[c]*Department of Industrial Engineering and Mathematics Sciences, Università Politecnica delle Marche, via Brecce Bianche 12, 60131 Ancona, Italy*
[d]*Department of Plant and Environmental Sciences, Norwegian University of Life Science, P.O. Box 5003, N–1432 Ås, Norway*

Abstract. Urban storm water runoff (USWR) often transports various trace metals, affecting the quality of receiving waters. Metal concentration and speciation in USWR were investigated by analysing data collected during a storm event on 19 July 2012. USWR samples were taken from a low–traffic intensity road, a pathway in a car–free pedestrian zone, and a galvanized metal roof located in the area of Vingio Park in the city of Vilnius, Lithuania. The metals were mostly in the insoluble form at all three sites, being more than 80% of the total amount measured, irrespective of the water quality characteristics and types of surface area. The partitioning coefficient (log Kd) of trace metals decreases in the order Zn > Cu ≈ Mn ≈ Pb >> Cd. The USWR types characterized by a higher pH or dissolved organic carbon (DOC) concentration produced a higher fraction of Cd, Mn, Pb, and Zn bound to fulvic acids (FA).

Keywords: storm water runoff, trace metals, macro elements, speciation, partitioning, water pollution, WHAM.

Introduction

Storm water runoff (SWR) from impervious surfaces can be considered a non–point source of water pollution (NPSWP) since the sources of pollution are diffuse and difficult to control. Traffic, industry and energy emissions (e.g. power stations), erosion of building materials and transportation are among the main sources of pollutants in SWR (Sabin *et al.* 2005; Göbel *et al.* 2007; Milukaitė *et al.* 2008).

Urban storm water runoff (USWR) is being recognized as a leading source of pollutants to receiving waters (Shaw *et al.* 2006). In urban areas, SWR represents a major environmental issue because it can be affected by elevated concentrations of trace metals (e.g. Cd, Cu, Ni, Pb and Zn) deteriorating the quality of water basins (Van Metre, Mahler 2003; Prestes *et al.* 2006; Brown, Peake 2006; Tiefenthaler *et al.* 2008). Of the trace metals typically found in SWR, Cd, Cu, Pb, and Zn are the major concern (Van Metre, Mahler 2003; Weiss *et al.* 2006; Kayhanian *et al.* 2008). Various studies have identified the corrosion of galvanized steel roofing materials and metal drainage elements (e.g. gutters, downspouts and steel sheets) as a main source of Cd, Pb, and Zn in SWR from roof surfaces (Van Metre, Mahler 2003; Chang *et al.* 2004; Rocher *et al.* 2004; Brown, Peake 2006; Lamprea, Ruban 2008; Ogburn *et al.* 2012). Vehicle exhaust emissions, the corrosion of galvanized safety barriers, wear and tear from tyres and brakes are the primary sources of Cd, Cr, Cu, Pb, and Zn in SWR from road surfaces (Legret, Pagotto 1999 cited in Rule *et al.* 2006; Adachi, Tainosho 2004).

Many parameters affect metal concentrations, behaviour and fate in SWR. Differences in land use and surface characteristics (e.g. surface roughness or material composition) result in a variety of trace metal concentrations and their presence in SWR (Dierkes *et al.* 2005; Gnecco *et al.* 2008; Wicke *et al.* 2012). The processes governing

Corresponding author: Enrico Mancinelli
E-mail: enrico.mancinelli@alice.it

metal partitioning depend on parameters such as surface characteristics (Göbel *et al.* 2007), rainfall pH, runoff residence time, and solid characteristics (e.g. particle size) (Sansalone, Ying 2008). Rainfall intensity and rainfall volume are important factors that influence the trace metal wash-off from impervious surfaces (Sonzogni *et al.* 1980 cited in Herngren *et al.* 2005). Wet atmospheric deposits are subject to significant evolutions due to transport and runoff from different urban surfaces. According to Pitt (1987 cited in Pitt *et al.* 1995), the relative importance of the contribution of organic and inorganic pollutants from source areas is a function of the area characteristics, pollutant wash-off potential, and the rainfall characteristics.

Municipalities should define and adopt the best available techniques (BAT) in order to minimize pollutant input to receiving water basins. In order to identify possible reduction measures or better treatment management it is necessary to define the sources, load and behaviour in the aquatic system of priority substances. The assessment of the partitioning of metals between the solid and liquid phases is an important factor in assessing the fate and bioavailability of the metals in receiving waters (Ciffroy *et al.* 2003). Regarding the toxicity of an element, its bioavailability is more important than its total concentration. Given the fact that the chemical form or speciation of metals affects bioavailability, and hence toxicity, it is important to determine the concentration of free metal ions, inorganic and organic complexes, and organometallic compounds rather than the total concentration of elements. For example, at the same concentration of total metal such as Cu, Cd, and Pb, the uncomplexed ionic forms are much more toxic than the complexed ones (Bricker 1999). In order to determine the real available trace metal fractions, the adoption of speciation modelling should involve the determination of trace metal concentration together with the physico-chemical properties of the water (e.g. major ions, organic metal-complexing agents, pH, etc.) (Peijnenburg, Jager 2003). According to Comber and Georges (2008), the bioavailability of Cu and Zn depends on pH, Ca concentration, and the presence of dissolved organic matter (DOM). If the physicochemical properties of the water are not determined, the evaluation of metal bioavailability and toxicity is hard to define. Compared to the extensive literature on trace metals in SWR from roof and road surfaces, the presence of major ions (e.g. Ca^{2+}, K^{+} and Mg^{2+}) has been poorly investigated (Göbel *et al.* 2007).

The study of SWR pollution is in its infancy in Lithuania and common water parameters (e.g. biochemical oxygen demand (BOD), suspended solids (SS), and pH) are usually the subject of research (Karlavičienė *et al.* 2009). To our knowledge, only Milukaitė *et al.* (2010) have investigated trace metal (e.g. Cd, Cu, Pb, Zn) concentrations in both the soluble and insoluble forms in SWR in the area of Vilnius city.

The purpose of the current field study was to collect data on dissolved and SS-bound trace metals (Cd, Cr, Cu, Mn, Ni, Pb and Zn) as well as macro elements (Ca, K, and Mg) that can compete in binding to humic substances in USWR from surfaces characterized by different material composition and usage. The information obtained was used to evaluate and compare metal concentrations and speciation at each site. We first hypothesize that pH and DOM concentration in USWR varies between the investigated sites. Moreover we hypothesize that a high pH and DOM concentration reduces the free metal ion activity of eroded trace metals, and hence increases their mobility as metal-DOM complexes. This study may have implications for the operational choices of storm water treatment best management practices (BMPs) for particle separation or treatment of the soluble fraction in order to avoid the water quality impairment of receiving waters.

1. Materials and methods

1.1. Sampling sites

The sampling sites, selected to represent an urban environment and to allow the assessment and comparison of the metal concentration, partition and speciation in USWR from different impervious surfaces, were located in the area of the town named "Vingio Park", situated on the North-Western side of the city of Vilnius, on a bend of the river Neris (54° 41′ 0″ N, 25° 14′ 23″ E) (Fig. 1). The soil parent material in the study area consists of Holocene alluvial deposits of fine sands (Zinkutė *et al.* 2011).

Given the fact that the location of the sampling sites is in a metropolitan area and not far from a heavily industrialized zone, atmospheric deposition is assumed to be an important trace metal source for USWR. In 2000, road transport was the largest source (97%) of the emissions into the atmosphere generated by mobile pollution sources in Lithuania (Baltrėnas *et al.* 2004). According to Zavadskas *et al.* (2007), vehicle emissions are mainly responsible for air pollution in the city of Vilnius. In addition, Milukaitė *et al.* (2008) have observed a close relationship between Cd, Cu, Pb, Zn and oil products in atmospheric deposition, suggesting that traffic represents the major source of pollution that can affect the chemical composition of atmospheric depositions in Vilnius. In the street zone, the deposition of Cd, Cu, Pb, and Zn is calculated to be about 20, 40, 29, and 25% of the total trace metals that deposit yearly in Vilnius.

USWR samples were collected from the downspouts of a roof surface of a building located in M.K. Čiurlionio Street No. 100, next to the amphitheatre of Vingio Park (i.e. sampling site number 1 in Fig. 1). It should be noted that the building was constructed a few years ago and the gutters, the downspouts and the roofing material are all made of galvanized metal.

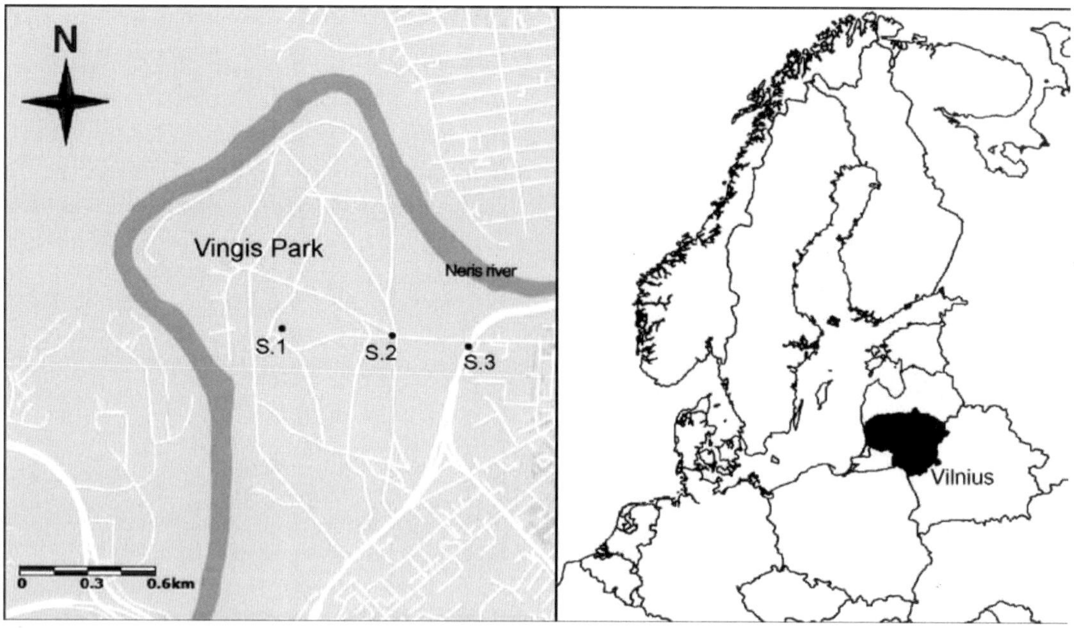

Fig. 1. Location of Lithuania in the Baltic Sea region. Location of the sampling sites in the area of Vingio Park, in Vilnius, Lithuania. The dots indicate the sampling points along M.K. Čiurlionio St.: S.1 is the roof site; S.2 is the pathway site; S.3 is the road site

Sampling site number 2 (Fig. 1) is a pathway in a car-free pedestrian zone inside the park, in proximity to the entrance from M.K. Čiurlionio Street. No point sources of trace metals pollution were identified in proximity to this sampling site.

The road USWR samples were taken in M. K. Čiurlionio Street (i.e. sampling site number 3 in Fig. 1), a low-traffic intensity road that is used only by people who go to Vingio Park, from the flyover that crosses over the heavily trafficked road named Geležinio Vilko Street, which is the main urban speedway in Vilnius, with a measured transport flow of about 140 thousand cars per 24 hours (Jarasuniene 2010). It should be noted that, in that area, Geležinio Vilko Street is a cut road (i.e. a road that is below ground level), thus a street canyon effect could be originated hence affecting the air circulation and limiting the dispersion of the smog.

1.2. Sampling of urban storm water runoff

The sampling was conducted during a storm event on 19 July 2012. The USWR sampling consisted in collecting SWR manually by taking grab samples directly from drainage flow (Spurlock 1999). All the USWR samples were collected by multiple sampling (about 5 L) in pre-cleaned polyethylene (PE) bottles; samples were collected until 25 L volume were obtained resulting in a single, composite sample for each site. After collection, the samples at each site were sealed in pre-cleaned PE bottles and stored at a temperature of 4 °C in darkness until analysis. The sampling technique, preservation and handling of the water samples were carried out in accordance with the following methodology: *ISO 5667-1: Water quality – Sampling – Part 1: Guidance on the design of sampling programmes and sampling techniques* and *ISO 5667-3: Water quality – Sampling – Part 3: Guidance on the preservation and handling of water samples.*

SWR measurements were taken at the time of sampling using the volumetric time-based Bucket method: a bucket and a stop-watch were used during the storm event to measure the runoff flow-rate as described in Spurlock (1999). The flow-rate of the runoff water showed a great variability between the three sampling sites, with the rate of fluid flow on the pathway (0.42 L s^{-1}) and the road surface (0.25 L s^{-1}) being, respectively, 8 and 5 times greater than on the roof surface (0.05 L s^{-1}).

Table 1. Daily precipitation in the period 13–19/07/12. The daily precipitation amount recorded when the urban storm water runoff samples were taken is highlighted in bold

Date	13–7	14–7	15–7	16–7	17–7	18–7	**19–7**
Daily precipitation (mm)	0	1.3	0.2	0	7.1	3.4	**6.6**

The meteorological data, from 13 to 19 July 2012, recorded the precipitation by means of the rain gauge of the Lithuanian Hydrometeorological Service; the rain gauge station is located in the district of Vilnius, in the town of Traku Vokė (54° 37′ 33.571″ N, 25° 6′ 25.430″ E). This is in proximity to the city of Vilnius and not far from the sampling sites. The storm event of 19 July 2012 was characterized by a dry period lasting less than 1 day (Table 1). The observed rainfall pattern during the previous days suggests that the length of the dry period was short for a consistent pollutant build-up due to the dry deposition of solids.

1.3. Physical and chemical preparation of the urban storm water runoff samples

Regarding the preparation of the USWR samples, a portion of the samples was thoroughly mixed and homogenized in 4 pre-cleaned PE buckets, with a volume of 10 L each, in order to ensure an appropriate homogeneity and create a single representative sample of USWR from each sampling site. Each well-mixed subsample was withdrawn from the bucket using a graduated cylinder.

Prior to analysis, the USWR samples were separated into solid and liquid phases. A measured volume of USWR sample was filtered in order to determine the metals adsorbed onto SS. The concentration of SS was determined by vacuum filtration through pre-weighted glass fibre Whatman GF/693 filters (Whatman International Ltd., Maidstone, UK) characterized by an average pore size of 1.2 µm, and a fast rate of filtration; after filtration, solid particles remaining in the filter paper were dried at a temperature of 105 °C and re-weighted to quantify the dry weight (DW) of SS.

The filtrate obtained in the first filtering process was subsequently filtered in order to detect the presence of dissolved metal ions and dissolved organic carbon (DOC) at each sampling site. For this purpose, sterile Acrodisc syringe filters with a Supor (hydrophilic polyethersulfone) membrane and pore size 0.45 µm, (Pall Corporation) were used.

Regarding the concentration of metals adsorbed onto SS, the insoluble fraction of USWR on the filter was digested prior to flame atomic absorption spectrophotometry (FAAS) or graphite furnace (GFAAS) analysis. In order to determine trace metal concentration, a rapid and efficient method of sample decomposition is microwave digestion (Peijnenburg, Jager 2003). This enables a more rapid extraction of the metals from the particles compared to traditional techniques; it prevents the loss of volatile elements and also avoids contamination from the environment. The insoluble fraction of the USWR on the filter was digested for metal analysis using an Ethos 900 Milestone Microwave digestion system. The digestion process consisted in mixing each sample with 3 mL of HNO_3 (65%) and 9 mL of HCl (37%); subsequently the samples were poured into special vessels and then placed in a Milestone ETHOS digester and heated for 43 minutes. The resulting solution was poured into a 50 mL flask and the volume was made up to 50 mL with deionized water (Butkus, Baltrenaitė 2007; Pundytė *et al.* 2011). In order to remove any possible presence of metal on the walls of the vessels, all the vessels were accurately rinsed with deionized water.

1.4. Chemical analysis of urban storm water runoff

As regards water chemistry characterization, laboratory tests of USWR samples were performed in the laboratories of the Research Institute of Environmental Protection (Vilnius Gediminas Technical University). The pollution parameters studied were: electric conductivity (EC), pH, SS, DOC, trace metals (Cd, Cu, Cr, Mn, Ni, Pb, and Zn) and macro elements (Ca, K, Mg).

EC and pH were determined by analysing unfiltered USWR samples. The pH of USWR samples was determined using a pH-meter with an electronic glass electrode (SevenMulti ion/pH/ORP module Mettler Toledo); before the determinations, the pH-meter was calibrated with buffer solutions at pH 4, 7, and 11 (Merck, Darmstadt, Germany); the resulting calibration curve was determined via a linear regression line. pH measurements were carried out in accordance with the methodology described in *ISO 10523: Water quality – Determination of pH*. EC was measured by a conductivity-meter (inoLab Cond 740 WTW) in accordance with the methodology of *ISO 7888: Water quality – Determination of electrical conductivity*.

The DOC in USWR samples filtered through a syringe (pore size 0.45 µm, Pall Corporation) was determined by employing a total organic carbon (TOC) analyser (Shimadzu-V CSN) at 900 °C. For each sampling site, three sub-samples of 40 mL volume were analysed. In this work, the DOC is equivalent to TOC because the USWR samples were first filtered; syringe filters with a pore size 0.45 µm (Pall Corporation) were used for this purpose.

The analysis of dissolved trace metals and macro elements in the solution phase was performed after a syringe filtration through a 0.45 µm Acrodisc filter. The concentration in solution was measured using flame atomic absorption spectrophotometry (FAAS) (210 VGP atomic absorption spectrophotometer Buck Scientific); when the concentration of metals such as Cd, Cu, Mn, and Pb was too low to be detected accurately by FAAS, a graphite furnace (GFAAS) was used (Baltrenaitė *et al.* 2010; Butkus, Baltrenaitė 2007). The detection limits of Ca, K, Mg, Mn, Ni, Zn, Pb, Cu and Cr were 0.3; 0.3; 0.05; 0.2; 0.2; 0.04; 0.2; 0.1 and 0.2 (mg L^{-1}) using FAAS; the GFAAS detection limit of Cd, Cu, Mn and Pb was 0.002 (mg L^{-1}). All the procedures were carried out in accordance with the methods mentioned in *ISO 8288: Water quality – Determination of cobalt, nickel, copper, zinc, cadmium and lead – Flame atomic absorption spectrometric methods; ISO 9174: Water quality – Determination of chromium – Atomic absorption spectrometric methods; ISO 9964-2: Water quality – Determination of sodium and potassium – Part 2: Determination of potassium by atomic absorption spectrometry* and *ISO 7980: Water quality – Determination of calcium and magnesium – Atomic absorption spectrometric method.*

The concentration of adsorbed metals in SS was measured using FAAS or GFAAS methods in accordance with *ISO 11047: Soil quality – Determination of cadmium, chromium, cobalt, copper, lead, manganese, nickel and zinc in acqua regia extracts of soil – Flame and electrothermal atomic absorption spectrometric methods*. This standard specifies ways to determine metal concentration in aqua regia extracts of soil through flame and electrothermal atomic absorption spectrometric methods. As regards the determination of macro elements adsorbed onto SS by FAAS, the samples were analysed under optimum analytical conditions setting the instrumentation according to Buck Scientific device description.

1.5. Quality assurance

Standard solutions of Ca, Cd, Cr, Cu, K, Mg, Mn, Ni, Pb and Zn (1000 ppm) (Buck Scientific, USA) were used for FAAS or GFAAS calibration. All the chemicals used in this study were of analytical grade and solutions were prepared with deionized water. Concentrated nitric (65%) and hydrochloric (35%) acid was employed for the digestion of the samples.

In order to monitor for possible contaminations resulting from the sample preparation procedure, a reagent blank was prepared for each of the digestion procedure, following the entire sequence of steps as described above for sample preparation in each case.

1.6. WHAM/Model VII, version for waters

For each sampling site, metal speciation in USWR was simulated using the chemical speciation program WHAM/Model VII based on WHAM (Tipping 1994). WHAM7 for waters includes different submodels such as the humic ion-binding Model VII, a surface complexation model simulating ion-binding to oxides of aluminum, iron, manganese and silicon, a model for inorganic solution chemistry speciation, and a model for cation exchange on clays. The model assumes that metal species are at chemical equilibrium and it does not consider kinetically-controlled reactions. It is based on the discrete-site/electrostatic formulation of the interactions of protons and metals with humic substances; in describing binding sites for metals it adopts a distributional approach.

1.7. WHAM modelling

The speciation of the detected trace metals (Cd, Cu, Mn, Pb, and Zn) and major elements (Mg, K, and Ca) in USWR was calculated using WHAM7 (Tipping *et al.* 2011) for waters. The values of pH, fulvic acids (FA), SS and metal concentration were input as measured in the USWR analysis (Table 2). As regards the solute type adopted for the modelling, the concentration in USWR of metals in soluble form (syringe filtered water, 0.45 μm pore size), determined with FAAS or GFAAS analysis, was identified as total dissolved concentration. A temperature of 15 °C and no CO_2 partial pressure were set for all USWR types.

Metal binding to DOM in WHAM assumes metal binding to acid groups on humic (HA) and fulvic acids (FA). We assumed that the DOM has the binding properties of "default" FA, and that the DOM is 50% made up of carbon. DOM was considered "colloidal" and, as suggested in the work by Dwane and Tipping (1998), it was assumed to consist of a fraction of 50% active fulvic acid (% aFA) for proton/metal binding and a fraction of 50% that is inert as regards ion-binding. The calculated metal concentrations in the "aqueous phase" as defined in WHAM, are hereafter reported as soluble. This includes those in true solution and metals in the diffuse layers of colloidal phases (here, the FA).

The Debye-Hückel equation for the activity correction of the solution equilibrium calculation and the default database of stability constants were adopted. The model outputs were analysed to provide the following results for each USWR type and metal: fractions of free metal ions relative to total dissolved metals, fraction of truly dissolved metal, fraction of metal bound to colloids (represented by FA).

1.8. Statistics and calculation

Statistical analysis was carried out using GNU PSPP Statistical Analysis Software (PSPP, n.d.). For the descriptive statistics of element concentrations in the USWR, values of the mean, median, standard deviation and coefficient of variance (CV), were calculated for each element detected from the three sites. In order to evaluate the correlation between the detected elements from each site and to determine if the means of concentrations were affected by

Table 2. WHAM modelling input data for pH, dissolved organic carbon (DOC), suspended solids (SS), and metal concentration as measured in the urban storm water runoff analysis. [] tot diss.: total dissolved concentration of trace metals and macro elements

	$[Mg]_{tot\ diss}$ (mol L^{-1})	$[K]_{tot\ diss}$ (mol L^{-1})	$[Ca]_{tot\ diss}$ (mol L^{-1})	$[Mn]_{tot\ diss}$ (mol L^{-1})	$[Cu]_{tot\ diss}$ (mol L^{-1})	$[Zn]_{tot\ diss}$ (mol L^{-1})	$[Cd]_{tot\ diss}$ (mol L^{-1})	$[Pb]_{tot\ diss}$ (mol L^{-1})	pH	SS (g L^{-1})	DOC (mg L^{-1})
Roof	4.06×10^{-6}	–	2.86×10^{-4}	2.71×10^{-7}	–	1.11×10^{-6}	2.4×10^{-8}	2.9×10^{-3}	6.9	0.038	3.36
Pathway	2.62×10^{-5}	4.53×10^{-5}	1.68×10^{-4}	3.84×10^{-7}	–	2.02×10^{-6}	1.87×10^{-8}	5.8×10^{-3}	7.3	0.074	6.4
Road	2.44×10^{-5}	1.24×10^{-5}	3.57×10^{-5}	3.39×10^{-7}	5.04×10^{-8}	8.31×10^{-7}	4.09×10^{-8}	4.6×10^{-3}	8.1	0.135	3.86

the different sampling site locations, Pearson's correlation coefficients were calculated; values of $p < 0.05$ and < 0.01 were regarded as significant.

The metal element partitioning between the dissolved phase and SS was determined in terms of the partition coefficient (Kd, L kg^{-1}) and was calculated according to the following equation:

$$Kd = \frac{Cs}{Cw}, \quad (1)$$

where: Cs (mg kg^{-1} DW) is the ratio between metal mass normalized to the dry mass of SS; Cw (mg L^{-1}) denotes the metal concentration in the solution phase.

Blank sample analysis showed that the mean concentration of elements such as Ca (0.70 mg L^{-1}), Mg (0.065 mg L^{-1}), Cu (0.052 mg L^{-1}), Cr (0.241 mg L^{-1}), Pb (0.011 mg L^{-1}), and Zn (0.156 mg L^{-1}) was above the detection limit. Therefore, for the computation of the concentration of metals in suspended solids C (mg kg^{-1} DW), a correction was applied by subtracting the mean concentration detected in the blank analysis B (mg L^{-1}) from the one obtained in the analysis of the digested filters A (mg L^{-1}), as follows:

$$C = \frac{(A - B) \cdot D \cdot V}{m}, \quad (2)$$

where: D is the dilution factor; V is the final volume (mL); m is the suspended solid mass (g).

It should be noted that a small mass of SS may lead to an overestimation of the metal concentration in SS. Comparisons of the roof USWR data showed that the sample with a mass of SS from 4 to 9 times smaller than the other three samples had much higher metal concentrations, up to one order of magnitude. Therefore, in order to have more homogeneous data and not to overestimate the concentration of metals in SS in the roof USWR, of the four samples analysed the one with the smallest mass of SS was not considered for the evaluation of metal mass concentration.

2. Results and discussion

2.1. Analysis of urban storm water runoff chemistry

The trace metal Maximum Allowable Concentrations (MACs) for the environment are provided by the Order of the Minister of the Environment of Lithuania No. D1–236 of 17 May 2006 on Wastewater Management Regulation and its subsequent amendments. Because of the ongoing restructuring of the European Community water policy, the Lithuanian Wastewater Management Regulation has been amended several times. As regards trace metals, the amendment made by the Order of the Minister of Environment of Lithuania No. D1–416 of 18 May 2010 complies with Directive 2008/105/EC. The trace metals (e.g. Cr, Cu, and Zn) that are not in the list of Directive 2008/105/EC have MAC limits that are set by Lithuanian legislation.

Compared with the Lithuanian MAC limits for the environment, none of the USWR samples collected from the three sites contained amounts of dissolved Cd, Cr, Cu, Pb, Ni, and Zn (Table 3) in harmfully high concentrations.

Table 3. The dissolved trace metal concentrations detected in the urban storm water runoff samples compared to the trace metal Maximum Allowable Concentration (MAC) limits for the environment according to the Order of the Minister of the Environment of Lithuania No. D1–236 of 17 May 2006 and its subsequent amendments based on Directive 2008/105/EC

Name of substance	Maximum Allowable Concentration (MAC) for the environment (µg L^{-1})	Sampling site		
		Road	Roof	Pathway
		(µg L^{-1})		
Chromium (total)	500	ND	ND	ND
Copper	500	3.2	ND	ND
Zinc	400	54.33	72.56	131.9
Nickel and its compounds	200	ND	ND	ND
Lead and its compounds	100	4.57	2.85	5.76
Cadmium and its compounds	40	4.64	2.72	2.06

Some of the investigated trace metals including Cr and Ni were not detected in the soluble phase while the concentrations of Cd, Cu, Pb and Zn were from 8 to 19 times, about one order of magnitude, up to 35 times and from 3 to 7 times lower than the MACs of these trace metals, respectively.

In the study by Milukaitė *et al.* (2010), chemical analysis of USWR from several areas of Vilnius revealed low quantities of trace metals (Cd, Cu, Pb, Zn) in the dissolved form with only Zn and Cu exceeding the limits set for the receiving waters in a few USWR samples. As regards the quality of USWR receiving water, in the river Neris trace metal (e.g. Cd, Cr, Cu, Ni, Pb, Zn) concentrations were found to be far lower than the MAC limits (Dudutytė *et al.* 2007).

The mean values of SS, DOC, EC, and pH for the three sampling sites are shown in Table 4. The measured pH was in the range from 6.9 to 8.1 in the samples taken from the roof and road surfaces, respectively. This is a similar pH range (7.0–8.5) to the one reported in Pitt *et al.* (1995) after the analysis of SWR from several source areas. This almost neutral to slightly alkaline medium reduces the concentration of free metal ions in SWR, inhibiting their desorption from SS (Bogacz 1993 cited in Milukaitė *et al.* 2010). The pathway USWR analysis showed a twofold higher concentration of DOC, possibly originating from the surrounding forest area.

Table 4. Summary of the data from the three sampling sites for the mean value of suspended solids (SS), dissolved organic carbon (DOC), electric conductivity (EC), and pH

	pH	EC (microS cm^{-1})	DOC (mg L^{-1})	SS (g L^{-1})
Road	8.1	75.3	3.86	0.135
Roof	6.9	22.7	3.36	0.038
Pathway	7.3	124.3	6.4	0.074

The measured SS average concentration (0.038 g L^{-1}) in USWR samples from the roof is in the range (0.005–0.092 g L^{-1}) measured by Pitt *et al.* (1995) in storm water roof runoff. The rooftop produced the lowest concentration of SS, being from 2 to 4 times lower than the one detected in road and pathway USWR, respectively. This can be explained by the low roughness and the comparatively greater inclination of the zinc sheet roof surface. Therefore the particles that were on the roof could easily have been washed off this surface during the previous precipitations. The fact that SS concentration in USWR samples from the road was much lower than the average values reported in scientific literature is likely to be due to the low-density traffic of M.K. Čiurlionio St.

At all the investigated sites, the soluble concentrations of the detected constituents were much lower than the insoluble concentrations (Table 5). Analyses of metals both in SS and the dissolved form showed the predominance of the SS-bound fraction of all the detected elements with more than 80% of the total amount measured, irrespective of the water quality characteristics (e.g. pH, DOC, SS) and types of surface area. These results harmonized with the observations reported by Milukaitė *et al.* (2010) for USWR in Vilnius. These authors reported that metals such as Cd and Pb were mostly in the insoluble form with more than 90% of the total amount measured. Figure 2 shows a comparison of log Kd for each metal detected in the USWR from the investigated surfaces. The highest log Kd values were observed for Zn (4.9–5.6) followed by Cu, Mn, and Pb reflecting high affinities of these trace metals to SS. The lowest log Kd values were found for Cd (3.1–3.5).

The ranking of adsorption potential for soil particles of the detected trace metals, with Zn having the highest potential, is as shown below:

$$Zn > Cu \approx Mn \approx Pb >> Cd.$$

The partitioning coefficient values of Cu, Cd, Pb, and Zn are consistent with values reported for surface waters by Allison, J. D. and Allison, T. L. (2005). As regards metal partitioning in the USWR from roof surfaces, there was a discrepancy between our findings and the ones reported in the study by Lamprea and Ruban (2008). These authors found that the partitioning of Cd, Cu, Pb and Zn was dominated by the dissolved phase due to low pH values

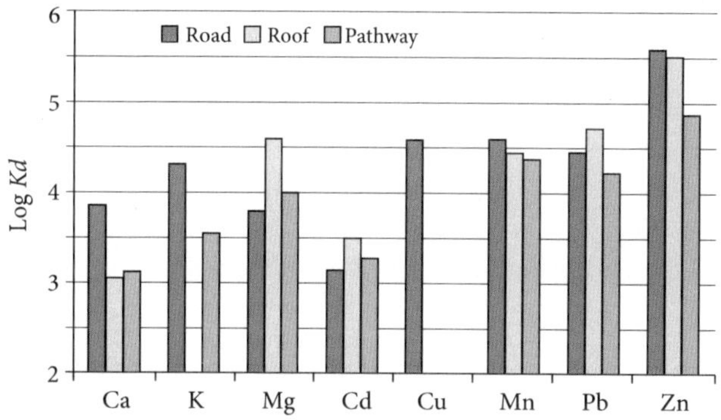

Fig. 2. Partition coefficients (log Kd) calculated for macro elements and trace metals in the urban storm water runoff samples from the investigated sites

(5.1–5.8) and low SS concentrations (0.004–0.016 mg L^{-1}). In the present study, a higher pH (6.9) and higher (up to one order of magnitude) SS concentrations are likely to have limited the amount of leaching in the dissolved phase and increased the tendency of metals to partition to SS. According to Herngren *et al.* (2005) the reduced solubility of trace metals in SWR and the insignificant release from SS can be connected to a high pH and low DOC concentration.

The data from the three sampling sites for the elements analysed in soluble and insoluble forms are summarized in Table 5. Of all the detected trace metals, Zn concentration was the highest both in the soluble and insoluble form, whereas Cd concentration was the lowest for all the investigated sites. The mass concentration of Zn was found at levels from 1 to 3 orders of magnitude higher than Pb and Cu, from 2 to 6 orders of magnitude higher than Cr and up to 50 times higher than Mn. In the soluble form, Zn was found from 10 to 25 times higher than Pb, from 2 to 6 times higher than Mn and up to 60 times higher than Cd.

Figure 3a shows that the concentrations of soluble metals in all samples from the investigated sites decreased in the following order, with Cu* detected only in the road USWR:

$$Zn > Mn > Pb > Cd > Cu^*.$$

The measured trace metal concentration (mg kg^{-1} DW) in the insoluble form (Fig. 3b) in the USWR from the different sampling sites decreased in the following order, with Ni* detected only in the pathway USWR:

$$Zn > Mn > Pb > Cu > Cr > Ni^* > Cd.$$

According to Taraškevičius *et al.* (2008), Zn and Pb were among the main contaminants identified in the topsoil of Vilnius city recreational areas. Based on wet and dry atmospheric deposition analyses, Milukaitė *et al.* (2008) identified Zn as the trace metal with the highest annual flux to the ground surface of Vilnius.

The USWR samples from the pathway surface exhibited the lowest concentration of Zn, Pb, Cu, and Cd bound to SS, as expected for typical pollutants associated with vehicular traffic (Legret, Pagotto 1999 cited in Rule *et al.* 2006; Göbel *et al.* 2007) that were investigated in the USWR from a car-free pedestrian zone. Conversely, road USWR analysis showed a high concentration of Zn, Pb, Cu, Cr and Mn in SS.

In a study of trace metal concentration in USWR from a district in Vilnius, the highest concentrations of Cu, Mn, and Zn were reported to be in the range 0.4–2.0 mg L^{-1}, 1.3–5.1 mg L^{-1}, and 0.9–12.4 mg L^{-1}, respectively (Vasarevičius *et al.* 2010). According to Kern *et al.*, Sansalone and Buchberger (1992; 1997 cited in Göbel *et al.* 2007), trace metals coming from traffic activities are mainly found as particulate matter in SWR. Wear and tear from tyres, from brakes and the corrosion of galvanized safety barriers are the primary sources of Zn (Legret, Pagotto 1999 cited in Rule *et al.* 2006); vehicle brake emissions are major sources of Cu, Cr and Pb (Adachi, Tainosho 2004). Correlations between the concentration of metals bound to SS such as Cd ($p < 0.05$), Cu ($p < 0.01$) and Pb as well as between Cu and Mn ($p < 0.01$) were found.

The soluble concentration of metals such as Mn, Pb, and Zn was found to be the highest in the USWR samples from the pedestrian zone. This is because of the higher concentrations of DOC compared to the ones measured in the USWR samples from the road and roof surfaces (Table 4). The speciation results reported below (Table 6) confirmed higher Mn, Pb and Zn complexation to FA in the modelling of USWR from the pathway compared to the road and roof USWR whereas the fractions of Mn–FA complexes were higher in the road USWR than the pathway one. It is unclear why Ni was found only in the USWR from the pathway and only bound to SS (22.19 mg kg^{-1} DW), whereas the concentration in the samples from the other sites was under the detection limit.

It should be noted that considerable amounts of trace metals were detected in the USWR from the pathway but no point sources of trace metal pollution were identified in proximity to the this sampling site. Therefore, atmospheric deposition was assumed to be a significant source of trace metal pollution for the area of Vingio Park. The observed similarity with the order of mass concentrations calculated for trace metal accumulation on the ground surface of a residential site in Vilnius in the work by Milukaitė *et al.* (2008) suggests that atmospheric deposition played a leading role in the pollutant build–up on the investigated surfaces. In addition, a strong relation was noticed between SS–bound metals such as Cu ($p < 0.01$) and Cd ($p < 0.01$) detected in the USWR from the roof and

Table 5. Summary of the urban storm water runoff data from the three sampling sites for trace metals and macro elements analysed in soluble and insoluble forms. 9 determinations in all cases. ND not detected as the result was less than the method detection limit

		Road				Roof				Pathway			
Macro elements		Mean	Median	Std.Dev.	CV,%	Mean	Median	Std.Dev.	CV,%	Mean	Median	Std.Dev.	CV,%
Ca	Soluble (mg L^{-1})	1.43	1.42	0.036	2.5	11.45	11.45	0.24	2.1	6.72	6.64	0.36	5.3
	Insoluble (mg kg^{-1}DW)	10611	12010	2732	25,7	12811	16038	4240	33.1	8333	9342	3124	37.5
K	Soluble (mg L^{-1})	0.48	0.50	0.105	21.7	ND				1.77	1.77	0.18	10.0
	Insoluble (mg kg^{-1}DW)	9983	11432	3003	30.1	10598	9668	2880	27.2	6190	7486	3363	54.3
Mg	Soluble (mg L^{-1})	0.59	0.59	0.014	2.4	0.1	0.098	0.004	3.7	0.64	0.65	0.05	8.1
	Insoluble (mg kg^{-1}DW)	3527	4080	1009	28.6	3571	3601	1723	48.3	6483	8603	4609	71.1
Trace metals		Mean	Median	Std.Dev.	CV,%	Mean	Median	Std.Dev.	CV,%	Mean	Median	Std.Dev.	CV,%
Cd	Soluble (μg L^{-1})	4.64	5.0	1.11	23.9	2.72	2.80	0.51	18.7	2.06	1.9	0.37	17.8
	Insoluble (mg kg^{-1}DW)	5.49	6.58	2.09	38.1	7.82	7.09	1.65	21.1	3.87	3.87	0.78	20.1
Cu	Soluble (μg L^{-1})	3.2	3.0	1.01	31.6	ND				ND			
	Insoluble (mg kg^{-1}DW)	117.1	111.3	10.8	9.3	60.3	52.7	33.2	55.0	58.4	71.3	28.2	48.3
Cr	Soluble (μg L^{-1})	ND				ND				ND			
	Insoluble (mg kg^{-1}DW)	44.7	45.6	11.8	26.4	36.0	52.0	29.1	81.0	41.7	52.1	24.9	59.7
Mn	Soluble (μg L^{-1})	18.61	19.3	2.92	15.7	14.88	14.7	2.29	15.4	21.1	20.8	4.5	21.3
	Insoluble (mg kg^{-1}DW)	650	746	171	26.3	429	448	329	76.8	504	603	241	47.8
Ni	Soluble (μg L^{-1})	ND				ND				ND			
	Insoluble (mg kg^{-1}DW)	ND				ND				22.2	22.2	1.1	5.0
Pb	Soluble (μg L^{-1})	4.57	3.5	3.06	66.9	2.85	3.05	1.69	59.3	5.76	6.10	1.14	19.8
	Insoluble (mg kg^{-1}DW)	135	150	34	25.2	138	142	69	50.0	99,1	127	51	51.3
Zn	Soluble (μg L^{-1})	54.33	58.0	16.55	30.5	72.56	85.0	43.54	60.0	131.9	138.0	19.91	15.1
	Insoluble (mg kg^{-1}DW)	20592	23798	6344	30.8	22373	20036	4265	19.1	9964	11605	5446	54.7

road surfaces as well as between Zn ($p < 0.01$) and Pb ($p < 0.01$) detected in the USWR from the road and the pathway surfaces.

For the insoluble form of trace metals such as Cd, Pb, and Zn the highest concentrations were measured in the USWR from the roof and road surfaces (from 30 to 50%, about 30% and 50% higher than Cd, Pb, and Zn detected in the USWR from the pathway, respectively); this may be attributed to the presence of galvanized steel materials such as safety barriers, gutters, downspouts and steel sheets that have been identified among the primary sources of Zn in roof and road SWR by many authors (Chang *et al.* 2004; Rocher *et al.* 2004; Brown, Peake 2006; Lamprea, Ruban 2008). Metals such as Cd and Pb are minor constituents of galvanized steel (Ogburn *et al.* 2012). Correlations ($p < 0.01$, $p < 0.05$) were found between the concentration of the metals in SS detected in the USWR from the road (e.g. Pb, Cd) and roof surfaces (e.g. Zn, Cd), between Zn and Cd in the USWR from the roof as well as between Cd and Pb in the USWR from the road.

A prolonged contact time with water can increase the weathering and deterioration of metal materials (Ogburn *et al.* 2012). The observed rainfall pattern (i.e. a previous dry period of less than one day) suggests that building materials may represent an important local source of metals in USWR from the roof and road surfaces since the prolonged wet weather conditions may have promoted erosion and weathering phenomena.

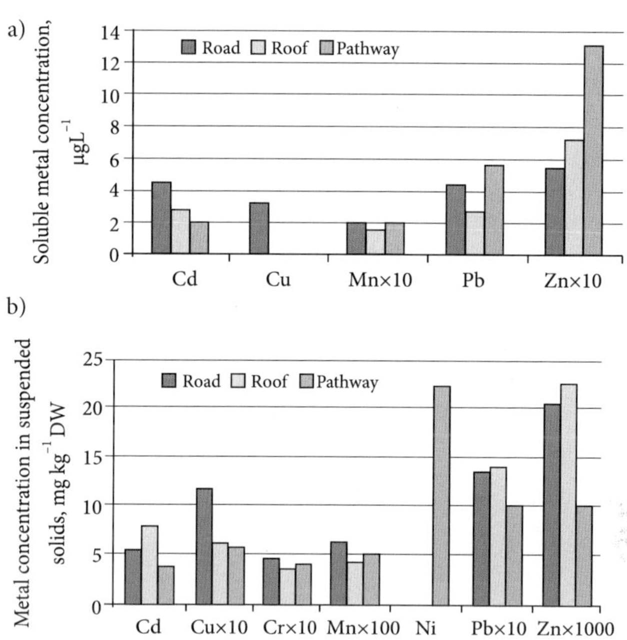

Fig. 3. Average concentration of the trace metals detected in soluble (a) and insoluble (b) forms in urban storm water runoff samples from the investigated sites

Table 6. Free ion activity and fraction bound to colloid (FA) of trace metals and macro elements in urban storm water runoff (USWR) from different surfaces, calculated by WHAM. [FA–M], concentration of M, trace metals and macro elements, bound to fulvic acids, FA. [ ] tot. diss., total dissolved concentration of trace metals and macro elements

	Mg				K			
	$[Mg^{2+}]$ (mol L^{-1})	[FA–Mg] (mol L^{-1})	$[Mg^{2+}]/[Mg]_{tot\ diss}$	$[FA–Mg]/[Mg]_{tot\ diss}$	$[K^{+}]$ (mol L^{-1})	[FA–K] (mol L^{-1})	$[K^{+}]/[K]_{tot\ diss}$	$[FA–K]/[K]_{tot\ diss}$
Roof	3.62×10^{-6}	4.79×10^{-8}	0.89	0.01	–	–	–	–
Pathway	2.36×10^{-5}	8.38×10^{-7}	0.90	0.03	4.42×10^{-5}	8.00×10^{-7}	0.97	0.02
Road	2.27×10^{-5}	1.61×10^{-6}	0.93	0.07	1.22×10^{-5}	5.42×10^{-7}	0.99	0.04
	Ca				Mn			
	$[Ca^{2+}]$ (mol L^{-1})	[FA–Ca] (mol L^{-1})	$[Ca^{2+}]/[Ca]_{tot\ diss}$	$[FA–Ca]/[Ca]_{tot\ diss}$	$[Mn^{2+}]$ (mol L^{-1})	[FA–Mn] (mol L^{-1})	$[Mn^{2+}]/[Mn]_{tot\ diss}$	$[FA–Mn]/[Mn]_{tot\ diss}$
Roof	2.55×10^{-4}	4.26×10^{-6}	0.89	0.01	2.55×10^{-4}	7.98×10^{-8}	0.64	0.29
Pathway	1.49×10^{-4}	6.86×10^{-6}	0.89	0.04	1.49×10^{-4}	2.10×10^{-7}	0.42	0.55
Road	3.25×10^{-5}	3.07×10^{-6}	0.91	0.09	3.25×10^{-5}	2.58×10^{-7}	0.24	0.76
	Cu				Zn			
	$[Cu^{2+}]$ (mol L^{-1})	[FA–Cu] (mol L^{-1})	$[Cu^{2+}]/[Cu]_{tot\ diss}$	$[FA–Cu]/[Cu]_{tot\ diss}$	$[Zn^{2+}]$ (mol L^{-1})	[FA–Zn] (mol L^{-1})	$[Zn^{2+}]/[Zn]_{tot\ diss}$	$[FA–Zn]/[Zn]_{tot\ diss}$
Roof	–	–	–	–	7.5×10^{-7}	2.78×10^{-7}	0.68	0.25
Pathway	–	–	–	–	9.93×10^{-7}	9.37×10^{-7}	0.49	0.46
Road	7.52×10^{-11}	5.01×10^{-8}	0.0	1.0	2.5×10^{-7}	5.53×10^{-7}	0.30	0.67
	Cd				Pb			
	$[Cd^{2+}]$ (mol L^{-1})	[FA–Cd] (mol L^{-1})	$[Cd^{2+}]/[Cd]_{tot\ diss}$	$[FA–Cd]/[Cd]_{tot\ diss}$	$[Pb^{2+}]$ (mol L^{-1})	[FA–Pb] (mol L^{-1})	$[Pb^{2+}]/[Pb]_{tot\ diss}$	$[FA–Pb]/[Pb]_{tot\ diss}$
Roof	1.91×10^{-8}	2.89×10^{-9}	0.80	0.12	9.36×10^{-10}	2.67×10^{-3}	0.0	0.92
Pathway	1.3×10^{-8}	4.72×10^{-9}	0.69	0.25	7.52×10^{-10}	5.6×10^{-3}	0.0	0.97
Road	2.27×10^{-8}	1.8×10^{-8}	0.55	0.44	2.36×10^{-10}	4.5×10^{-3}	0.0	0.98

2.2. Metal speciation in USWR using WHAM

WHAM modelling results showed a variability in the speciation behaviour of trace metals (Cd, Mn Pb, and Zn) between the sites, whereas major cation (Ca, Mg, and K) speciation was not substantially affected by the different USWR types (Table 6). In all the analysed USWR types, the major cations (Ca, Mg, K) were characterized by the predominance of free metal ions forms, with a value of free ion activity calculated in the range 89–99% of the total dissolved elements (Table 6). Speciation model results for Pb and Cu showed their affinity to form complexes. In all the USWR types, the predominant Pb species (with more than 90% of the total dissolved Pb) is the Pb–fulvic acid complex. In addition, compared to the total dissolved concentration, the concentration of free Pb^{2+} activity is negligible. In road USWR modelling Cu–FA were found to be the dominant complexes, being up to 99% of the total dissolved Cu.

Of all the modelled water-types, road USWR showed the highest ratio between the free major cation activity and total dissolved concentration; conversely the lowest ratio of free Cd^{2+} and Zn^{2+} activity was noticed with a value of 55 and 30% of total dissolved Cd and Zn, respectively. Roof USWR modelling showed the highest ratio of free Mn^{2+}, Zn^{2+}, and Cd^{2+} activity with a value of 64, 68 and 80% of total dissolved Mn, Zn and Cd, respectively. The highest fraction of Pb and Zn bound to FA was found in the USWR from the pathway. The Pb–FA and Zn–FA complexes were calculated to be from 19.6 to 52.3% and from 41 to 70.3% higher than Pb–FA and Zn–FA complexes in the road and roof USWR modelling, respectively. The metal fulvic acid complexes of Mn were 62% higher in the modelling of USWR from the pathway compared to the roof USWR whereas they were 18.6% less than the road USWR.

Comparisons of USWR modelling results showed that the USWR types characterized by a higher pH or DOC concentration produced a higher fraction of metals bound to FA. A higher fraction of Cd (38.8%), Mn (62%), Pb (52.3%), Zn (70.3%) bound to FA was calculated in the modelling of USWR from the pathway compared to the roof USWR. The pathway USWR exhibited a slightly higher pH and a higher DOC (about 47%) concentration than the roof ones. In the road USWR modelling, the Cd, Mn, Pb, Zn–FA complexes were from about 40 to 80% higher than the ones calculated for the roof USWR. USWR from the road had a higher pH (about 14%) and an almost equal DOC concentration compared to the one from the roof. Investigating the complexation of Cd, Ni, and Zn as a function of DOC concentrations in groundwater for a range of pH values, Christensen, J. B. and Christensen, T. H. (2000) found that DOC–metal complexing increased with increasing pH.

Conclusions

1. The presence of trace metals in the USWR from the investigated sites was likely to be due to non-point source pollution such as atmospheric deposition as well as local pollution sources such as the erosion of building materials and traffic activities. The high concentrations of Cd, Pb, and Zn detected in the USWR from the roof and road surfaces may be attributed to the corrosion of galvanized steel materials (e.g. steel sheets, downspouts, gutters and safety barriers). Correlations ($p < 0.01$, $p < 0.05$) between these metals were found.

2. Analyses of metals both in SS and the dissolved form showed the predominance of the SS–bound fraction of all the detected elements with more than 80% of the total amount measured, irrespective of the water quality characteristics (e.g. pH, DOC, SS) and types of surface area. The partitioning coefficient (log Kd) of trace metals decreases in the order Zn > Cu ≈ Mn ≈ Pb >> Cd.

3. The concentration of trace metals (Cd, Cr, Cu, Pb, Ni, and Zn) in the dissolved form did not exceed the Lithuanian Maximum Allowable Concentration (MAC) limits in all the USWR samples. Trace metals such as Cr and Ni were not detected in the soluble phase. The concentration of Cd, Cu, Pb and Zn in the dissolved form was far lower (up to one order of magnitude) than the MAC limits of these metals for the environment.

4. WHAM modelling results showed a variability in the speciation behaviour of trace metals (Cd, Mn, Pb, and Zn) between the investigated sites, whereas major cation (Ca, Mg, K) speciation was not substantially affected by the different USWR types. Ca, Mg, and K were characterized by the predominance of free metal ion forms, being from 89 to 99% of the total dissolved elements. USWR types characterized by a higher pH or DOC concentration produced a higher fraction of trace metals bound to FA. A higher fraction of Cd (38.8%), Mn (62%), Pb (52.3%), and Zn (70.3%) bound to FA was calculated in the modelling of USWR from the pathway compared to the roof USWR. In the road USWR modelling, the Cd, Mn, Pb, Zn–FA complexes were about 40 to 80% higher than the ones calculated in the roof USWR.

5. Since the studied metals were mainly related to SS, the removal of SS can be an effective method for reducing metal concentrations in USWR. For example, the adoption of best management practices (BMPs) that employ sedimentation or filtration (e.g. porous pavement, wet and dry detention ponds) may be efficient in abating the flow of metals loads to receiving water basins.

6. Since roof USWR analysis exhibited the highest concentrations of Zn and Cd in SS (from 10 to 55% and 30 to 50% higher than the road and pathway levels, respectively) and the speciation modelling showed the highest ratio of free ion activity to total dissolved zinc (68%) and

cadmium (80%), if a change in water chemical parameters occurs, the release of metals from SS may pose a problem because of the strong bioavailability of the dissolved metal fraction.

Acknowledgements

We gratefully acknowledge the help of our colleagues from the Department of Environmental Protection at Vilnius Gediminas Technical University for the chemical analyses, Justina Vaiciute for the legal advice about environmental legislation, as well as Dr. Rodolfo Venerucci for his assistance in statistical analysis. The data and information provided by the Lithuanian Hydrometeorological Service are greatly acknowledged.

References

Adachi, K.; Tainosho, Y. 2004. Characterization of heavy metal particles embedded in tire dust, *Environment International* 30(8): 1009–1017. http://dx.doi.org/10.1016/j.envint.2004.04.004

Allison, J. D.; Allison, T. L. 2005. *Partition coefficients for metals in surface water, soil, and waste.* Prepared for U.S. EPA Office of Research and Development: Washington, DC, EPA/600/R-05. 74 p.

Baltrenaitė, E.; Butkus, D.; Booth, C. A. 2010. Comparison of tree-ring sampling methods for trace metal analysis, *Journal of Environmental Engineering and Landscape Management* 18(3): 170–178. http://dx.doi.org/10.3846/jeelm.2010.20

Baltrėnas, P.; Kaziukonienė, D.; Kvasauskas, M. 2004. Air pollution at parking-lots of Vilnius, *Journal of Environmental Engineering and Landscape Management* 12(1): 38–43. http://dx.doi.org/10.1080/16486897.2004.9636813

Bricker, O. P. 1999. An overview of the factors involved in evaluating the geochemical effects of highway runoff on the environment, *U.S. Geological Survey Open File Report 98–630.* 28 p.

Brown, J. N.; Peake, B. M. 2006. Sources of heavy metals and polycyclic aromatic hydrocarbons in urban stormwater runoff, *Science of the Total Environment* 359(1): 145–155. http://dx.doi.org/10.1016/j.scitotenv.2005.05.016

Butkus, D.; Baltrenaitė, E. 2007. Accumulation of heavy metals in tree seedling from soil amended with sewage sludge, *Ekologija* 53(4): 68–76.

Chang, M.; McBroom, M. W.; Scott, B. R. 2004. Roofing as a source of nonpoint water pollution, *Journal of environmental management* 73(4): 307–315. http://dx.doi.org/10.1016/j.jenvman.2004.06.014

Christensen, J. B.; Christensen, T. H. 2000. The effect of pH on the complexation of Cd, Ni and Zn by dissolved organic carbon from leachate-polluted groundwater, *Water Research* 34: 3743–3745. http://dx.doi.org/10.1016/S0043-1354(00)00127-5

Ciffroy, P.; Garnier, J. M.; Benyahya, L. 2003. Kinetic partitioning of Co, Mn, Cs, Fe, Ag, Zn and Cd in fresh waters (Loire) mixed with brackish waters (Loire estuary): experimental and modelling approaches, *Marine Pollution Bulletin* 46(5): 625–641. http://dx.doi.org/10.1016/S0025-326X(02)00517-9

Comber, S.; Georges, K. 2008. Tiered approach to assessment of metal compliance in surface waters, *Environment Agency* [online], [cited 9 June 2013]. Available from Internet: http://www.wfduk.org/sites/default/files/Media/Environmental standards/Tiered approach metal compliance.pdf

Dierkes, C.; Göbel, P.; Lohmann, M.; Coldewey, W. G. 2005. Development and investigation of a pollution control pit for treatment of stormwater from metal roofs and traffic areas, *Urban Drainage* 54(6): 291–298.

Dudutytė, Z; Manusadžianas, L.; Ščeponavičiūtė, R. 2007. Report on dangerous substances in the aquatic environment of Lithuania. Prepared within the project *Screening of dangerous substances in the aquatic environment of Lithuania.* Vilnius, Lithuania. 53 p.

Dwane, G. C.; Tipping, E. 1998. Testing a humic speciation model by titration of copper-amended natural waters, *Environmental International* 24(5/6): 609–616. http://dx.doi.org/10.1016/S0160-4120(98)00046-4

European Commission (EC). 2008. Directive 2008/105/EC of the European Parliament and of the Council of 16 December 2008 on environmental quality standards in the field of water policy, *Official Journal of the European Union* L348.

Gnecco, I.; Sansalone, J. J.; Lanza, L. G. 2008. On the kinetics of pollutants in storm water runoff, in *11 ICUD – 11th International Conference on Urban Drainage,* 31 August – 5 September, 2008, Edinburgh, Scotland, UK [online], [cited 2 May 2014]. Available from Internet: http://web.sbe.hw.ac.uk/staffprofiles/bdgsa/11th_International_Conference_on_Urban_Drainage_CD/ICUD08/pdfs/587.pdf

Göbel, P.; Dierkes, C.; Coldewey, W. G. 2007. Storm water runoff concentration matrix for urban areas, *Journal of Contaminant Hydrology* 91: 26–42. http://dx.doi.org/10.1016/j.jconhyd.2006.08.008

Herngren, L.; Goonetilleke, A.; Ayoko, G. A. 2005. Understanding heavy metal and suspended solids relationships in urban stormwater using simulated rainfall, *Journal of Environmental Management* 76(2): 149–158. http://dx.doi.org/10.1016/j.jenvman.2005.01.013

ISO 10523:2008. Water quality – Determination of pH. International Organization for Standardization, Geneva, Switzerland, 2008. 13 p.

ISO 11047:1998. Soil quality – Determination of cadmium, chromium, cobalt, copper, lead, manganese, nickel and zinc in acqua regia extracts of soil – Flame and electrothermal atomic absorption spectrometric methods. International Organization for Standardization, Geneva, Switzerland, 1998. 18 p.

ISO 5667-1:2006. Water quality – Sampling – Part 1: Guidance on the design of sampling programmes and sampling techniques. International Organization for Standardization, Geneva, Switzerland, 2006. 31 p.

ISO 5667-3:2006. Water quality – Sampling – Part 3: Guidance on the preservation and handling of water samples. International Organization for Standardization, Geneva, Switzerland, 2006. 42 p.

ISO 7888: 1985. Water quality – Determination of electrical conductivity. International Organization for Standardization, Geneva, Switzerland, 1985. 6 p.

ISO 7980:1986. Water quality – Determination of calcium and magnesium – Atomic absorption spectrometric method. International Organization for Standardization, Geneva, Switzerland, 1986. 3 p.

ISO 8288:1986. Water quality – Determination of cobalt, nickel, copper, zinc, cadmium and lead – Flame atomic absorption spectrometric methods. International Organization for Standardization, Geneva, Switzerland, 1986. 11 p.

ISO 9174:1998. Water quality – Determination of chromium – Atomic absorption spectrometric methods. International Organization for Standardization, Geneva, Switzerland, 1998. 10 p.

ISO 9964-2:1993. Water quality – Determination of sodium and potassium. Part 2: Determination of potassium by atomic absorption spectrometry. International Organization for Standardization, Geneva, Switzerland,1993. 4 p.

Jarasuniene, A. 2010. The solutions to vehicle traffic organisation problems in Vilnius, in *Proceedings of the 10th International Conference "Reliability and Statistics in Transportation and Communication" (RelStat'10)*, 20–23 October 2010, Riga, Latvia, 79–83.

Karlavičienė, V.; Švedienė, S.; Marčiulionienė, D. E.; Randerson, P.; Rimeika, M.; Hogland, W. 2009. The impact of storm water runoff on a small urban stream, *Journal of Soils and Sediments* 9(1): 6–12. http://dx.doi.org/10.1007/s11368-008-0038-9

Kayhanian, M.; Stransky, C.; Bay, S.; Lau, S. L.; Stenstrom, M. K. 2008. Toxicity of urban highway runoff with respect to storm duration, *Science of the Total Environment* 389(2): 386–406. http://dx.doi.org/10.1016/j.scitotenv.2007.08.052

Lamprea, K.; Ruban, V. 2008. Micro pollutants in atmospheric deposition, roof runoff and storm water runoff of a suburban Catchment in Nantes, France, in *11 ICUD – 11th International Conference on Urban Drainage*, 31 August – 5 September, 2008, Edinburgh, Scotland, UK [online], [cited 2 May 2014]. Available from Internet: http://web.sbe.hw.ac.uk/staffprofiles/bdgsa/11th_International_Conference_on_Urban_Drainage_CD/ICUD08/pdfs/194.pdf

Milukaitė, A.; Sakalys, J.; Kvietkus, K.; Vosyliene, M. Z.; Kazlauskiene, N.; Karlaviciene, V. 2010. Physico-chemical and ecotoxicological characterizations of urban storm water runoff, *Polish Journal of Environmental Studies* 19(6): 1279–1285.

Milukaitė, A.; Valiulis, D.; Sakalys, J. 2008. Fluxes and loading of heavy metals, benzo[a]pyrene and oil products in Vilnius city, *Water Air Soil Pollution* 8: 485–494. http://dx.doi.org/10.1007/s11267-008-9181-y

Ogburn, O.; Pitt, R.; Clark, S. 2012. Pollutant releases from gutter and piping materials into urban stormwater runoff, in *Proceedings of World Environmental and Water Resources Congress 2012: Crossing Boundaries,* Albuquerque, 20–24 May 2012, New Mexico, US, 3674–3685. http://dx.doi.org/10.1061/9780784412312.369

Peijnenburg, W. J. G. M.; Jager, T. 2003. Monitoring approaches to assess bioaccessibility and bioavailability of metals: matrix issues, *Ecotoxicology and Environmental Safety* 56(1): 63–77. http://dx.doi.org/10.1016/S0147-6513(03)00051-4

Pitt, R.; Field, R.; Lalor, M.; Brown, M. 1995. Urban storm-water toxic pollutants: assessment, sources, and treatability, *Water Environment Research* 67(3): 260–275. http://dx.doi.org/10.2175/106143095X131466

Prestes, E. C.; dos Anjos, V. E.; Sodrè, F. F.; Grassi, M. T. 2006. Copper, lead and cadmium loads and behaviour in urban stormwater runoff in Curitiba, Brazil, *Journal of Brazilian Chemical Society* 17(1): 53–60. http://dx.doi.org/10.1590/S0103-50532006000100008

Pundytė, N.; Baltrenaitė, E.; Pereira, P.; Paliulis, D. 2011. Anthropogenic effect on heavy metals and macronutrients accumulation in soil and wood of Pinus sylvestris L., *Journal of Environmental Engineering and Landscape Management* 19(1): 33–43. http://dx.doi.org/10.3846/16486897.2011.557473

Rocher, V.; Azimi, S.; Gasperi, J.; Beuvin, L.; Muller, M.; Moilleron, R.; Chebbo, G. 2004. Hydrocarbons and metals in atmospheric deposition and roof runoff in central Paris, *Water, Air, and Soil Pollution* 159(1): 67–86. http://dx.doi.org/10.1023/B:WATE.0000049165.12410.98

Rule, K. L.; Comber, S. D. W.; Ross, D.; Thomton, A.; Makropoulos, C. K.; Rautiu, R. 2006. Diffuse sources of heavy metals entering an urban wastewater catchment, *Chemosphere* 63(1): 64–72. http://dx.doi.org/10.1016/j.chemosphere.2005.07.052

Sabin, L. D.; Lim, J. H.; Stolzenbach, K. D.; Schiff, K. C. 2005. Contribution of trace metals from atmospheric deposition to stormwater runoff in a small impervious urban catchment, *Water Research* 39(16): 3929–3937. http://dx.doi.org/10.1016/j.watres.2005.07.003

Sansalone, J.; Ying, G. 2008. Partitioning and granulometric distribution of metal leachate from urban traffic dry deposition particulate matter subject to acidic rainfall and runoff retention, *Water Research* 42(15): 4146–4162. http://dx.doi.org/10.1016/j.watres.2008.06.013

Shaw, S. B.; Walter, M. T.; Steenhius, T. S. 2006. A physical model of particulate wash-off from rough impervious surface, *Journal of Hydrology* 327: 618–626. http://dx.doi.org/10.1016/j.jhydrol.2006.01.024

Spurlock, F. 1999. DPR SOP FSWA008.00 Sampling for surface water runoff in agricultural fields [online], [cited 2 January 2013]. Available from Internet: http://www.cdpr.ca.gov/docs/emon/pubs/sops/fswa008.pdf

Taraškevičius, R.; Zinkutė, R.; Jankauskaitė, M. 2008. Differences of Vilnius topsoil contamination in Neris River valley due to anthropogenic factors, *Geologija 50* 3(63): 135–142.

Tiefenthaler, L. L.; Stein, E. D.; Schiff, K. C. 2008. Watershed and land use-based sources of trace metals in urban storm water, *Environmental Toxicology and Chemistry* 27(2): 277–287. http://dx.doi.org/10.1897/07-126R.1

Tipping, E. 1994. WHAM – a chemical equilibrium model and computer code for waters, sediments and soils incorporating a discrete-site/electrostatic model for ion-binding by humic substances, *Computers and Geosciences* 20: 973. http://dx.doi.org/10.1016/0098-3004(94)90038-8

Tipping, E.; Lofts, S.; Sonke, J. E. 2011. Humic ion-binding model VII: a revised parameterisation of cation-binding by humic substances, *Environmental Chemistry* 8: 225–235. http://dx.doi.org/10.1071/EN11016

Van Metre, P. C.; Mahler, B. J. 2003. The contribution of particles washed from rooftops to contaminant loading to urban streams, *Chemosphere* 52(10): 1727–1741. http://dx.doi.org/10.1016/S0045-6535(03)00454-5

Vasarevičius, S.; Mineikaite, A.; Vaitiekunas, P. 2010. Investigation into heavy metals in storm wastewater from Vilnius Zirmunai district and pollutantits spread model in Neris river, *Journal of Environmental Engineering and Landscape Management* 18(3): 242–249. http://dx.doi.org/10.3846/jeelm.2010.28

Weiss, J.; Hondzo, M.; Biesboer, D.; Semmens, M. 2006. Laboratory of heavy metal phytoremediation by three wetland macrophytes, *International Journal of Phytoremediation* 8: 245–259. http://dx.doi.org/10.1080/15226510600846798

Wicke, D.; Cochrane, T. A.; O'Sullivan, A. 2012. Build-up dynamics of heavy metals deposited on impermeable urban surfaces, *Journal of Environmental Management* 113: 347–354. http://dx.doi.org/10.1016/j.jenvman.2012.09.005

Zinkutė, R.; Taraškevičius, R.; Želvys, T. 2011. Major elements as possible factors of trace element urban pedochemical anomalies, *Central European Journal of Chemistry* 9(2): 337–347. http://dx.doi.org/10.2478/s11532-011-0012-z

Zavadskas, E. K.; Kaklauskas, A.; Šaparauskas, J.; Kalibatas, D. 2007. Vilnius urban sustainability assessment with an emphasis on pollution, *Ecology* 53: 64–72.

Enrico MANCINELLI. Master of Science in Civil Engineering (Università Politecnica delle Marche, 2013), Bachelor of Science in Civil Engineering (UNIVPM 2009). Research interests: trace metals in surface water, environmental protection.

Edita BALTRĖNAITĖ. Dr, Assoc. Prof. at the Department of Environmental Protection, Vilnius Gediminas Technical University (VGTU). She gained her Environmental Engineering and Landscape Management PhD in 2007, the diploma of the Associate Professor in 2010 from Vilnius Gediminas Technical University (Lithuania). Publications: author/co-author of approx. 38 scientific papers, participant of about 20 international conferences. She is a deputy member of the Management Committee of COST TD1107 Biochar as Option for Sustainable Resource Management and FA0905 Mineral-Improved Crop Production For Healthy Food and Feed. Main research visits: Latvian State Institute of Wood Chemistry under FP7 project The Implementation of Research Potential of the Latvian State Institute of Wood Chemistry in the European Research Area (WOOD-NET) (FP7-203459)); University of Ljubljana (Slovenia) under FP7 project SPIRIT Support of Public and Industrial Research Using Ion Beam Yechnology; short-term scientific mission at Norwegian University of Life Sciences, in the framework of COST 859 action Phytotechnologies to Promote Sustainable Land Use and Improve Food Safety (COST-859-STSM-04824). Research interests: biogeochemistry, phytogeochemistry, bioindication.

Pranas BALTRĖNAS. Dr Habil, Prof. and Head of the Department of Environmental Protection, Vilnius Gediminas Technical University (VGTU). Doctor Habil of Science (air pollution), Leningrad Civil Engineering Institute (Russia), 1989. Doctor of Science (Air Pollution) Ivanov Textile Institute (Russia), 1975. Employment: Professor (1990), Associate Professor (1985), Senior Lecturer (1975), Vilnius Civil Engineering Institute (VISI, now VGTU). Publications: author of 13 monographs, 24 study-guides, over 320 research papers and 67 inventions. Honorary awards and membership: prize-winner of the Republic of Lithuania (1994), a corresponding Member of the Ukrainian Academy of Technological Cybernetics, a full Member of International Academy of Ecology and Life Protection. Placements in Germany and Finland. Research interests: air pollution, pollutant properties, pollution control equipment and methods.

Dainius PALIULIS. Assoc. Prof. at the Department of Environmental Protection, Vilnius Gediminas Technical University (VGTU). Doctor of Science (air pollution) Vilnius Gediminas Technical University, 2000. Employment: Associate Professor (2006), Assistant (1997), Vilnius Gediminas Technical University. Publications: author of 1 monograph, 2 study-guides, over 50 research papers and 2 inventions. Research interests: air pollution, pollution control equipment and methods.

Giorgio PASSERINI. Assoc. Prof. at the Engineering Faculty of Marche Polytechnic University. MD in Electronic Engineering, PhD in Applied Physics. Employment: Associate Professor (2003), Researcher (1998) Marche Polytechnic University. Publications: 2 authored books, 4 edited books, over 90 papers. Research interests: air pollution; global warming assessment; energy conversion and management.

Åsgeir R. ALMÅS, Senior Research Scientist at the Norwegian University in Life Sciences, the Department of Environmental Sciences. Since 1999, he has been appointed as Post Doc, research scientist and project leader in several projects related to metal speciation in soil, sediments and fresh water and the impact of metal speciation on the soil biology. During most of his career he has used several techniques to investigate metal speciation in environmental samples using i.e. anodic stripping voltammetry (visiting scientist at Cornell University) WHAM/Model VI (Windermere Humic Aqueous Model, visiting scientist at the Centre of Ecology and Hydrology, UK) and DGT (diffusive gradients in thin films) technique. He has supervised a number of PhD and MSc students and collaborated with partners around the world. This has resulted in several peer-reviewed articles and contributions in conferences. His citation index at the moment is 9.

2

REMOVAL OF AMMONIUM IONS FROM DIGESTED SLUDGE FUGATE BY USING NATURAL ZEOLITE

Aušra MAŽEIKIENĖ, Marina VALENTUKEVIČIENĖ

Department of Water Engineering, Vilnius Gediminas Technical University, Saulėtekio al. 11, 10223 Vilnius, Lithuania

Abstract. Wastewater treatment loadings with total nitrogen can increased by introducing fugate that is saturated with ammonium ions, generated when dewatering wastewater treatment sludge. In this article the possibility to reduce the concentration of ammonium ions in the fugate by the use of natural zeolite (Transcarpatian clinoptilolite) has been analysed. Lab-scale experiments were carried out with different particle sizes zeolite: 0.8–1.6 mm, 1.6–2.5 mm and 2.5–3.2 mm. At the first stage of the experiments, zeolite particles were mixed with fugate and left to settle until a 99% efficiency of ammonium ion removal was achieved. At the second stage, ammonium ions were removed from the fugate by filtering this liquid through zeolite where the 70–92% efficiency was achieved. Zeolite saturated with ammonium ions possibly can be used in agriculture as nitrogenous fertilizers. Obtained fertilizer grade of zeolite saturated with fugate can also be produced from the high ammonium nitrogen concentration side-streams separated from wastewater flow from sludge digestion processes.

Keywords: sludge fugate, ammonium ion concentration, removal efficiency, zeolite, wastewater treatment.

Introduction

Urban wastewater treatment is a high priority environmental issue of and a very important task in the area is the reduction of nitrogen and phosphorus loads (Farkas *et al.* 2005; Cheng *et al.* 2011). The removal of mentioned pollutants in wastewater treatment units is affected by the composition of the initial wastewater, i.e. the higher the concentrations of nitrogen and phosphorus in the wastewater, the more complicated it is to meet the required threshold values (Fig. 1). The loading of wastewater treatment units with total nitrogen can be increased by using fugate enriched with ammonium, which has been generated when dewatering sludge by means of centrifuges that have been returned to the starting link of the wastewater treatment unit (Constantine 2006; Thornton *et al.* 2007). The ammonium concentration in the sludge liquor is often between 500–1500 mg NH_4^+-N/L, e.g. high sludge thickening prior to digestion increases the ammonium concentration Gustavsson 2010).

In this case the ammonium ions that are present in the sludge fugate can increase the concentration of the total amount of nitrogen to be treated by more than 25% (Thornton *et al.* 2007). However, the pollution of the influent wastewater with ammonium ions burdens the processes of biological removal of both the nitrogen and phosphorus from the wastewater (Vabolienė *et al.* 2007). In order to reduce the higher concentration of NH_4^+, an older activated sludge (approximately 20 days) is needed,

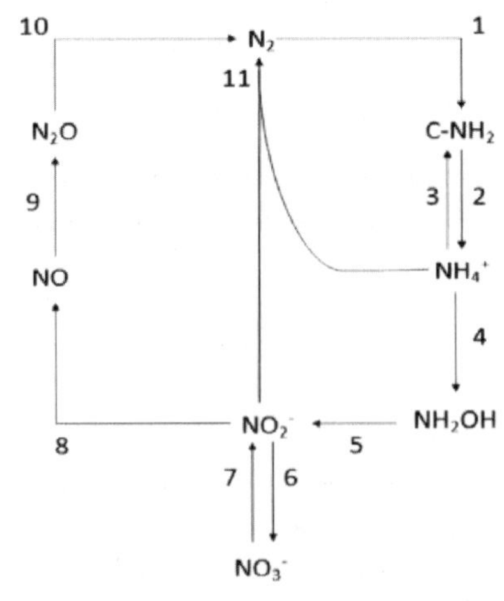

Fig. 1. The main nitrogen compounds transformation processes at WWTP: 1 – gaseous nitrogen fixation, 2 – decomposition of organic material, 3 – assimilation, 4, 5 – nitrification, 6 – nitrite oxidation, 7–10 – denitrification (Gustavsson 2010)

Corresponding author: Aušra Mažeikienė
E-mail: ausra.mazeikiene@vgtu.lt

but this does not provide favourable conditions for the removal of phosphorus. High concentrations of NH_4^+, in bioreactors and digestion systems inhibit the activity of methanogenic bacteria and prolong the period of digestion (Calli *et al.* 2005; Tada *et al.* 2005; Wang *et al.* 2010).

Vilnius wastewater treatment unit treats 100,000 m^3 of wastewater, which equals 660,000 inhabitants equivalent. In addition wastewater treatment sludge is generated, which is thermally treated, digested and dewatered by centrifuges. Technological processes of biological wastewater and sludge treatment are presented in Figure 2.

In scientific literature it is recommended that sludge liquor generated by dewatering the digested sludge should be treated separately before it is returned to the wastewater treatment link (Gustavsson 2010). Sludge liquor can be treated by employing sorption or biological (ANAMMOX) methods. Biological ANAMMOX processes require complex treatments and depend on the period of adaptation as well as on the variability of temperature (Monballiu *et al.* 2013; Zekker *et al.* 2014; Bao-Shan Xing *et al.* 2015), and due to which it is suggested that sorption processes are more appropriate (Vassileva, Voikova 2009; Karapinar 2009).

The aim of the experimental study described in this article, is to analyse the concentration of the fugate saturation with ammonium ions of the dewatered sludge in the Vilnius wastewater treatment plant and to reduce the pollution by either stirring or filtering the sludge liquid through the natural zeolite particles of various sizes. The zeolitic materials are often used in scientific investigations, mostly due to their crystallinity, thermal stability, catalytic, sieve and ion exchange properties, as well as the well-defined cage structure of molecular size (Rožič *et al.* 2000; Kurama *et al.* 2002; Du *et al.* 2005; Miladinovic, Weatherley 2008; Sileg *et al.* 2010). The most widely used natural zeolite is clinoptilolite (simplified formula $(Na, K)_6Si_{30}Al_6O_{72}{\cdot}24H_2O$), whose K and Na forms are highly selective for ammonium ions, independent of NH_4^+ concentration (Lebedynets *et al.* 2004; Inglezakis 2005, Sileg *et al.* 2010).The advantages of the clinoptilolite are: low cost, high selectivity towards the ammonium ions in the presence of other concurrent cations in the wastewaters and wide spreading in nature (Jafarpour *et al.* 2010; Ivanova *et al.* 2010).

The sorption process was chosen because the zeolite that absorbed ammonium ions can be used as fertilizer in agriculture (Rehaková *et al.* 2004).

1. Materials and methods

The samples of sludge liquid after dewatering (25 L of total samples volume) from Vilnius biological wastewater treatment plant were transported to the laboratory of the Water Management Department of Vilnius Gediminas Technical University (Lithuania). The temperature of the liquor samples, pH, and the concentrations of ammonium

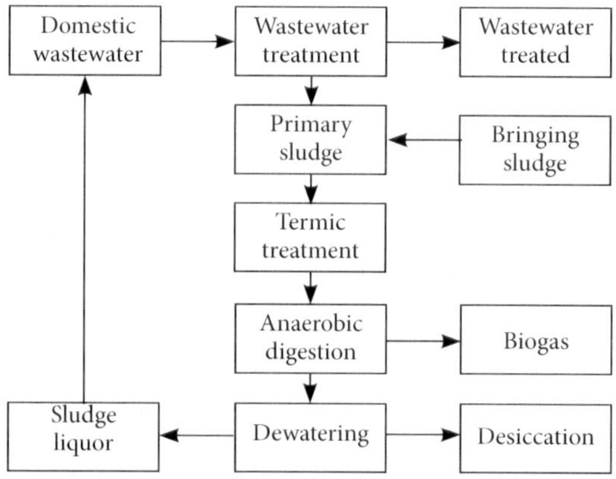

Fig. 2. Conventional technological processes of biological wastewater and sludge treatment

ions and of suspended solids were measured in the laboratory by using of standard approved methods (ISO, EN and LST). The concentration of TSS (total suspended solids) was determined by using a filtering method described in the regulatory document LAND 46–2007 "Water quality. Determination of suspended solids. A filtering method using a fibreglass filter". Vacuum filtering equipment; fibreglass filter with pores of 0.45 μm; drying unit with a temperature of 105 °C; analytical scales (precision 0.1 mg) were used. The concentration of total suspended solids was determined 3 times in each sample. To determine the concentration of ammonium nitrogen, "MerckSpectroquant, 1.00683.0001 Ammonium test" test which determination limits are 5.0–150 mg/L NH_4^+-N, and a spectrophotometer "Genesys 10 Vis" (wave length 190–1100 nm) were used. Samples with bigger than 150 mg/L NH_4^+-N were diluted with distilled water (1:10; 1:100 and 1:1000 dilution) and the concentration of ammonium nitrogen was determined accordingly. The studies of NH_4^+-N concentrations in the samples were carried out by using natural zeolite (Table 1), a Transcarpatian clinoptilolite (its mineral content was found to be about 87%) of aluminium–silicate origin, which adsorbs high quantities of ammonium (Sakalauskas, Valentukevičienė 2003; Sprinsky *et al.* 2005; Mažeikienė *et al.* 2008; Valentukevičienė, Rimeika 2007; Saltah *et al.* 2007).

The zeolites contain channels, interconnected cavities or voids of a discrete size (0.4–0.6 nm). The micropores that exist between particles and accompanying materials in the natural zeolites can be occupied by different ions. Previous work in Lithuania observed that this method is an efficient way to achieve ammonium sorption at low cost (Valentukevičienė, Jankauskas 2004). Zeolite fractions (particles of 0.8–1.6; 1.6–2.5; 2.5–3.2 mm size) were selected using calibrated sieves. Before the filtration run all zeolite was washed with distilled water in order to remove small particles and dried in an oven at a temperature not exceeding 105 °C. The drying took 2 hours.

Table 1. The physico-chemical properties of natural zeolite used in this study

Physical properties		Chemical content	
Index		Component	%
Porosity	34%	SiO_2	71.5
		Al_2O_3	13.1
Density	2.37 kg/dcm^3	Fe_2O	0.9
		FeO	0.78
Humidity	4%	TiO_2	0.2
		MnO	0.03
Conditional surface	160 m^2/g	CaO	2.1
		MgO	1.07
Mechanical resistance	150 kg/cm^2	P_2O_5	0.03
		K_2O	2.85
Attrition	4 %	Na_2O	2.25
		Loss on ignition	5.19

Quality parameters – composition of elements in dry samples of fugate and zeolite (before and after experiments) were analyzed. Each zeolite and fugate sample was homogenized to achieve a uniform appearance, and a mixed sample from each sampling procedure was taken and placed in a ceramics container to access all surface drying. All samples were dried for 11 hours at the temperature of 110 °C and disintegrated using a porcelain pestle and mortar. Obtained samples were sieved through a 2.0 mm and 250 μm stainless steel sieves, as required for X-Ray Fluorescence analysis. Sieved samples were analyzed using a Thermo Scientific Niton® XL2 X-Ray fluorescence spectrometer. Analytical method of X-ray fluorescence spectrometry and specifically Niton XL2 device for analysis of substances composition was used.

STAGE I. Three 100 and 200 g samples of three zeolite fractions were mixed with 1 L of sludge liquor in an automatic mixer for 30 min at rotation 100 rpm and left to set in for 30 min. All samples were filtered through glass fibre filters with a pore size of 1.0 μm. The concentration of ammonium ions was measured in the filtered samples.

STAGE II. The experiment to reduce ammonium ion concentrations in the fugate was carried out by filtering it through zeolite filler. The fugate was filtered through constructed columns filters. The columns were 46 mm in diameter and 1.4 m high. The filter columns were filled with 0.8–1.6 mm or 1.6–2.5 mm zeolite fractions. The hight of zeolite filler was either 0.5 or 1.0 m. The fugate was supplied into the column and filtered at 2 m/h filtration rate. After filtered 1 L of fugate in each case, fugate samples were collected and the concentration of ammonium ions measured. The samples were filtered through glass fibre filters (47 mm in diameter) where mesh size was 1.0 μm. In order to avoid random errors and to obtain more precise results, the concentrations of suspended materials as well as ammonium ions were determined 5 times for each sample.

All results obtained are presented as the arithmetic mean of five replicates. Significant differences ($p < 0.05$) were removed from the experiment estimations and the measurements were analysed once again. The average concentration was calculated:

$$\overline{c} = \frac{1}{n}\sum_{i=1}^{k} c_i m_i \,, \tag{1}$$

where: c_i – concentration of substances at typical points; m_i – probability at the occurrence of concentration; n – number of trials; k – number of different values of the concentration.

The average concentrations of the variables substances, mentioned above, at the characteristic point were calculated as well. The standard statistical estimation error was evaluated.

2. Results and discussion

The characteristics of sludge liquid are described in Table 2.

Table 2. Main characteristics of sludge liquid

Samples	pH	NH_4^+ (mg/L)	TSS (mg/L)	T (°C)	BOD (mg/L)	COD_t (mg/L)
1	7.2	1025	1040	20	165	2475
2	7.4	1750	1135	24	176	2550
3	7.4	1720	1220	23	197	2498
4	7.3	1420	1093	22	155	2464
5	7.4	1450	1090	22	205	2567

It was determined that the dewatering of digested sludge in the Lithuanian wastewater treatment unit generates a liquid fraction, i.e. fugate, where the ammonium concentration varied from 1025 to 1750 mg/L. The concentration of total COD in fugate was about 2500 mg/L (BOD–180 mg/L) because of degradation of COD and dissolution of ammonium in the digesters (table 2). Clinoptilolite does not affect COD removal, since organic carbon such as COD neither has anionic or cationic charge to adsorb on zeolite, nor could exchange with its cations (Rezaei *et al.* 2014). The ammonium ions concentrations were approximately two times higher than indicated in the studies done by other researchers (Constantine 2006; Thornton *et al.* 2007). The pH values measured for this study were not considerably different and varied between 7.2 and 7.4. In the digesting mechanisms, when anaerobic organic decomposition of organic materials is taking place, the process can be expressed with the formulas below:

$$C_5H_7O_2N + 3H_2O + H^+ \rightarrow 2.5CO_2 + 2.5CH_4 + NH_4^+ \,. \tag{2}$$

The equation shows that the destruction of organic materials leads to an increase of ammonium ions concentration. Sludge liquids contain especially high concentrations of TSS (1040–1220 mg/L). Due to the thermal decomposition of the sludge, which is carried out in the Lithuanian wastewater unit – suspended solids in the fugate are particularly small. Therefore, when sludge is dewatered by using conventional centrifuges the suspended materials cannot be suspended effectively in the dry fraction of the sludge and a high proportion of it obtained into the fugate. TSS concentration is an important characteristic of fugate as it influences either the ion exchange or the sorption process (Hedstrom 2001; Rahmani *et al.* 2004). Suspended and colloid particles can be attached to the surface of zeolite particles and block the micro pores, thus slowing down the ion exchange.

The results from five laboratory mixers run can be seen in Table 3.

Table 3. Batch test (I) experiment results

No.	Zeolite weight, g	Zeolite particle size, mm	Removed NH_4^+ conc., mg/L	Removal efficiency, %
1	100	2.5–3.2	40±4	2±0.5
2	200	2.5–3.2	85±9	5±0.4
3	100	1.6–2.5	410±40	24±2
4	200	1.6–2.5	885±85	51±5
5	100	0.8–1.6	1164±115	68±7
6	200	0.8–1.6	1540±150	90±9

As can be seen in Table 3, ammonium ions were most efficiently removed by the smallest zeolite fraction when a larger amount of it was used (200 g). Zeolite particle fraction sized 2.5 to 3.2 mm proved to be inefficient with an efficiency of only 2–5% of ammonium ions removed from the samples. Therefore, this fraction size was no longer used for further research. Two smaller zeolite fractions sized 0.8 to 1.6 mm and 1.6–2.5 mm were used for the reduction of ammonium ions concentration in other fugate samples and Langmuire isotherm was obtained. The following formula was used for the calculations:

$$q_e = (C_0 - C_e) \cdot \frac{V}{m_z}, \qquad (3)$$

where: q_e – equilibrium adsorption capacity, mg/g; C_0 – initial NH_4^+ concentration in the sample, mg/L; V – volume of the sample, L; m_z – the adsorbent weight, g.

This calculation can be carried out when q_e, C_0, C_e and proportion of C_e/q_e are known or estimated from the experimental results (Table 4).

Table 4. Experimental results of equilibrium adsorption capacity

C_0, mg/L	C_e, mg/L	q_e	C_e/q_e
1720	1310	2.05	639
1420	1210	1.05	1152
1450	1165	1.37	850
1720	835	2.21	378
1420	720	1.75	411
1450	740	2.11	350

All experimental results were under statistical verification and the following numbers were estimated to plot the adsorption equilibrium presented in Figure 3.

Figure 3 shows the results of the equilibrium adsorption of ammonium ions on zeolite. The experimental data were fitted through the isoterm of Lagmuir, often applied for adsorption equilibrium description for NH_4^+ removal from water solutions (Vassileva, Voikova 2009; Valentukevičienė, Jankauskas, 2004; Huang *et al.* 2010). It can be observed that the adsorption equilibrium will be established when C_e/q_e is in the range between 0.6–0.8. A variety of researchers (Huang *et al.* 2010; Ivanova *et al.* 2010; Rožič *et al.* 2000; Vassileva, Voikova, 2009; Sprinsky *et al.* 2005) describe in their articles the capacity of zeolite to absorb ions from natural water or artificial solutions. The results received in the present study are nearly similar, when compared with such experiments as were carried out with natural zeolite obtained from North-Eastern Rhodope Mountain (Ivanova *et al.* 2010). An artificial solution made from distilled water and NH_4Cl was used in the experiments carried out by E. Ivanova and others. Concentration of ammonium ions reached 800 mg/L and 1 g of zeolite absorbed up to 9 mg ammonium ions. In the present study 1 g of zeolite can remove up to 2.2 mg of ammonium ions only (Table 4). This is due to high pollution of the sludge liquid (concentration of ammonium ions – up to 1750 mg/L) and high TSS concentrations in the fugate (up to 1220 mg/L).

At a second stage of the experiments 0.8 to 1.6 mm sized zeolite particles were tested for the filtration of sludge liquor through zeolite filter media.

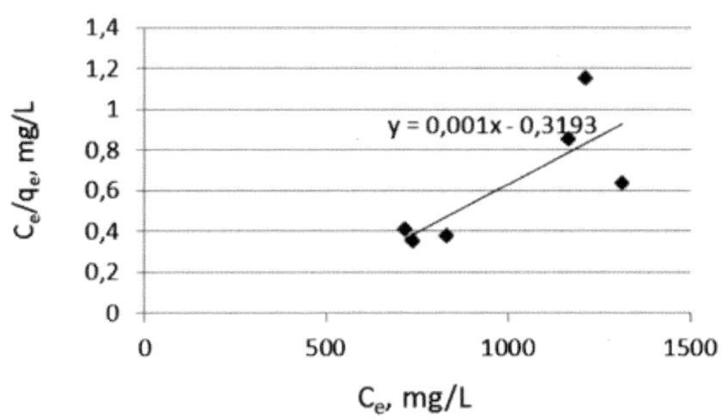

Fig. 3. Langmuir isotherm, when initial NH_4^+ concentrations in the sludge liquid were in range between 1420–1720 mg/L

Once it was completed (the height of filter media 1.5 m; the volume 2.3 L) only 1 L of the sample could be filtered. TSS materials in the sludge liquor soon blocked the filter media. The efficiency of ammonium ions removal was in this case very high, i.e. 99.9%; however, under laboratory conditions further filtering was impossible during the experiment due to a small surface area of the filter (0.045 m), where a layer of sediments appeared and the filter became hydraulically impermeable. Therefore, further research of the filter was carried out by reducing the height of the filter media to 0.5–1 m and choosing 1.6 to 2.5 mm sized zeolite particles. The results of ammonium ion concentration reduction in samples by filtering them through 1 m height filler are presented in Table 5. The primary NH_4^+ concentration in the sludge liquid was 1025 mg/L.

Table 5. Ammonium ions removal, using column test (zeolite particles size 1.6–2.5 mm)

Filtered volume, L	Filtration run, min	Removed NH_4^+ concentration, mg/L	Removal efficiency, %
1	5	937	91
1.5	9	857	84
2	14.5	795	78
2.5	21.5	760	74
3	30	752	73

As can be seen from Table 5, the efficiency of ammonium ion removal from the samples decreased from 91 to 73% after half an hour. Liquor pollution had a considerable influence on that. Having noticed that the filter media can easily be blocked by suspended materials from the sludge liquid, the concentrations of pollutants were later reduced and the samples were left in a sedimentation unit for one day. The results of sedimentation experiment are reflected in Table 6.

Table 6. Experimental results of sedimentation process, initial NH_4^+ concentration in the sludge liquid was 1450 mg/L, TSS –1090 mg/L

Sampling depth, m	*TSS, mg/L	*NH_4^+ concentration, mg/L
0.50	920	1400
1.00	870	1320
1.50	1150	1540

**Note*: Triple samples were analysed, from different sedimentation depth

Table 6 shows that the precipitation of sludge liquid was not efficient because the particles of over digested sludge were floating due to gas released bubbles. Suspended materials in the fugate are particularly small due to thermal sludge decomposition, which is carried out in the Lithuanian sludge treatment plant. When sludge is dewatered after digesting it by using conventional centrifuges – a considerable part of TSS is obtained into the fugate. A part of the disintegrated materials precipitated, due to which the highest concentrations of TSS were obtained at the bottom of the sedimentation unit, lower concentrations were observed on the surface and the minimum in the middle. As the sedimentation experiment was insufficient, new alternatives to reduce the concentration of suspended materials in the fugate more efficiently need to be found.

The volume of 13 L of sludge liquid from the sedimentation unit was filtered through the filler layer of 0.5 m height (0.8–1.6 mm zeolite fraction). The results are presented in Figure 4.

As can be seen from Figure 4, the efficiency of ammonium ion removal from sludge liquor decreases from 65 to 9% in 50 minutes of filtering. During the given period 13 L of sludge liquor were filtered. The speed of filtering was ~10 m/h. In order to achieve better results of ammonium ion removal from the sludge liquor, the period in which the sludge liquid was stored in the filtering media need to be increased. The Figure 5 shows the results of ammonium ion removal from sludge liquor, where a sample of 2 L was filtered several times through the same filter media of 0.5 m height (0.8–1.6 mm zeolite particles size) by using total retention time of 90 min.

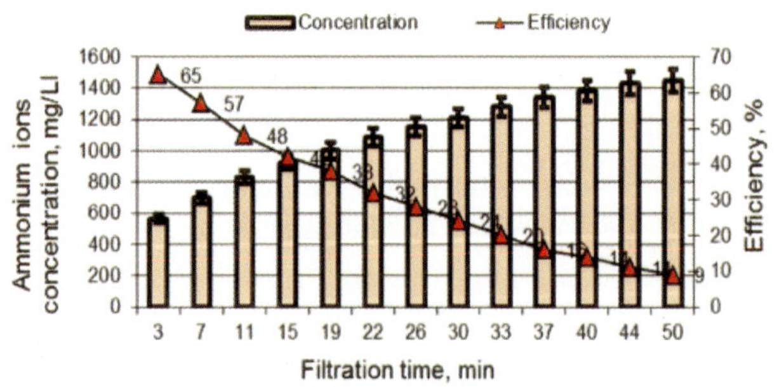

Fig. 4. Experimental results of ammonium ions removal efficiency (filtration run 50 min)

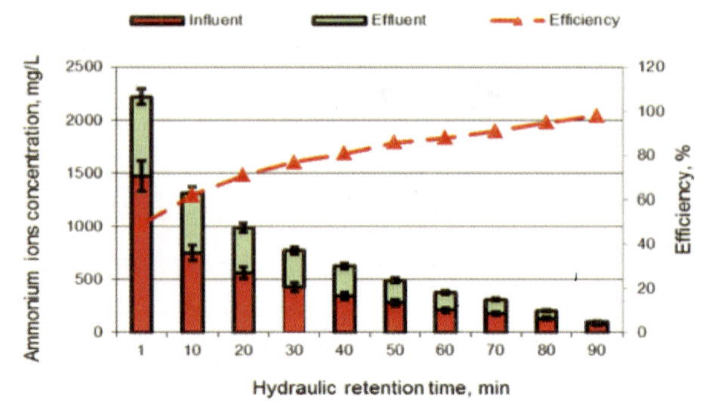

Fig. 5. Experimental results of ammonium ions removal efficiency (filtration run 90 min)

It can be seen that higher efficiency of ammonium ion removal is achieved by prolonging the period in which sludge liquid remained in the filter. Once the sample had remained in the filter media for 90 minutes, the efficiency reached 98%, where in the beginning (1 min retention run) it only reached 49% (Fig. 5). In order to achieve maximum ammonium ions sorption a period of 1.5 h is needed while other parameters of the experiment were remained the same. During the lab-scale experiment it turned out that the filter media of 1 L volume (0.5 m height, 0.8–1.6 mm fraction) would be sufficient to treat 2.5 L sludge liquid, where the primary concentration of ammonium ions were 1470 mg/L.

It was observed that doing filtration of fugate, the zeolite do not aggregates into monolith mass, and nor layering. When using zeolite for fugate treatment, after the ammonium ions sorption, the zeolite can be reutilized. As it is pointed out in scientific literature (Vesela *et al.* 2003; Montemurro *et al.* 2007; Rehaková *et al.* 2004; Beler-Baykal, Allar 2008; Lija *et al.* 2012), zeolite saturated with ammonium ions could be successfully used in agriculture as it is a nitrogen fertilizer.

Results of the elements X-Ray Fluorescence analysis in samples (average values) are presented in Table 7.

As can be seen from Table 7, the almost all average values of heavy metals (As, Pb, Zn, Cu) in samples of fugate are higher or in samples of natural zeolite. Only Cr (10 times) and Hg (1,5 times) in zeolite exceeds the values of fugate. Used zeolite (after sorption) is not contaminated with Rb, Se, Fe. Cd, Ag, Pb, Ni, Co, Cr, V, Sn, Pd, Au, W concentrations are below the level of detection. Results of the obtained elements X-Ray Fluorescence analysis presented in Table 7, can be compared along with maximum allowable concentrations for the wastewater sludge of I-III category (LAND 20-2005). Almost all average values of heavy metals in zeolite fugate samples do not exceed the concentrations that are determined for the I sludge category. The amounts of Cd and Ni were below detection limits and Hg in all samples exceeds the requirements for I sludge category. Fugate and zeolite samples were attributed to the II category sludge which means that sediments are slightly polluted with heavy metals and can be used with restrictions that are specified in Lithuanian environmental normative document (LAND 20-2005).

One of the most common methods for assessing quantifying metal enrichment is to compare identified concentrations with a background level or with a concentration posing serious environmental risks (e. g. maximum concentration defined in legal environmental regulations and guidelines). Amount of Ca, K, S in zeolite samples increased significantly after sorption process. The following elements are known to be essential for plant growth. Calcium, potassium and sulphur normally derived from the soil in the form of inorganic salts. Potassium with nitrate and phosphates in high amount are required to all plants and microorganisms nutrition. Saturated zeolite can be used as excellent fertilizer because the plants can assimilate valuable materials step by step, and those materials can be assimilated from the zeolite only by plants roots (Ahmed *et al.* 2010). Additional and well balanced material transportation to prior- roots zone is going on. Therefore we cannot over-fertilise the soil. Plants roots start dissolving reactions because of absorption, and needed nutrients are accessible. Thereafter the zeolite is saturated with excess materials again (Vesela *et al.* 2003). Nitrogen compounds are not washed out to the ground water and the zeolite particles improve the structure of soil (Montemurro *et al.* 2007). This material fulfils soil with mineral compounds. However, according to (Rehaková *et al.* 2004) such zeolite is not harmful to nature as the sludge of wastewater in the wastewater treatment unit is treated in the thermal hydrolysis phase, where pathogenic bacteria and helminthic eggs are deactivated. Therefore zeolite could be used to

Table 7. Composition of elements in substances

Element	Fugate	Natural zeolite before sorption	Zeolite after sorption
		ppm	
As	5.15	<LOD*	4.55
Hg	2.45	3.62	3.72
Cd	<LOD	<LOD	<LOD
Ba	<LOD	<LOD	<LOD
Sb	<LOD	<LOD	<LOD
Ag	<LOD	<LOD	<LOD
Zr	10.88	96.14	101.4
Sr	135.26	325.32	346.45
Rb	16.53	101.08	92.36
Pb	8.19	<LOD	<LOD
Se	2.42	3.03	2.65
Zn	934.52	31.44	114.62
Cu	164.18	<LOD	16.24
Ni	<LOD	<LOD	<LOD
Co	<LOD	48.64	<LOD
Fe	5149.52	7368.63	6268.46
Mn	96.06	241.72	254.27
Cr	<LOD	10.83	<LOD
V	<LOD	<LOD	<LOD
Ti	2198.14	998.07	1018.78
Sc	495.52	62.82	107.54
Ca	117548.32	15277.6	38713.44
K	40505.78	21749.2	22560.45
S	12055.89	<LOD	674.48
Sn	<LOD	<LOD	<LOD
Pd	<LOD	<LOD	<LOD
Au	<LOD	<LOD	<LOD
W	<LOD	<LOD	<LOD

reduce the use of N, P and K fertilizers on acid soils (Lija *et al.* 2012). The mostly useful fraction to the plants roots is 2–5 mm (Rehaková *et al.* 2004).

In the present study the smallest zeolite fraction (0.8–16 mm) proved to be the most efficient to reduce ammonium ions concentration in the sludge liquor. When used in production, the hydraulic permeability of filter can be decreased, as there is a number of TSS in sludge liquor. During lab-scale experiments, the height of filter media should not exceed 50 cm; however, the same or several filters gradually following one another can be used several times. Having assessed high NH_4^+-N concentrations and a poor precipitation of suspended materials (when the fugate is kept in the sedimentation unit for one day), it can be presumed that sludge under digestion was not fully stabilized in the course of the experiment. The fact that ammonium ions are still generated in sludge liquid after one day of storage shows further decomposition of organic materials.

This fact only proves the need for the reduction of ammonium ions concentration in the fugate before it is returned to the wastewater treatment stage. The present studies are valuable as the concentration of ammonium ions in the fugate can be reduced considerably by means of simple filtration through the zeolite filter media with no chemical reagent employed.

Conclusions

The results showed a high concentration of ammonium ions (1025–1750 mg/L) in the samples of digested sludge liquor collected from the Lithuanian wastewater treatment plant. These results confirm the hypothesis that the dewatering of sludge in centrifuges leads to higher ammonium concentrations in fugate, as predicted in scientific literature. To reduce the NH_4^+-N concentration of fugate the samples were mixed with zeolite fractions and an efficiency of 90% was reached. In the present study 1 g of zeolite can remove up to 2.2 mg of ammonium ions only. This is due to high pollution of the sludge liquid (concentration of ammonium ions – up to 1750 mg/L) and high TSS concentrations in the fugate (up to 1220 mg/L).

The NH_4^+-N concentrations in digested sludge liquor can be reduced up to ~ 98% when the fugate is filtered through the zeolite filler of 0.5 m height (0.8–1.6 mm particles size zeolite) several times.

It was observed that doing filtration of fugate, the zeolite do not aggregates into monolith mass, and nor layering. This sorbent is interfering with the water not only as filtration material, but also like adsorbent and as ion exchange carrying out material. Zeolite saturated with ammonium ions possibly can be used in agriculture as nitrogenous fertilizers. To gain deeper knowledge of this, research can be done to investigate cultivation in practice. Obtained fertilizer grade of zeolite saturated with fugate can also be produced from the high ammonium nitrogen concentration side-streams separated from wastewater flow from sludge digestion processes by sorption on zeolite filter media.

Acknowledgements

We would like to thank the administration of JSC "Vilniaus vandenys" (Vilnius, Lithuania) and colleagues for the possibility to carry out the experiments with fugate that is generated at wastewater treatment plant.

References

Ahmed, O. H.; Braine, C. H; Muhamad, A. M. 2010. Minimizing of ammonia loss through mixing with zeolite and acid sulphate soil, *International Journal of Physical Sciences* 5(14): 2198–2202.

Bao-Shan Xing; Qiong Guo; Guang-Feng Yang; Zheng-Zhe Zhang; Peng Li; Li-Xin Guo; Ren-Cun Jin. 2015. The properties of anaerobic ammonium oxidation (anammox) granules: Roles of ambient temperature, salinity and calcium, *Separation and Purification Technology* 147: 311–318. http://dx.doi.org/10.1016/j.seppur.2015.04.035

Beler-Baykal, B.; Allar, A. D. 2008. Upgrading fertilizer production wastewater effluent quality for ammonium discharges through ion exchange with clinoptilolite, *Environmental Technology* 29(6): 665–672. http://dx.doi.org/10.1080/09593330801987004

Calli, B.; Mertoglu, B.; Inanc, B.; Yenigun, O. 2005. Effects of high free ammonia concentrations on the performances of anaerobic bioreactors, *Process Biochemistry* 40(3–4): 1285–1292. http://dx.doi.org/10.1016/j.procbio.2004.05.008

Cheng, G. V., *et al.* 2011. A study on the ammonia removal mechanisms for the A/O biological nitrogen removal process with the enchancement of the zeolite, *Advanced Materials Research* 322: 52–56. http://dx.doi.org/10.4028/www.scientific.net/AMR.322.52

Constantine, T. 2006. North American experience with centrate treatment technologies for ammonia and nitrogen removal, *Proceedings of the Water Environment Federation 2006*: 5271–5281. http://dx.doi.org/10.2175/193864706783763291

Du, Q.; Liu, S.; Cao, Z.; Wang, Y. 2005. Ammonia removal from aqueous solution using natural Chinese clinoptilolite, *Separation and Purification Technology* 44: 229–234. http://dx.doi.org/10.1016/j.seppur.2004.04.011

Farkas, A.; Rozic, M.; Barbaric-Mikocevic, Z. 2005. Ammonium exchange in leakage waters of waste dumps using natural zeolite from the Krapina region, Croatia, *Journal of Hazardous Materials* 117: 25–33. http://dx.doi.org/10.1016/j.jhazmat.2004.05.035

Jafarpour, M.; Foolad, A.; Mansouri, M.; Nikbakhsh, Z.; Saeedizade, H. 2010. Ammonia removal from nitrogenous industrial waste water using Iranan natural zeolite of Clinoptilolite type, *Engineering and Technology* 70: 939–945.

Gustavsson, D. 2010. Biological sludge liquor treatment at municipal wastewater treatment plants – a review, *Journal of Water Management and Research* 66: 179–192.

Hedstrom, A. 2001. Ion exchange of ammonium in zeolites: a literature review, *Journal of Environmental Engineering* 127(8): 673–681. http://dx.doi.org/10.1061/(ASCE)0733-9372(2001)127:8(673)

Huang, H.; Xiao, B.; Yan, L. 2010. Ammonium removal from aqueous solutions by using natural Chinese (Chende) zeolite as adsorbent, *Journal of Hazardous Materials* 175: 247–252. http://dx.doi.org/10.1016/j.jhazmat.2009.09.156

Inglezakis, V. J. 2005. The concept of "capacity" in zeolite ion-exchange systems, *Journal of Colloid and Interface Science* 281: 68–79. http://dx.doi.org/10.1016/j.jcis.2004.08.082

Ivanova, E.; Karsheva, M.; Koumanova, B. 2010. Adsorption of ammonium ions onto natural zeolite, *Journal of the University of Chemical Technology and Metallurgy* 45(3): 295–302.

Karapinar, N. 2009. Application of natural zeolite for phosphorus and ammonium removal from aqueous solutions, *Journal of Hazardous Materials* 170(2–3): 1186–1191. http://dx.doi.org/10.1016/j.jhazmat.2009.05.094

Kurama, H.; Poetzschke, J.; Haseneder R. 2002. The application of membrane filtration for the removal of ammonium ions from potable water, *Water Research* 36(12): 2905–2909. http://dx.doi.org/10.1016/S0043-1354(01)00531-0

LAND 20–2005. *Water quality. Use of sewage sludge as fertilizer requirements.* Lithuanian Ministry of Environment, 2015, D1–867.

LAND 46–2007. *Water quality. Determination of suspended solids. Filtration through glass fiber filters.* Lithuanian Ministry of Environment, 2007, D1–412.

Lebedynets, M.; Sprinskyy, M.; Sakhnyuk, I.; Zbytniewsky, I.; Golembiewsky, R.; Buszewsky, B. 2004. Adsorption of ammonium ions onto a natural zeolite: Transcarpathian clinoptilolite, *Adsorption Science & Technology* 22(9): 731–741. http://dx.doi.org/10.1260/0263617043026541

Lija, W. B. M.; Ahmed, O. H.; Kasim, S. 2012. Reducing ammonia volatilization from compound fertilizers amended with zeolite, *African Journal of Biotechnology* 11: 13903–13906.

Mažeikienė, A.; Valentukevičienė, M.; Rimeika, M.; Matuzevičius, A. B.; Dauknys, R. 2008. Removal of nitrates and ammonium ions from water using natural sorbent zeolite (clinoptilolite), *Journal of Environmental Engineering and Landscape Management* 16(1): 38–44. http://dx.doi.org/10.3846/1648-6897.2008.16.38-44

Miladinovic, N.; Weatherley, L. R. 2008. Intensification of ammonia removal in a combined ion-exchange and nitrification column, *Chemical Engineering Journal* 135(1–2): 15–24. http://dx.doi.org/10.1016/j.cej.2007.02.030

Monballiu, A.; Desmidt, E.; Ghyselbrecht, K.; De Clippeleir, H.; Van Hulle, S. V. H.; Verstraete, W.; Meesschaert, B. 2013. Enrichment of anaerobic ammonium oxidizing (Anammox) bacteria from OLAND and conventional sludge: features and limitations, *Separation and Purification Technology* 104: 130–137. http://dx.doi.org/10.1016/j.seppur.2012.10.046

Montemurro, F.; Maiorana, M.; Lacertosa, G. 2007. Plant and soil nitrogen indicators and performance of tomato grown at different nitrogen fertilization levels, *Journal of Food, Agriculture & Environment* 5(2): 143–148.

Rahmani, R., *et al.* 2004. Investigation of ammonia removal from polluted waters by Clinoptilolite zeolite, *International Journal of Environmental Science & Technology* 1(2): 125–133. http://dx.doi.org/10.1007/BF03325825

Rehaková, M.; Cuvanová, S.; Dzivak, M.; Rimarb, J.; Gavalovac, Z. 2004. Agricultural and agrochemical uses of the clinoptilolite type, *Current Opinion in Solid State and Material Science* 8(6): 397–404. http://dx.doi.org/10.1016/j.cossms.2005.04.004

Rezaei, M.; Mehrnia, M. R.; Sarrafzadeh, M. H. 2014. Effect of clinoptilolite addition on nutrient removal in a membrane bioreactor, *Desalination and Water Treatment* 54(11): 2920–2927. http://dx.doi.org/10.1080/19443994.2014.913203

Rožič, M.; Cerjan–Stefanovic, Š.; Kurajica, S.; Vančina, V.; Hodžic, E. 2000. Ammoniacal nitrogen removal from water by treatment with clays and zeolites, *Water Research* 34(14): 3675–3681. http://dx.doi.org/10.1016/S0043-1354(00)00113-5

Sakalauskas, A.; Valentukevičienė, M. 2003. Investigation into the influence of natural powered zeolite on drinking water treatment at Druskininkai waterworks III, *Journal of Environmental Engineering and Landscape Management* 11(4): 69–178.

Saltah, K., *et al.* 2007. Removal of ammonium ion from aqueous solution by natural Turkish (Yildizeli) zeolite for environmental quality, *Journal of Hazardous Materials* 141: 258–263. http://dx.doi.org/10.1016/j.jhazmat.2006.06.124

Sileg, M.; Folgar, L.; Kukucka, M. 2010. The groundwater ammonium sorption onto Croatian and Serbian clinoptilolite, *Journal of Hazardous Materials* 178: 572–577. http://dx.doi.org/10.1016/j.jhazmat.2010.01.123

Sprinsky, M.; Lebedynets, M.; Terzyk, A. P.; Kowalczyk, P.; Namiesnik, J.; Buszewsky, B. 2005. Ammonium sorption from aqueous solutions by the natural zeolite Transcarpathian Clinoptilolite studied under dynamic conditions, *Journal of Colloid and Interface Science* 284: 408–415. http://dx.doi.org/10.1016/j.jcis.2004.10.058

Tada, Ch., *et al.* 2005. Effect of natural zeolite on methane production for anaerobic digestion of ammonium rich organic sludge, *Bioresource Technology* 96: 459–464. http://dx.doi.org/10.1016/j.biortech.2004.05.025

Thornton, A.; Pearce, P.; Parsons, S. A. 2007. Ammonium removal from digested sludge liquors using ion exchange, *Water Research* 41: 433–439. http://dx.doi.org/10.1016/j.watres.2006.10.021

Vabolienė, G.; Matuzevičius, A. B.; Valentukevičienė, M. 2007. Effect of nitrogen on phosphate reduction in biological phosphorus removal from wastewater, *Ekologija* 53 (1): 80–88.

Valentukevičienė, M.; Jankauskas, J. 2004. *The invention of using natural adsorbents for the treatment of natural water.* Lithuanian Patent Nr. 5104 B. CO2F 1/28.

Valentukevičienė, M.; Rimeika, M. 2007. Development of a fluidized batch process using natural powdered zeolite (clinoptilolite), *Polish Journal of Environmental Studies* 16(2): 283–288.

Vassileva, P.; Voikova, D. 2009. Investigation on natural and pretreated Bulgaran clinoptilolite for ammonium ions removal from aqueous solutions, *Journal of Hazardous Materials* 170(2–3): 948–957. http://dx.doi.org/10.1016/j.jhazmat.2009.05.062

Vesela, M.; Drdak, M.; Sandara, S. 2003. Relation between free amino acids and the biogenic amines contents in green tomatoes inoculated with Lactobacillus plantarum, *Czech Journal of Food Science* 21: 51–58.

Wang, F.; Ding, Y.; Ge, L.; Ren, H.; Ding, L. 2010. Effect of high-strength ammonia nitrogen acclimation on sludge activity in sequencing batch reactor, *Journal of Environmental Sciences* 22(11): 1683–1688. http://dx.doi.org/10.1016/S1001-0742(09)60306-5

Zekker, I.; Rikmann, E.; Tenno, T.; Kroon, K.; Seiman, A.; Loorits, L.; Fritze, H.; Tuomivirta, T.; Vabamäe, P.; Raudkivi, M.; Mandel, A.; Taavo, T. 2014. Start-up of low-temperature anammox in UASB from mesophilic yeast factory anaerobic tank inoculum, *Environmental Technology* 35(24): 1–12.

Aušra MAŽEIKIENĖ. Dr., Assoc. Prof., Dept of Water Management, Vilnius Gediminas Technical University (VGTU), Lithuania. Doctor of Science (environmental engineering), VGTU, 2005. Author of more than 25 research papers. Research interests: water treatment, ecotechnologies for water management.

Marina VALENTUKEVIČIENĖ. Dr., Assoc. Prof., Dept of Water Management, Vilnius Gediminas Technical University (VGTU), Lithuania. Doctor of Science (environmental engineering), VGTU, 2003. Author of more than 30 research papers. Research interests: water treatment, ecotechnologies for water management.

3

ENVIRONMENTAL POLLUTION BY WASTEWATER FROM BROWN COAL PROCESSING

Arndt Wiessner[a], Jochen A. Müller[a], Peter Kuschk[a], Uwe Kappelmeyer[a], Matthias Kästner[a], Yong-Jun Liu[b], Ulrich Stottmeister[a]

[a]*UFZ-Helmholtz Centre for Environmental Research, Permoserstrasse. 15, D-04318 Leipzig, Germany*
[b]*Key Laboratory of Northwest Water Resource, Ecology and Environment, Ministry of Education, Xi'an University of Architecture & Technology, Xi'an, China*

Abstract. The large scale of the contamination by the former carbo-chemical industry in Germany requires new and often interdisciplinary approaches for performing an economically sustainable remediation. For example, a highly toxic and dark-colored phenolic wastewater from a lignite pyrolysis factory was filled into a former open-cast pit, forming a large wastewater disposal pond. This caused an extensive environmental pollution, calling for an ecologically and economically acceptable strategy for remediation. Laboratory-scale investigations and pilot-scale tests were carried out. The result was the development of a strategy for an implementation of full-scale enhanced *in situ* natural attenuation on the basis of separate habitats in a meromictic pond. Long-term monitoring of the chemical and biological dynamics of the pond demonstrates the metamorphosis of a former highly polluted industrial waste deposition into a nature-integrated ecosystem with reduced danger for the environment, and confirmed the strategy for the chosen remediation management.

Keywords: disposal pond, environmental pollution, lignite pyrolysis, remediation strategy, wastewater.

Reference to this paper should be made as follows: Wiessner, A.; Müller, J. A.; Kuschk, P.; Kappelmeyer, U.; Kästner, M.; Liu, Y.-J.; Stottmeister, U. 2014. Environmental pollution by wastewater from brown coal processing – a remediation case study in Germany, *Journal of Environmental Engineering and Landscape Management* 22(1): 71–83.

Introduction

Extent of brown coal mining and processing in Central Germany

The mining of lignite and the subsequent processing for the chemical industry has a long history in several countries of the world. These activities have produced relics of different quality and magnitude. In contrast to engineering reports in technical bulletins, there are only few scientific publications about the remediation of contaminated sites by this industry in peer-reviewed journals (Stottmeister *et al.* 2009; Kubal *et al.* 2011). Because of the increasing price for oil, carbon/lignite chemistry will probably experience a renaissance (Volkmann, Borner 2012; Zhou *et al.* 2012). This assumption is supported by an increasing number of publications about the processing of lignite for the chemical industry (Tokarska 2009; Ciahotny *et al.* 2011). This article not only summarizes the history of environmental problems caused by this industry using the example of Germany with the intention of providing environmental engineers with an idea of consequences when wastes are not handled in a proper way, but also includes suggestions of how to manage these problems.

From the middle of the nineteenth century until the reunification of Germany in 1990 the central regions of Germany were characterized by extensive brown coal mining. The areas particularly around Böhlen, Espenhain, Altenburg, Zeitz, Bitterfeld, Wolfen, and Leuna-Buna near the cities of Leipzig and Halle were dominated by large-scale open-cast mining (Stottmeister *et al.* 1997).

Massive exhausting (subrosion of anhydrite) during the Tertiary Period caused the formation of mighty basins within the marine Zechstein and Sandstone from the early Triassic epoch. In the Eocene and Oligocene epochs, these basins were filled with brown coal, forming altogether four seams of different mightiness and quality of the lignite (Eissmann 1994). The oldest seam in the south of Leipzig was

Corresponding author: Peter Kuschk
E-mail: peter.kuschk@ufz.de

raised by sedimentation of a hydrodynamic bog and the lignite was found to have a high content of bitumen suitable for coking and applications in chemical industries. Newer seams are characterized by a high load of xylite suitable for energy production.

Brown coal mining started in the middle of the nineteenth century near the cities of Weißenfels and Zeitz. From the middle of the 1930s, on the basis of lignite carbon-chemical industries were growing in Central Germany. The German Democratic Republic (GDR) became the world's largest producer of lignite due to a lack of alternative raw materials for the production of energy and chemicals (Stottmeister *et al.* 1997, 2009; Bilkenroth, Snyder 1998). In 1985, approximately 310 million tons or 28% of the worldwide total brown coal production was mined in 33 open-cast mines in the GDR (Stottmeister *et al.* 1997). Nearly 70% of the lignite was directly burned for energy production (electrical power) and 30% was used for the production of briquettes (Stottmeister *et al.* 2002). About 92% of the briquettes were used for house heating and 8% for the carbon-chemical industries (Stottmeister *et al.* 2002). The main processes of the carbo-chemical industries were carbonization, coking, gasification and hydrogenation for producing coke, synthesis gas, fuel, tar, oil, wax and several basic chemicals, such as benzene, ammonia and sulphuric acid (Ringpfeil *et al.* 1988).

The demand for brown coal drastically decreased in the years after the reunification of Germany. Nowadays lignite is of subordinated importance but is still used, to a lesser extent, by modernized or modern power stations for the production of electrical power. The economically inefficient carbo-chemical industries of the former GDR were completely discontinued in the Central German territories.

1. Environmental consequences of long-term brown coal mining and processing in the Central German territories

1.1. Extent of environmental pollution

The extensive brown coal mining in the Central German regions caused enormous environmental damages, such as landscape destruction and pollution, exhaust gas pollution, as well as pollution of rivers and lakes from wastewaters and solid wastes (Stottmeister *et al.* 2002).

Most of the technical equipment installed in the 1920s and the 1930s in the industrial centres went out of date years ago, but was still used until their closure in the early 1990s. In some cases, provisional reconstruction of massive damages during World War II remained the state of the art. Dangerous wastes from industrial processing, such as tars, wastewaters from lignite pyrolyses, and so on, were usually released untreated and often uncontrolled into the environment (Stottmeister *et al.* 1997, 2002).

The whole territory of the lignite deposits of Central Germany was characterized by many open-cast mines. For example, an area with a length of approximately 90 km north/south and 30 km east/west enclosing the city of Leipzig was extensively used for brown coal mining. Nearly one-third of the territory south of Leipzig was considered to be a so-called *post-mining area* (Stottmeister *et al.* 1997). Former open-cast mines and other areas transformed by extensive mining were normally used for the disposal of all kinds of wastes from mining, industries and households, resulting in an imminently dangerous contamination of groundwater and/or air of the related districts (Stottmeister *et al.* 1999).

Finally, after the termination of extensive mining and processing in the early 1990s, a total of 570 suspicious areas with dangerous wastes from the past were found in these regions. Exactly 327 of them were remediated by 2002, and for 128 no treatment was considered necessary. However, in 2002, 115 suspicious areas were found to be still untreated and 30 of them were regarded dangerous and in need of urgent remediation (Stottmeister *et al.* 2002).

In general, the single depots of dangerous wastes of the past typically have unique site characteristics and require defined remediation strategies related to the prevailing pollution. Irrespective of the particular site characteristics, focus is on prevention of gas emissions into the air and protection of the water resources, and so on.

1.2. Environmental pollution due to wastewater from lignite pyrolysis

The extensive lignite pyrolysis, such as coking, gasification, hydrogenation or low-temperature carbonization, had led to a widespread generation of highly polluted wastewater. This wastewater resulted from the heating of the brown coal, whereby large amounts of water steam originated from the thermal reactions and the release of the bounded water from within the coal. As an example, the mass balance of coking briquettes for producing gas, oil and formed coke is presented in Table 1.

In this way, nearly one-fourth of the initial mass was transformed into wastewater. The water steam

Table 1. Mass balance of coking brown coal briquettes (Stottmeister 2008)

Coking of 1000 kg of brown coal briquettes resulted in the formation of:	
Coke	431.4 kg
Gas	262.6 kg
Tar	58.1 kg
Reaction water	91.9 kg
Inherent water	156.0 kg

produced during these processes contained low-molecular phenolics and fatty acids, as well as hydrogen sulphide and ammonia. The composition and the amounts typical for the wastewater generated from brown coal of the Central German mining territories are given in Table 2.

Among the polyhydric phenols (polyphenolic compounds) were detected catechol, resorcinol, hydrochinone and pyrogallol and their derivatives (Stottmeister 2008). Usually, the condensed waters from the generated steam were of light red color and the pH was in the range of 7.5 to 10 (Ringpfeil *et al.* 1988). Caused by the air oxygen, the color of the condensate turned to dark brown over time due to autoxidation of aromatic compounds, as seen with phenolic derivatives forming polymeric molecules (Kuschk *et al.* 2010). These generated artificial fulvic/humic substances are known to be highly resistant to microbial degradation (Wießner *et al.* 1994; Kopinke *et al.* 1995a, b; Pörschmann *et al.* 1998a). The polymerization by autoxidation in such wastewater also involves nitrogen and sulphur compounds, but does not involve any aliphatic fatty acids (Stottmeister 2008). In this way, the long-term disposal of such wastewater in abandoned open-cast mines resulted in the formation of colloid structures and finally in suspended solid particles formed by the aggregation of the humic macromolecules (Wießner *et al.* 1993; Stottmeister 2008). The reactivity of the dissolved and colloidal or solid suspended humic substances in the deposited wastewater is partly comparable to soils in respect to the incorporation of contaminants, masking of smaller molecules, and so on (Pörschmann *et al.* 1996, 1997, 1998a, b; Mackenzie *et al.* 2002). The disappearance of low-molecular fatty acids and the unchanged high ammonia concentrations were usually observed in wastewater from lignite pyrolysis being stored in derelict open-cast-mines for a long time. Importantly, the toxicity of the water probably reduced the long-term efficiency of microbial degradation of contaminants (Wießner *et al.* 1993; Stottmeister *et al.* 1997).

However, the insufficiently treated (Table 3) and, in occasional cases, even untreated wastewater was released into the environment such as rivers, lakes, unused fallow land, and former open-cast mines for disposal or seepage (Stottmeister *et al.* 1999; Eccarius *et al.* 2001). In some cases, the wastewater was pumped into the deep layers of the Upper Permian Zechstein (Eccarius 1998). The long-term input of the wastewater into abandoned open-cast mines resulted in the formation of large ponds with a significant potential environmental danger particularly for groundwater and the human population of the surrounding areas (Eccarius 1998).

1.3. Perspectives of large-scale environmental problems and sustainable development

The importance of coal mining and processing is still high nowadays due to the drastically increasing needs of energy and raw materials worldwide (Biswas *et al.* 2010; Thiruvenkatachari *et al.* 2011; Zhang *et al.* 2011). Particularly in countries such as China, India, Brazil, and Australia with a rapidly accelerating energy demand and enhanced coal mining/industries, the related environmental problems are still acute and have sparked a search for new solutions regarding wastewater problems (Domazetis *et al.* 2010; Huang *et al.* 2010). Those novel approaches target environmental issues associated with classical coal processing such as combustion, coking, and gasification, as well as new coal applications, such as pyrolysis of sludges and extraction of rare elements (Folgueras, Diaz 2010; Zhang *et al.* 2011; Wu *et al.* 2012;). Similar treatment technologies may be applicable to wastewater problems caused by hydrothermal dewatering of peat, wet air oxidation of black liquor, petroleum refineries and other industries (Garg *et al.* 2008; Tomei, Annesini 2008; Mursito *et al.* 2010).

Moreover, increasing traffic worldwide causes large-scale environmental problems (Baltrenas, Kazlauskiene 2009; Valentukeviciene, Ignatavicius 2011). In general, all of these environmental waste problems may concern grounds and diverse water bodies such as groundwater, rivers, lakes and stored wastewater inside abandoned coal mines, ponds and lakes and research

Table 2. Composition of phenolic waters from brown coal carbonization, gasification, and tar hydrogenation (Ringpfeil *et al.* 1988)

	Carbonization					
	Gas liquor (g/l)	Liquor (g/l)	Tar distillation (g/l)	Washing oil distillation water (g/l)	Gasoline distillation water (g/l)	Phenosolvan weak gas liquor (g/l)
Monohydric phenols	4.2	9.9	18.7	2.0	4.7	
Polyhydric phenols	4.6	8.3	4.0	7.5	6.5	2.2
Fatty acids	2.0	9.9	3.8	3.5	4.4	5.6
H_2S	0.3	0.5	0.2	8.5	6.0	
NH_3	4.6	5.1	1.8	19.8	13.5	4.9

Table 3. Efficiency of the treatment of wastewater from brown coal processing using a bio-tank reactor (Ringpfeil *et al.* 1988)

	Inlet concentration (mg/l)	Degradation efficiency (%)
BOD_5	2000–12,000	85
NH_4^+–N	500–3.000	35
Volatile phenols	50–600	99
Volatile fatty acids	1000–2,000	90
Formaldehyde	1000–3,000	45

for sustainable ecological development is in progress and has to be intensified (Pathiratne *et al.* 2007; Baltrenas, Kazlauskiene 2009; Xu *et al.* 2010; Qi *et al.* 2011; Valentukeviciene, Ignatavicius 2011). Research and technological expertise as documented in this paper may be helpful to find solutions for the remediation of environmental pollution by coal and related industries.

2. Large-scale remediation of a phenolic wastewater disposal pond – a case study

2.1. Characterization of the wastewater pond

The so-called "phenol lake" Vollert-Süd was a former open-cast lignite mine in which highly polluted wastewater was discharged from a near low-temperature coking plant between 1950 and 1968. The disposal pond is located in the state of Saxony-Anhalt in Germany (N51°06′15.8″, E12°03′28.0″, WGS84). It has an oval surface shape, a surface area of nearly 9 ha, a maximum depth of 27 m and contains approximately 2 million m^3 water (Wießner *et al.* 1993; Kuschk *et al.* 1994; Stottmeister *et al.* 1998, 1999). The geometrical data of the pond were measured by echograph (Table 4).

The sediment layer at the bottom of the pond was measured to be 0.1 m thick at the edges and had a maximum of 1.2 m in the centre. The transparency of the deeply dark brownish colored water, which was free of dissolved oxygen, was only about 3 cm (Wießner *et al.* 1993; Stottmeister *et al.* 2002).

Table 4. Relation between the depth and the resulting theoretical surface area of the disposal pond (Stottmeister *et al.* 2009)

Depth (m)	Theoretical surface area (%)
0	100
10	40
15	22
20	22
24	11
25	5

Phenolic compounds and ammonium were found to be the highest concentrated contaminants of the water dependent on the depth of the pond, as shown in Figure 1.

The phenol, ammonium and COD concentrations increased up to a depth of approximately 15 m before reaching a constantly high level in the lower layers of the water body. The composition of the specific phenolic loads, *n*-alkanes and further alcohols were analyzed in the water from a depth of 24 m. The results are listed in Table 5.

Altogether, more than 100 different organic compounds were detected using thermoanalytical methods. In this way, several polycyclic aromatic hydrocarbons (PAHs) were determined in the water body and the sediment of the pond. For instance, the naphthalene concentration in the water amounted to 0.13 μg/l on the surface and 210 μg/l at a depth of 24 m. In the sediment at the centre of the disposal pond, the concentration reached even 30 mg/kg (Wießner *et al.* 1993).

Water from the surface layer and at a depth of 24 m was investigated by micro- and ultrafiltration in relation to the dissolved organic carbon (DOC) to evaluate the molecular size distribution of the organic compounds (Fig. 2; Table 5).

The total organic load of the deep water was found to be higher compared to the surface water. Approximately one half of the total DOC-load of this deep water was in the low-molecular range of <0.5 kDa and the remainder was almost similarly distributed in the higher molecular, colloidal or particular ranges. In contrast, in the surface water, the organics were similarly distributed involving all ranges and only 16% of the total amount was found to be in the low-molecular range of <0.5 kDa. These results reveal the efficiency of auto-oxidative polymerization of lower molecules up to certain ranges forming artificial fulvic/humic compounds (Pörschmann, Stottmeister 1993; Wießner *et al.* 1994; Kopinke *et al.* 1995a, b; Stottmeister 2008). Furthermore, the removal of the organic load by physical/chemical and biological processes

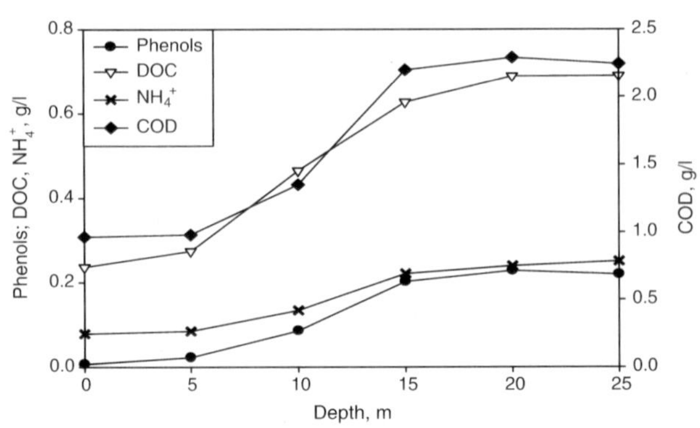

Fig. 1. Analytical data of the disposal pond water depending on the depth before remediation (Wießner *et al.* 1993)

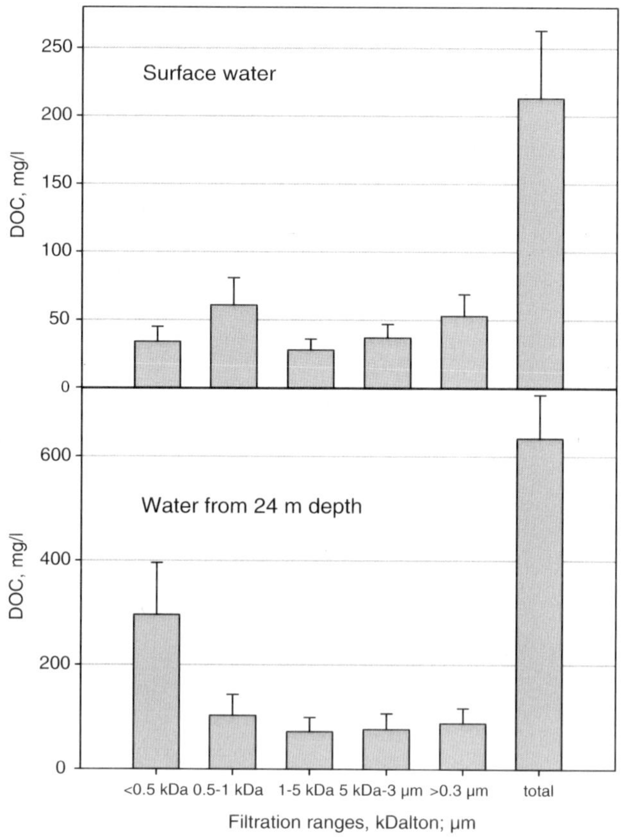

Fig. 2. Molecular size distribution of the DOC in the wastewater disposal pond "Vollert-Süd" before remediation

within the upper layer of the disposal pond (by auto-oxidation, sedimentation, degradation) was confirmed. Toxicity of water from different depths and for all filtration ranges was estimated using the Photobacterium-test (Wießner *et al.* 1993; Stottmeister *et al.* 1997). Accordingly, the surface water of the pond was found to be fivefold more toxic than the water contaminated with pure phenol of a similar DOC concentration. The highest toxicity was estimated for the fraction of <0.5 kDa of the surface water. The results indicate a comparably higher toxic efficiency of alkylphenols and/or higher molecular auto-oxidation products of phenolic compounds. Nevertheless, high concentrations of microorganisms were detected at all depths of the water body. The mean numbers of living bacterial cells per ml were found to range from 4×10^4 in the water at a depth of 24 m to 2×10^6 in the surface water (Stottmeister *et al.* 2002). Apparently, the relevant microorganisms were well adapted to these conditions.

Table 5. Analytically detectable main monomeric compounds of the water of the disposal pond at a depth of 24 m (Wießner *et al.* 1993)

Group	Compound	Concentration (mg/l)
Phenols	Phenol	135
	m-Cresol	22
	p-Cresol	57
	Sum of C_2-phenols	53
	Sum of C_3-phenols	20
	Sum of C_4-phenols	7
n-Alkanes	$C_{10}\ldots\ldots C_{42}$	210

The removal of organics from the upper layer of the disposal pond should be facilitated by atmospheric oxygen. Consequently, the seasonal dynamics of the water body was investigated, and the disposal pond was found to behave as a meromictic water body with seasonal changes in the water density, the stratification, and the mixing due to wind and waves influenced by the specific geometry of the pond (Table 4). The water body below a depth of 8 m was found to be not directly affected by perturbations of the surface zone (Stottmeister *et al.* 2002; Stottmeister 2008). The long-term efficiency of auto-oxidation, polymerization, sedimentation and biodegradation caused the resulting concentration decrease of the organics inside the upper water layer. Additionally, the immobilized low-molecular contaminants could be transferred into the deeper layer by precipitated organic particles. In this way, a natural sub-hydric deposition in the ground layer of the pond was formed. A forecast to 2030 showed no direct danger to the groundwater by the lake water. The phenol plume was unlikely to exceed a length of 250 m (Eccarius *et al.* 2001).

However, the results of the physicochemical analysis of the deposed water and the emissions of volatile phenolic compounds and hydrogen sulphide indicated a high potential of danger to the health of humans in nearby settlements, particularly in summer time. It was concluded that remediation was urgently necessary. At the time, no experience was available for treatment of a wastewater pond of this dimension and contaminant composition. Consequently, further investigations were necessary to evaluate the physicochemical and biological treatment possibilities in order to design and implement an efficient remediation strategy.

2.2. Biological degradability and physicochemical removal of contaminants

The aerobic and anaerobic microbial degradability of wastewater components were investigated in several laboratory-scale experiments (Wießner *et al.* 1994). Water from a depth of 24 m inoculated with different aerobically enriched bacterial cultures was treated in continuous mode in a laboratory-scale aerobic bioreactor (Fig. 3).

Firstly, the results suggested the potential for the aerobic microbial degradability of the organic load. Secondly, the extent of ammonium oxidation was estimated with nearly complete nitrification using related inoculates. Specifically, 50–70% of the totally

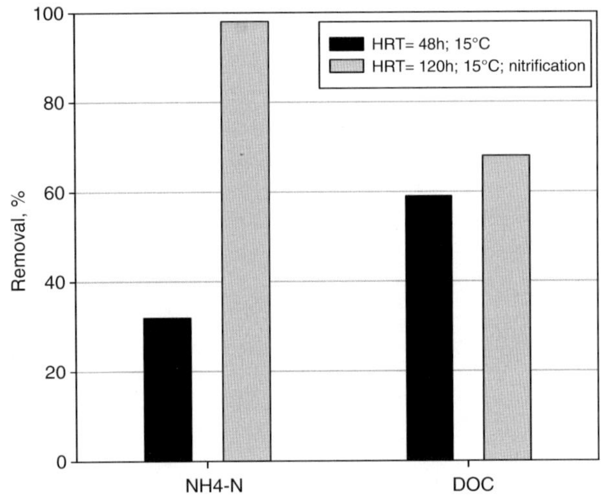

Fig. 3. Aerobic microbial degradability of water contaminants from a depth of 24 m of the wastewater disposal pond in a fermenter of continuous mode in dependence on the hydraulic retention time (HRT) at 15 °C

removed DOC was found to be low-molecular compounds (<0.5 kDa), 8–13% were high-molecular compounds and colloid immobilized by the biomass, and 2–8% could be determined as transformed from the low molecular range to the range of predominantly <0.5–5 kDa. Simultaneously, the water was decolorized by 30–60% as measured as changes in light absorption at a wave length of 436 nm.

For querying the potential toxicity of the contaminants to anaerobic degradation, the influence of the deep water on the efficiency of an enrichment culture of acetoclastic methanogenic bacteria was measured (Fig. 4).

A nearly total inhibition of methanogenesis was observed when untreated deep water was used. In contrast, the removal of the high-molecular parts of the organics by physicochemical pre-treatment methods, such as H_2O_2-oxidation, flocculation/precipitation and ultrafiltration, enabled a highly efficient methanogenesis by the enrichment culture. These results were in accordance with previous findings investigating the toxicity by using the Photobacterium test indicating comparably high toxicity of high molecular autooxidation products of phenolic compounds. The most efficient transformation of organics by the methanogenic enrichment culture was found for deep water pretreated by ultrafiltration and flocculation/precipitation (Fig. 5). Accordingly, about 60% to more than 80% of the DOC were removed and transformed into methane. Additionally, all physicochemical pre-treatments improved the efficiency of decolorization of the deep brownish colored water (Table 6).

The main findings from these pre-investigations can be outlined as follows:

(1) The removal of high-molecular-weight, colloidal, and particulate organic contaminants was necessary for an enhanced biological degradation.
(2) High-molecular-weight, colloidal, and particulate organics can be removed efficiently by using physicochemical methods and
(3) low-molecular-weight organics were efficiently biodegraded using aerobic and anaerobic bacteria.

In addition, the efficiency of the wetland technology to treat the water of the disposal pond was evaluated. A pilot-scale subsurface horizontal-flow constructed wetland with an area of 125 m^2 was established on the shore of the wastewater disposal

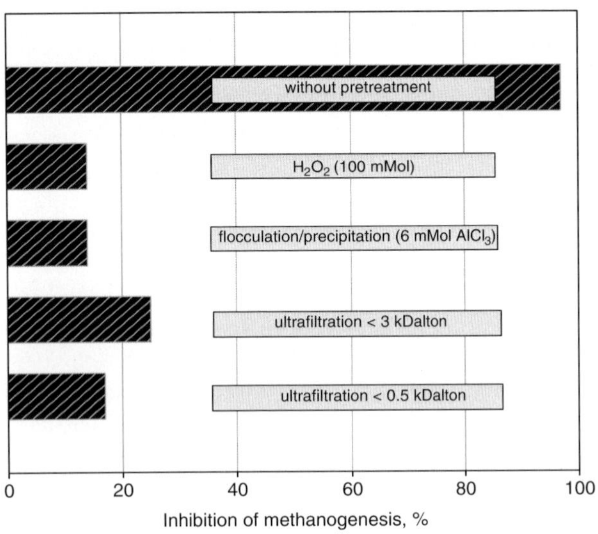

Fig. 4. Relative inhibition of methane formation of an acetoclastic methanogenic enrichment culture after different pre-treatments of water from a depth of 24 m of the wastewater disposal pond

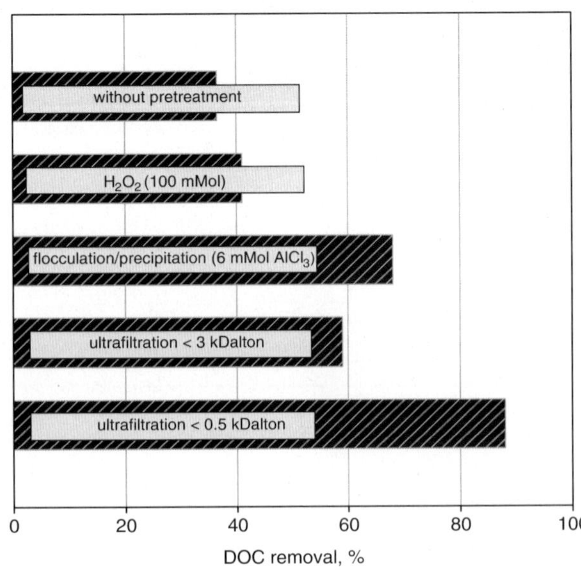

Fig. 5. The degradability of DOC of different pre-treated water from a depth of 24 m of the wastewater disposal pond in a methanogenic enrichment culture

Table 6. Decolorization efficiencies of several physicochemical treatment methods of water from a depth of 24 m of the wastewater disposal pond (Wießner *et al.* 1994)

Method	Decrease in light absorption at a wave length of 436 nm (%)
Ultrafiltration (<3 kD)	91.7
Ultrafiltration (<0.5 kD)	98.2
Oxidation with H_2O_2 (3.5%)	93.3
Flocculation/precipitation with $AlCl_3$	99.2

pond (Wießner *et al.* 1999; Kuschk *et al.* 2003). The constructed wetland was operated continuously for several years and its efficiencies for the removal of contaminants were determined.

The monitoring results suggest a long-term stable removal of nitrogen at high levels. Concerning ammonium-nitrogen, mean removal rates in the range of 0.51 to 0.59 g/m^2 d resulted in a 35–52% eliminating efficiency (Kuschk *et al.* 2003). Simultaneously, DOC was found to be removed by up to 47% (Wießner *et al.* 1999).

These results demonstrated that large-scale removal of ammonium-nitrogen in a subsurface horizontal-flow constructed wetland would be feasible. A constructed wetland with an area of 0.65 ha was considered necessary for continuous treatment of the water from the upper layer of the pond (Wießner *et al.* 1999). However, lack of suitable land close to the disposal pond prevented the construction of a sufficiently large treatment wetland.

2.3. Remediation strategy and in situ pilot-scale investigations

A particular remediation strategy for the disposal pond ensuring maximum treatment success with minimum expenditure and energy requirements had to be developed. Taking into account the depth distribution of the contaminants, the hydrodynamics of the pond, the results of the physicochemical and biological investigations and the land required for a treatment wetland, an *in situ* remediation strategy incorporating the principle of enhanced natural attenuation was developed. All stakeholders involved had to agree to this strategy.

The aim of the long-term strategy was to transform this meromictic disposal pond into a safe ecosystem containing the following two separate compartments:

(1) A subhydric deposition of contaminants in the undisturbed zone of the water body below a depth of approximately 8–10 m and
(2) a hydrodynamic water body in the depth of approximately 0–8 m free of high-molecular-weight, colloidal, and particulate organics to be cleaned continuously over years by biological and chemical removal of the low-molecular-weight contaminants (mainly organics and ammonium).

Firstly, it was concluded that a short-term initial treatment was necessary to facilitate the long-term strategy as follows:

(1) The physicochemical treatment of the water body to remove high-molecular-weight, colloidal, and particulate organics, and
(2) the initiation of microbiological activity to achieve long-term efficiency.

Secondly, flocculation/sedimentation was selected to be the most advantageous physicochemical method to remove the high-molecular-weight and aggregated organics considering all technological (amounts of treatment chemicals, procedures, efficiency) as well as economic (specific treatment costs) aspects of the decision. In particular, the following treatment steps were decided to be necessary:

(1) Flocculation using iron(III)-chloride at pH 4–5, sedimentation of most of the high-molecular-weight organics into the bottom layer of the pond and
(2) the dosage of lime solution for neutralization of the pond water to pH 7–7.5 and the dosage of phosphate for nutrition, in order to enhance the activity of the inherent microbiota.

With the intention to test the efficiencies of *in situ* flocculation and sedimentation, the long-term stability of the resulting state and the influences of seasonal changes of water temperature and wind, a pilot-scale experiment was necessary prior to starting full-scale treatment of the disposal pond. An enclosure of 33 m in diameter and 25.5 m in depth (with a volume of about 20,000 m^3) was placed at the deepest location almost in the centre of the disposal pond (Stottmeister 2008; Stottmeister *et al.* 2009). This artificial basin, which was completely separated from the surrounding water body, was filled with pond water from a depth of about 10 m. Defined amounts of $FeCl_3$ solution and lime suspension were added in succession and the main parameters such as DOC, transparency and water-steam-volatile phenols were observed over a total period of 700 days (Stottmeister *et al.* 2009). By this means, the DOC immediately decreased by about 50% and the pH was found to be adjusted to a value of 5 after introducing a dosage of the $FeCl_3$ solution (Fig. 6). Simultaneously, the transparency increased

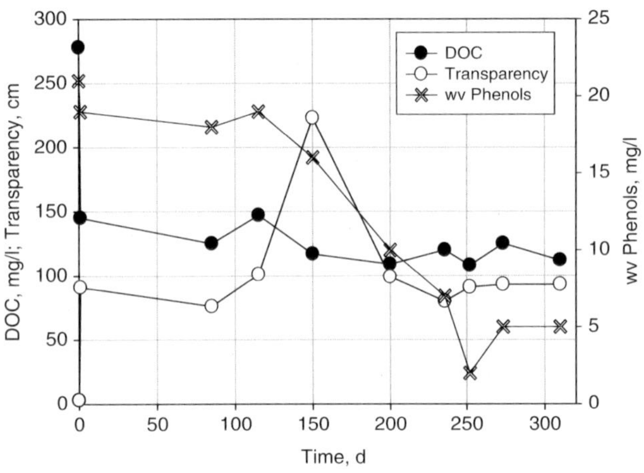

Fig. 6. Dynamics of DOC, watersteam-volatile (wv) phenols and transparency inside the pilot-scale *in situ* enclosure after flocculation on day zero (14–15 December 1994) and neutralization on days 80/81 (adapted from Stottmeister *et al.* 2009)

from the initial 3 cm to about 80 cm caused by the settling of the generated humate flocs.

During the following acidic phase of about three months winter time, an independent secondary flocculation was observed. On days 80 to 81 after flocculation, lime suspension was dosed to neutralize the water body at a pH of 7.5 (Stottmeister *et al.* 2009). Subsequently, the steam-volatile phenols decreased to a concentration of nearly 2 mg/l (from initially 21.5 mg/l) simultaneously with the seasonal increase in water temperature. The transparency improved up to 2 m, and in the upper layer of the enclosure, concentrations of about 1 mg/l of dissolved oxygen were measured. Furthermore, microbial degradation of several organic compounds such as phenols, cresols, xylenols and ethylphenols increased significantly only 5 months after the removal of the high-weight molecules, colloids and particulates. During the following period of approximately 450 days, the DOC decreased slowly but steadily, and the availability of low-molecular-weight aromatic compounds, ammonium and oxygen was found to have enhanced the intensity of microbiological re-colonization after the initial decrease in bacterial numbers as a consequence of the iron (III) flocculation (Becker 1999; Becker *et al.* 1997, 1998; Becker, Stottmeister 1998; Stottmeister *et al.* 2002). Microbial activities were further enhanced by counteracting iron phosphate precipitation with additions of phosphoric acid (75%, 1:200 dilution) to raise the mean phosphate concentration to about 1.4 mg/l inside the enclosure (Stottmeister *et al.* 2002).

The results and the experience of the enclosure experiment confirmed the treatment strategy in principal and encouraged the full-scale remediation of the disposal pond.

2.4. Large-scale in situ treatment of the disposal pond and long-term remediation efficiency

Based on the results of the laboratory-scale and enclosure experiments, the necessary concentrations and amounts of iron(III)-chloride, lime suspension and phosphoric acid for full-scale treatment of the disposal pond were determined, and operations started in October 1996 (Stottmeister *et al.* 2009). The acidic $FeCl_3$ solution (3200 m^3 40% $FeCl_3$) was injected layer by layer into the disposal pond within a period of about 9 weeks using an engine-driven raft along a grid system (Stottmeister 2008; Stottmeister *et al.* 2002, 2009). The neutralization with 2200 m^3 of 20% calcium carbonate suspension started in late April 1997 and was terminated at the end of August 1997. Phosphoric acid (three times 0.8 m^3 0.75% phosphoric acid) was added in April and August 1998 and once more in May of 1999. The first visible success of the remediation process was the increased transparency ranging from 60 to 200 cm depending on the season. At this treatment stage, the calculated 300 tons of macro-molecular humic-like organic material were precipitated. Simultaneously, large amounts of low-molecular-weight contaminants from the upper parts of the water body were removed by adsorption to the high-weight molecules and the aggregated particles. The characteristic bad odor disappeared entirely, and the pond had attracted water birds.

The short-term and long-term efficiency of the remediation were evaluated by analyzing the characteristic contaminants and key parameters such as DOC, watersteam-volatile phenols and ammonium for different water depths (Stottmeister 2008; Stottmeister *et al.* 2009). The results of the investigations are presented in Figure 7.

Acidic flocculation and neutralization immediately affected all of the analyzed parameters. Accordingly, the DOC decreased very efficiently in all depths of the disposal pond, and the watersteam-volatile phenols and ammonium in the zones deeper than 5 m were found to be significantly removed up to the lower levels of the surface-near layers, as shown in Figure 6. During the next two years, the DOC slowly continued to decrease, the watersteam-volatile phenols completely disappeared from the water body, and the concentration of ammonium was found to have stabilized in the range of 80–120 mg/l at a depth below 10 m and to a level of about 50 mg/l at a depth of 5 m. After these treatment steps, the targeted long-term stability of the meromictic stratification of the disposal pond was proven, most recently in 2008, 12 years after iron(III) flocculation. The concentrations of DOC and ammonium in the upper part of the disposal pond had decreased to levels of about 30 and 37 mg/l, respectively.

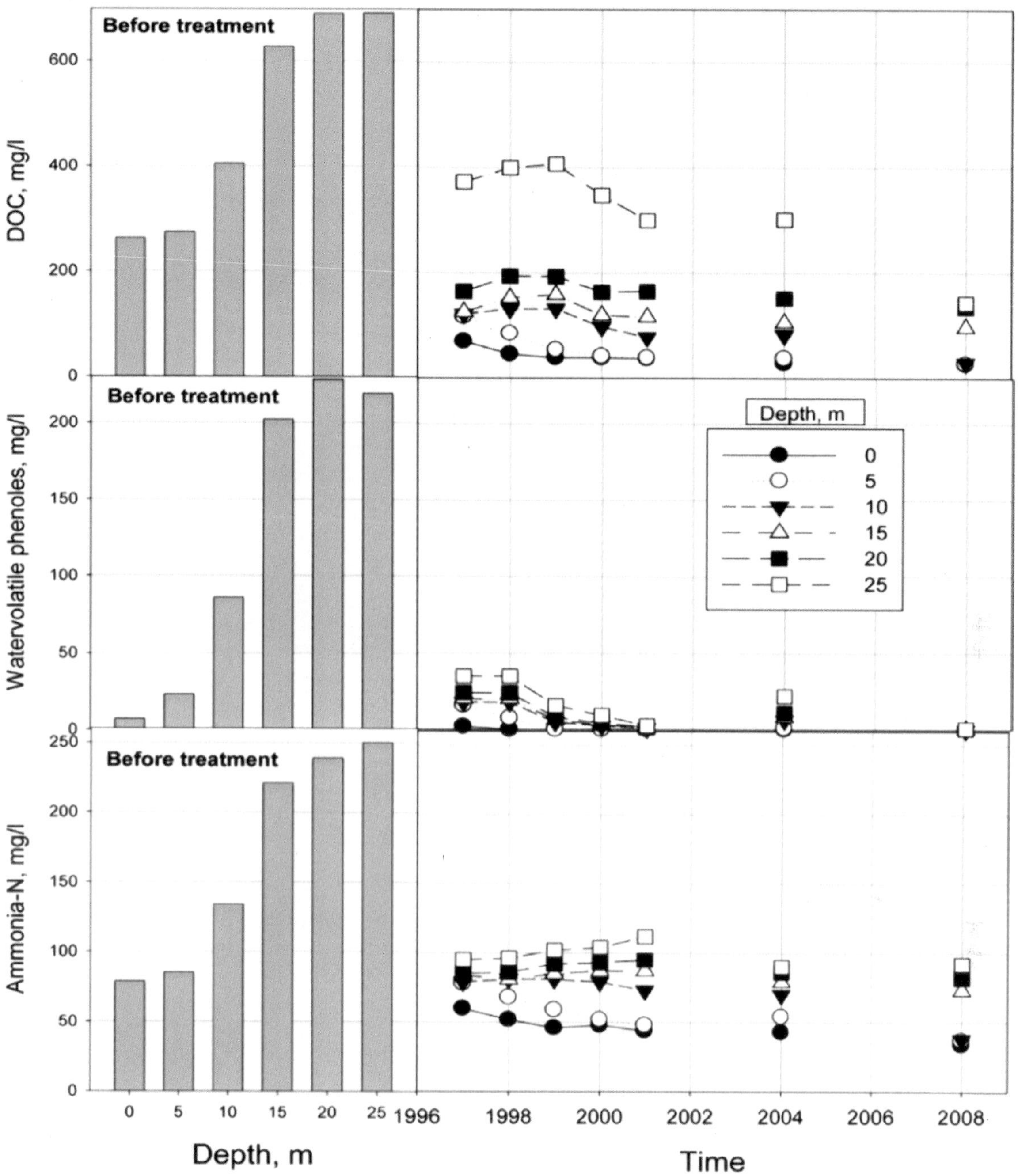

Fig. 7. DOC, watersteam-volatile phenols and ammonia-N before remediation and long-term monitoring of the parameters during and after the *in-situ* remediation of the disposal pond

In principle, a further long-term decrease of the contaminant concentrations can be postulated.

The dynamics of the different biological systems inside the pond were monitored during treatment and long-term remediation in addition to the physical-chemical analyses (Stottmeister *et al.* 2002). After neutralization, an enhanced growth of algae was observed. The density of bacteria and colorless flagellates increased as well, and the first multicellular organisms such as wheel animals (rotifers) were identified. Furthermore, the larvae of the mosquitoes *Culex* sp. and *Chironomidae* sp. were found. The observed dominance of bacteria and heterotrophic flagellates is typical for humic-rich lakes and lakes highly loaded with organics from external sources (Stottmeister *et al.* 2002). During the period until the dosage of the phosphoric acid, the low microbial activity to remove pollutants was caused by phosphorous limitation. Dissolved oxygen could be measured to a depth of 2 m and the concentrations increased over time due to algae metabolisms and input from the air. However, because of the estimated lack of available phosphate (>3 µg/l), a rich plankton community did not develop despite oxygen availability and nutrition by bacteria and flagellates.

The dosage of phosphoric acid finally enhanced the growth of algae. In spring 1998, algae biomass was found to double daily from an initial 0.01 mm^3/l before the phosphate dosage and finally to 77 mm^3/l only one month later (Stottmeister *et al.* 2002). The oxygen concentration increased and the contaminant removal processes continued mainly by plankton metabolism and sorption.

The ongoing removal of pollutants during the next three years affected a nutrient decrease and the settlement of zooplankton such as ciliates (*Halteria grandinella*), wheel animals (*Brachionus calyciflorus f. anuraeiformis*) and cladocerans (*Moina macrocopa* and *Daphnia pulex*) (Stottmeister *et al.* 2002). The algae concentrations decreased simultaneously and a nearly fully developed food chain (except fish), including all trophic levels, was found to be established (Stottmeister *et al.* 2002). Finally, highly sensitive organisms such as the larvae of the top predator phantom midges (*Chaoboridae*) were observed in high densities, and algae almost disappeared completely.

Conclusions

1. Wastes from the carbo-chemical industry in Central Germany have been deposed off in open-cast mines or former industrial areas for many decades. The large scale of the contamination and the chemical composition of the wastes urgently required new and often case-specific methods for an economically sustainable remediation.
2. A highly toxic and dark-colored phenolic wastewater from a lignite pyrolysis factory was filled into a former open-cast pit, forming a wastewater disposal pond. This caused extensive environmental pollution, requiring an ecologically and economically acceptable strategy for remediation.
3. Laboratory-scale investigations and pilot-scale tests were necessary on the basis of an interdisciplinary cooperation between chemists, microbiologists, hydrologists, biologists and technologists.
4. A strategy for a full-scale enhanced *in situ* natural attenuation on the basis of separate habitats inside this meromictic pond was developed. A stable anaerobic layer in the depth, including an undisturbed subhydrical deposition with a low influence on the adjacent groundwater layer, and a biological dynamic upper layer to a depth of approximately 10 m was created.
5. Long-term monitoring of the chemical and biological dynamics of the pond shows the metamorphosis of a former highly polluted industrial waste deposition to a nature-integrated ecosystem with reduced danger for the environment.

Acknowledgments

The authors would like to thank E. Weissbrodt for the management of the project and her dedication as well as I. Richter and R. Schumann for technical assistance. The technical field work was financed and executed by the Lausitzer und Mitteldeutsche Bergbau-Verwaltungsgesellschaft mbH (LMBV). The authors are grateful to the staff of LMBV, and, in particular, they would like to thank W. Kletzander for the fruitful cooperation.

References

Baltrenas, P.; Kazlauskiene, A. 2009. Sustainable ecological development reducing negative effects of road maintenance salts, *Technological and Economic Development of Economy* 15: 178–188.

Becker, P. M. 1999. About the order in heterotrophic microbial communities from hydrocarbon contaminated sites, *International Biodeterioration and Biodegradation* 43: 135–146. http://dx.doi.org/10.1016/S0964-8305(99)00042-6

Becker, P. M.; Stottmeister, U. 1998. General (Biolog GN) versus site-relevant (pollutant-dependent) sole-carbon-source utilization patterns as a means to approaching community functioning, *Canadian Journal of Microbiology* 44: 913–919. http://dx.doi.org/10.1139/w98-084

Becker, P. M.; Wand, H.; Martius, G. G. S.; Weissbrodt, E.; Stottmeister, U. 1998. Functional and structural successions in arbitrary samples of heterotrophic bacteria during aerobic treatments of lignite-carbonization wastewater in *in-situ* enclosures, *Canadian Journal of Microbiology* 44: 211–220. http://dx.doi.org/10.1016/S0045-6535(97)00001-5

Becker, P. M.; Wand, H.; Weißbrodt, E.; Kuschk, P.; Stottmeister, U. 1997. Distribution of contaminants and the self-purifying potential for aromatic compounds in a carbonization waste water deposit, *Chemosphere* 34: 731–748.

Bilkenroth, K.-D.; Snyder, D. O. 1998. *Brown coal mining in Central Germany – history, presence, and future MIBRAG – Mitteldeutsche Braunkohlengesellschaft mbH*, Theißen. Germany (in German).

Biswas, R.; Bagchi, S.; Urewar, C.; Gupta, D.; Nandy, T. 2010. Treatment of wastewater from a low-temperature carbonization process industry through biological and chemical oxidation processes for recycle/reuse: a case study, *Water Science and Technology* 61: 2563–2573. http://dx.doi.org/10.2166/wst.2010.181

Ciahotny, K.; Kusy, J.; Kolarova, L.; Safarova, M.; Andel, L. 2011. Brown coal and rape cakes copyrolysis, *Chemicke Listy* 105: 879–884.

Domazetis, G.; Barilla, P.; James, B. D. 2010. Lower emission plant using processed low-rank coals, *Fuel Processing*

Technology 91: 255–265.
http://dx.doi.org/10.1016/j.fuproc.2009.10.006

Eccarius, B. 1998. Groundwater monitoring and isotope investigation of contaminated wastewater from an open pit mining lake, *Environmental Geosciences* 5: 156–161. http://dx.doi.org/10.1046/j.1526-0984.1998.08022.x

Eccarius, B.; Christoph, G.; Ebhardt, G.; Gläßer, W. 2001. Grundwassermodellierung zur Gefährdungsabschätzung eines phenolverseuchten Tagebaurestsees [Groundwater modelling for risk assessment of a phenol contaminated open pit mining lake], *Grundwasser* 2: 61–70 (in German). http://dx.doi.org/10.1007/PL00010392

Eissmann, L. 1994. *From the earth-archives of Central Germany*. Beucha, Germany: Sax-Verlag.

Folgueras, M. B.; Diaz, R. M. 2010. Influence of $FeCl_3$ and lime added to sludge-coal pyrolysis, *Energy* 35: 5250–5259. http://dx.doi.org/10.1016/j.energy.2010.07.040

Garg, A.; Mishra, I. M.; Chand, S. 2008. Catalytic oxidative treatment of diluted black liquor at mild conditions using copper oxide cerium oxide catalyst, *Water Environment Research* 80: 136–141.
http://dx.doi.org/10.2175/106143007X220752

Huang, L. P.; Yang, X. H.; Quan, X.; Chen, J. W.; Yang, F. L. 2010. A microbial fuel cell-electro-oxidation system for coking wastewater treatment and bioelectricity generation, *Journal of Chemical Technology and Biotechnology* 85: 621–627. http://dx.doi.org/10.1002/jctb.2320

Kopinke, F.-D.; Pörschmann, J.; Remmler, M. 1995a. Sorption behavior of anthropogenic humic matter, *Naturwissenschaften* 82: 28–30.
http://dx.doi.org/10.1007/BF01167866

Kopinke, F.-D.; Pörschmann, J.; Stottmeister, U. 1995b. Sorption of organic pollutants on anthropogenic humic matter, *Environmental Science & Technology* 29: 941–950.
http://dx.doi.org/10.1021/es00004a014

Kubal, M.; Podhola, M.; Patocka, T.; Ciahotny, K.; Kuras, M. 2011. Treatment of ammonia-polluted groundwater in North Bohemian brown coal mining region – Feasibility study, *Desalination and Water Treatment* 33: 36–43.
http://dx.doi.org/10.5004/dwt.2011.2616

Kuschk, P.; Stottmeister, U.; Liu, Y.-J.; Wiessner, A.; Kästner, M.; Müller, R.-A. 2010. Batch methanogenic fermentation experiments of wastewater from a brown coal low-temperature coke plant, *Journal of Environmental Sciences* 22: 192–197.
http://dx.doi.org/10.1016/S1001-0742(09)60092-9

Kuschk, P.; Wießner, A.; Kappelmeyer, U.; Weißbrodt, E.; Kästner, M.; Stottmeister, U. 2003. Annual cycle of nitrogen removal by a pilot-scale subsurface horizontal flow in a constructed wetland under moderate climate, *Water Research* 37: 4236–4242.
http://dx.doi.org/10.1016/S0043-1354(03)00163-5

Kuschk, P.; Wießner, A.; Martius, G.; Weißbrodt, E.; Stottmeister, U. 1994. Untersuchungen zur Sanierung wäßriger Altlasten der Braunkohlepyrolyseindustrie [Investigations on remediation of wastewaters from lignite pyrolysis industries], *Energieanwendung, Energie- und Umwelttechnik* 43: 309–312 (in German).

Mackenzie, K.; Georgi, A.; Kumke, M.; Kopinke, F.-D. 2002. Sorption of pyrene to dissolved humic substances and related model polymers. 2. Solid-phase microextraction (SPME) and fluorescence quenching technique (FQT) as analytical methods, *Environmental Science & Technology* 36: 4403–4409. http://dx.doi.org/10.1021/es010310x

Mursito, A. T.; Hirajima, T.; Sasaki, K.; Kumagai, S. 2010. The effect of hydrothermal dewatering of Pontianak tropical peat on organics in wastewater and gaseous products, *Fuel* 89: 3934–3942.
http://dx.doi.org/10.1016/j.fuel.2010.06.035

Pathiratne, K. A. S.; De Silva, O. C. P.; Hehemann, D.; Atkinson, I.; Wei, R. 2007. Occurrence and distribution of polycyclic aromatic hydrocarbons (PAHs) in Bolgoda and Beira Lakes, Sri Lanka, *Bulletin of Environmental Contamination and Toxicology* 79: 135–140.
http://dx.doi.org/10.1007/s00128-007-9092-z

Pörschmann, J.; Kopinke, F.-D.; Balcke, G.; Mothes, S. 1998a. Pyrolysis pattern of anthropogenic and natural humic organic matter, *Journal of Microcolumn Separations* 10: 401–411.
http://dx.doi.org/10.1002/(SICI)1520-667X(1998)10:5<401::AID-MCS3>3.0.CO;2-K

Pörschmann, J.; Kopinke, F.-D.; Pawliszyn, J. 1998b. Solid-phase microextraction for determining the binding state of organic pollutants in contaminated water rich in humic organic matter, *Journal of Chromatography* A 816: 159–167. http://dx.doi.org/10.1016/S0021-9673(98)00525-1

Pörschmann, J.; Kopinke, F.-D.; Remmler, M.; Mackenzie, K.; Geyer, W.; Mothes, S. 1996. Hyphenated techniques for characterizing coal wastewaters and associated sediments, *Journal of Chromatography* A 750: 287–301.
http://dx.doi.org/10.1016/0021-9673(96)00403-7

Pörschmann, J.; Stottmeister, U. 1993. Methodical investigation of interaction between organic pollutants and humic organic material in coal wastewaters, *Chromatographia* 36: 207–211. http://dx.doi.org/10.1007/BF02263865

Pörschmann, J.; Zhang, Z.; Kopinke, F.-D.; Pawliszyn, J. 1997. Solid phase microextraction for determining the distribution of chemicals in aqueous matrices, *Analytical Chemistry* 69: 597–600.
http://dx.doi.org/10.1021/ac9609788

Qi, W. X.; Liu, H. J.; Qu, J. H.; Hu, C. Z.; Lan, H. C.; Berg, M.; Ren, H. M.; Xu, W. 2011. Polycyclic aromatic hydrocarbons in effluents from wastewater treatment plants and receiving streams in Tianjin, China, *Environmental Monitoring and Assessment* 177: 467–480.
http://dx.doi.org/10.1007/s10661-010-1648-4

Ringpfeil, M.; Stottmeister, U.; Behrens, U.; Martius, G.; Bürger, G.; Wenige, L. 1988. Aerobic treatment of sewage from lignite (brown coal) processing, in Wise, D. L. (Ed.). *Biotreatment Systems*. Boca Raton, Florida: CRC Press.

Stottmeister, U. 2008. Altlastensanierung mit Huminstoffsystemen [Remediation of wastes from the past using humic substances]. *Chemie in unserer Zeit* 42: 24–41 (in German). http://dx.doi.org/10.1002/ciuz.200800439

Stottmeister, U.; Glässer, W.; Klapper, H.; Weissbrodt, E.; Eccarius, B.; Kennedy, C.; Schultze, M.; Wendt-Potthoff, K.; Frömmichen, R.; Schreck, P.; Strauch, G. 1999. Strategies

for remediation of former opencast mining areas in eastern Germany, in Azcue, J. M. (Ed.). *Environmental Impacts of Mining Activities: Emphasis on Mitigation and Remediation*. Berlin, Heidelberg, New York: Springer-Verlag, 263–296.
http://dx.doi.org/10.1007/978-3-642-59891-3_16

Stottmeister, U.; Kuschk, P.; Wießner, A. 2009. Full-scale bioremediation and long-term monitoring of a phenolic wastewater disposal lake, *Pure & Applied Chemistry* 82: 161–173. http://dx.doi.org/10.1351/PAC-CON-09-05-05

Stottmeister, U.; Kuschk, P.; Wießner, A.; Weißbrodt, E.; Martius, G.; Becker, P. M.; Eismann, F.; Kotte, H. 1997. Altlastenprobleme der Carbochemie in Nordwestsachsen: Entstehung, Ausmaß und erste Sanierungskonzepte [Wastes from the past of the carbo-chemistry in North-Western Saxony: formation, size, and first remediation strategies], in Knorr, C.; Schell, T. (Eds.). *Mikrobieller Schadstoffabbau*. Braunschweig, Wiesbaden, Germany: Vieweg, 357–375 (in German).

Stottmeister, U.; Weißbrodt, E.; Becker, P. M.; Pörschmann, J.; Kopinke, F.-D.; Martius, G. G. M.; Wießner, A.; Kennedy, C. 1998. Analysis, behaviour and fate of a lignite pyrolysis wastewater deposit, in *Contaminated Soil 98, Sixth International FZK/TNO Conference on Contaminated Soil*. London: Thomas Telford Publishing, 113–121.

Stottmeister, U.; Weißbrodt, E.; Tittel, J. 2002. From contamination of the past to a lake [Von der Altlast zum See], *Biologie in unserer Zeit* 32: 276–285.
http://dx.doi.org/10.1002/1521-415X(200209)32:5 <276::AID-BIUZ276 >3.0.CO;2-G

Thiruvenkatachari, R.; Younes, M.; Su, S. 2011. Coal mine site investigation of wastewater quality in Australia, *Desalinations and Water Treatment* 32: 357–364.
http://dx.doi.org/10.5004/dwt.2011.2722

Tokarska, A. 2009. Preliminary studies on brown coals liquefaction, *Rocznik Ochrona Srodowiska* 11: 1095–1104.

Tomei, M. C.; Annesini, M. C. 2008. Biodegradation of phenolic mixtures in a sequencing batch reactor – a kinetic study, *Environmental Science and Pollution Research* 15: 188–195.
http://dx.doi.org/10.1065/espr2007.12.470

Valentukeviciene, M.; Ignatavicius, G. 2011. Analysis and evaluation of the effect of the solids from road surface runoff on the sediments of river bed, *Ekologija* 57: 39–45.
http://dx.doi.org/10.6001/ekologija.v57i1.1308

Volkmann, N.; Borner, K. 2012. The Deutsches Energie Rohstoff-Zentrum Freiberg: advanced prospects for applied organic petrology, focused on the generation of coal based chemical products, *International Journal of Oil, Gas and Coal Technology* 5: 399–409.
http://dx.doi.org/10.1504/IJOGCT.2012.048984

Wießner, A.; Kuschk, P.; Martius, G.; Eismann, F.; Zehnsdorf, A.; Weißbrodt, E.; Stottmeister, U. 1994. Abbaubarkeit von Schadstoffen im Braunkohlepyrolyse-Altwasser [Degradability of contaminants in disposed wastewater from lignite pyrolysis], *Wasser Abwasser Praxis* 4: 44–47 (in German).

Wießner, A.; Kuschk, P.; Stottmeister, U.; Struckmann, D.; Jank, M. 1999. Treating a lignite pyrolysis wastewater in a constructed subsurface flow wetland, *Water Research* 33: 1296–1302.
http://dx.doi.org/10.1016/S0043-1354(98)00310-8

Wießner, A.; Kuschk, P.; Weißbrodt, E.; Stottmeister, U.; Pörschmann, J.; Kopinke, F.-D. 1993. Charakterisierung des Wassers und des Sedimentes einer Braunkohle-Schwelwasserdeponie [Characterization of the water and the sediment of a wastewater disposal from lignite pyrolysis], *Wasser Abwasser Praxis* 6: 375–379 (in German).

Wu, Y. B.; Fu, J.; Zeng, S. S.; Sun, P.; Yan, Y. C. 2012. A thermal analysis calculation of pharmaceutical wastewater sludge, *Journal of Thermal Analysis and Calorimetry* 108: 389–395.
http://dx.doi.org/10.1007/s10973-011-1683-6

Xu, Z. X.; Wei, Z.; Yin, H. L.; Huang, L. H. 2010. Optimized design of natural ecological wastewater treatment system based on water environment model of dynamic mesh technique, *Journal of Hydrodynamics* 22: 1–8.
http://dx.doi.org/10.1016/S1001-6058(09)60021-4

Zhang, Y. M.; Bao, S. X.; Liu, T.; Chen, T. J.; Huang, J. 2011. The technology of extracting vanadium from stone coal in China: History, current status and future prospects, *Hydrometallurgy* 199: 116–124.
http://dx.doi.org/10.1016/j.hydromet.2011.06.002

Zhou, L.; Hu, S. Y.; Li, Y. R.; Jin, Y.; Zhang, X. L. 2012. Modeling and optimization of a coal-chemical eco-industrial system in China, *Journal of Industrial Ecology* 16: 105–118.
http://dx.doi.org/10.1111/j.1530-9290.2012.00447.x

Arndt WIESSNER. Dr, senior scientist, PhD (1991), 28 publications indexed in the ISI Web of Science. Research interests: redox reactions in the root zone of constructed wetlands, sulphur cycling in freshwater wetlands.

Jochen Ait MÜLLER. Dr, senior scientist, PhD (2000), 12 publications indexed in the ISI Web of Science. Research interests: anaerobic transformation of organic pollutants, genomics of sulphur-oxidizing bacteria.

Peter KUSCHK. Dr, head of the group of Ecological Water Treatment Technologies, PhD (1991), 68 publications indexed in the ISI Web of Science. Research interests: microbial anaerobic processes in constructed wetlands, hygienization of domestic sewage.

Uwe KAPPELMEYER. Dr, research engineer, PhD (2001), 14 publications indexed in the ISI Web of Science. Research interests: molecular biological techniques for the characterization of microbial cenoses, simulation of biotechnological processes.

Matthias KÄSTNER. Professor, head of the Department of Environmental Biotechnology, PhD (1989), more than 100 publications indexed in the ISI Web of Science. Research interests: bioremediation processes, metabolic flow in environmental systems, C and N turnover in soil, genesis of soil organic matter.

Yong-Jun LIU. Professor, head of the group of Biological Wastewater Treatment Technologies, PhD (2004), 21 publications indexed in the ISI Web of Science. Research interests: bioremediation of environmental pollution, characterization of biological toxic effects for environmental pollutants.

Ulrich STOTTMEISTER. Professor emeritus, PhD (1968), 71 publications indexed in the ISI Web of Science. Research interests: technology assessment, biological wastewater treatment, microbial product formation.

A HYBRID MADM ANALYSIS IN EVALUATING PROCESS OF CHEMICAL WASTEWATER PURIFICATION REGARDING TO ADVANCE OXIDATION PROCESSES

Mohamad Reza KHODADADI[a], Sarfaraz HASHEMKHANI ZOLFANI[b], Morteza YAZDANI[c], Edmundas Kazimieras ZAVADSKAS[d]

[a]Department of Nanotechnology, Semnan University, P.O Box 19111-35131, Semnan, Iran
[b]Department of Management, Science and Technology, Technology Foresight Group, Amirkabir University of Technology (Tehran Polytechnic), P.O. Box 1585-4413, Tehran, Iran
[c]Department of Business Management, Faculty of Social Sciences, Universidad Europea de Madrid, 28670 Madrid, Spain
[d]Research Institute of Smart Building Technologies, Vilnius Gediminas Technical University, Saulėtekio al. 11, LT-10223 Vilnius, Lithuania

Abstract. Organic components, widely spread in water by industrial sources become emerging Contaminants in the water. Due deficient removal in conventional wastewater treatments, advanced treatment should be considered for industrial wastewaters removing. Advance oxidation processes (AOPs) can be used as an effective subset of chemical treatment for degradation of toxic from water in industrial processing plant. Two Multiple Criteria Decision Making (MCDM) methods based on hybrid SWARA-WASPAS are organized to investigate efficiency of different sort of AOPs: ozonation, fenton, electrochemical Oxidation, UV/Photo-catalysis, UV/H_2O_2. Based on MCDM, Fenton is the most possible AOPs for application in wastewater treatment areas. In this study, SWARA is deployed in evaluating criteria that were identified in literature review, and WASPAS is used for evaluating and ranking the methods. The main point of this study is based on real data.

Keywords: wastewater, wastewater purification, AOPs, MCDM, SWARA, WASPAS.

Introduction

Population growth and industrialization caused more pollution of hydrosphere with organic and inorganic compounds (Oppenländer 2003). Sanitation is one of the essential practices in developing infrastructures in each government while encompasses wastewater treatment in each region of a country to fight for pollution and control it. Regarding recent UNICEF reports of 780 million people without access to safe drinking water, improvement in sanitation and purification methods needs critical attention. In fact purification is done to remove chemicals, biological and physical contaminants. Water purification is a process to reuse and recycle wastewaters for a specific purpose such as industrial usage.

In the literature four main sources of wastewaters are (1) domestic sewage, (2) industrial wastewaters, (3) agricultural runoff and (4) storm water and urban runoff. Although agricultural and urban runoff wastewaters have less importance, however domestic and industrial wastewater treatments are becoming increasingly critical due to rapid growth of industrialization and mechanization. Therefore all of the chemical compounds, materials and energy which are discharged into the water can potentially influence hydrosphere, global health and environment. This discharge can finally lead to transaction of energies and materials within an ecosystem's hydrology and can be a strong source of pollutant (contaminant). Role of AOPs as an effective chemical treatment is to degrade toxics and minimize the pollutants from the water that is investigated in this section. Almost all of the industries as cosmetics, wood and food and textile industries which contain chemical nature and processes can potentially produce water pollution.

Corresponding author: Sarfaraz Hashemkhani Zolfani
E-mail: sa.hashemkhani@gmail.com

Different treatment technologies are used to remove contaminants as environmental disinfectants and wastewater purification methods like chemical, physical, and biological operations. Wastewater purification is a complex process that needs to combine physical–chemical methods. Particularly, elimination of organic contaminations such as phenols, solvents, dye, etc., from wastewater is hard. Physical and mechanical treatment including filtration and membrane technique leads to secondary wastewater and higher cost. None of these treatment methods is effective enough to produce water with acceptable levels of organic compounds (Leyva-Díaz *et al.* 2015). Also, organic substances are stable compounds to be decomposed by common biological treatment. Literatures show that biological methods employed for degradation of organic compounds are ineffective and combined chemical- biological methods are suggested (Mantzavinos, Psillakis 2004). For instance, the biological treatments cannot be effective to remove most of complex poly-aromatic dye in textile industry. Chemical processes cannot be neglected and are highly recommended in decision making. Chemical treatment of wastewater may also be applied where there is limited access to the biological treatment.

Recent developments in the domain of chemical water treatment have led to an improvement in oxidative degradation procedures to remove organic materials in wastewater by oxidation through catalytic and photochemical methods. They are generally referred to advanced oxidation processes (AOPs) which has a strong impact on simple sanitation infrastructure decision and economical feasible approach. AOPs as chemical processes are promising methods to treat the most organic compounds from polluted water (Sievers 2011). It is a powerful tool for degradation and mineralization of pollutants in water and wastewater treatment, including hydrogen peroxide, ozone or special catalysts such as titanium dioxide. The use of a strong oxidizing agent can result in a high degree of wastewater treatment, including the breakdown of chemical recalcitrant and toxic compounds.

The aim of the present paper is to investigate the efficiency of different AOPs in wastewater treatment plant for wastewater management, namely Ozonation, Fenton, electrochemical Oxidation, UV/Photo-catalysis, UV/H_2O_2, and various integrated processes, regarding (i) efficiency and capacity of each technology for mineralization organic materials loads, (ii) comparable low cost and high cost technologies related economic and operational feasibility (iii) important chemical indicators to rank the possibility of techniques for removing effluents' toxicity. In this way, policy and decision making processes related to these issues are complicated. Multiple Criteria Decision Making (MCDM) (Liu *et al.* 2016; Liu, Shi 2015; Khan, Samadder 2015; Liu, Jin 2012; Turskis *et al.* 2012; Kucas 2010) frameworks can be a good choice for making an appropriate decision. Significance and contribution of multi criteria decision tools was applied in wide range of real life applications (Liu *et al.* 2011, 2012). Two MCDM methods are deployed in this study for making sound decisions. At first, SWARA is deployed for evaluating and weighting the criteria of this research. Must important criteria are identified from literature review. At next step, WASPAS method is deployed for evaluating different considerable alternatives of AOPs.

1. Literature review

In the literature review section, the significance of AOPs and its important parameters which have direct effect on term of AOPs wastewater treatment are discussed in Table 1. Several researches have reported applying different sort of AOPs and comparing them with conventional chemical methods. The high efficiency of AOPs in degradation of chemical toxin and pollution from water has been mentioned.

Sharma *et al.* (2012) compared advance and conventional oxidation processes for removing Microcystins (MCs) from water. It has been reported that AOPs are more effective and friendly to the environment than conventional chemical processes to convert a great variety of organic compounds into CO_2 and mineral acids. Study showed rate of pollution degradation with AOPs is faster than chlorine and chlorine dioxide as conventional methods.

Ozonation technology as initial AOPs has been scrutinized. Gómez-Pacheco *et al.* (2011) evaluated Ozonation process and showed a reasonable decrease in total organic carbon (TOC) performance with degradation of tetracyclines (TCs) from water. The result showed a rapid mineralization of organic compound and direct relation between operation variables and pollution degradation.

Several parameters are included to investigate different sort of AOPs, which make it hard to approach a prospective strategic decision making. Canizares *et al.* (2009a) compared three different AOPs such as: ozonation, electrochemical oxidation and Fenton. They illustrated that the efficiencies of these three technologies were critically depended on concentration of pollutant. Also, the operation cost analysis showed Fenton oxidation process had lower expenses than two others methods. In contrast, the cost calculation for two important oxidation processes, ozonation and Fenton reaction in a laboratory and pilot scale have been presented. This investigation showed Fenton reagent treatment was more expensive when compared to ozonation that is due to additional chemical cost for adjusting pH value of Fenton process (Krichevskaya *et al.* 2011).

Chang *et al.* (2008) took into account chemical and economic indicators for ozonation experiment. They investigated pollution and organic carbon removing were pH- dependent and the results showed the ozonation could not remove the pollutants entirely. The results declare a short ozone contact time has a better cost savings. Also, Fenton process has been under total cost and mineralization processes analyzes (Carra *et al.* 2013). They evaluated cost reduction has a direct correlation with the level of mineralization that higher mineralization levels cause significant decrease in financial terms.

Tizaoui *et al.* (2011) reported photo catalytic as an available AOPs technique for removing wastewater pollution by comparison with UV/H_2O_2 and ozonation processes. The results of this study successfully showed photo catalytic process was the most effective way for degradation pollution from water. Also, several researches tried to analyze the efficiency of different catalysts such as TiO_2 and ZnO for removing chemicals from water. They evaluated the optimum amount of catalyst, value of pH and the electrical energy cost as important indicators for reducing pollution from water (Lizama *et al.* 2002).

Exploring different advance oxidation processes (ozonation and its relatives, Fenton and photo catalysis) have been performed with ultra violet or solar source as energy consumption indicators (Esplugas *et al.* 2002). In addition, the research showed the ratio of photocatalysis process was five times lower than UV/H_2O_2 while the ability of recycling of photocatalyst powder could decrease the total cost of this process.

Cabral da Silva *et al.* (2013) used electrochemical oxidation to produce fresh water. Through this method they analyzed different parameters such as pH, total organic carbon (TOC) and energy consumption and mentioned electrochemical treatment method could be a feasible pretreatment process for the petrochemical industry. Also, Canizares *et al.* (2009a, 2009b) compared electrochemical oxidation method with other oxidation process. They believed electrolyte salt as a chemical parameter can improve efficiency and decrease total amount of pollution.

Autin *et al.* (2013) used an annular reactor to compare photocatalyst and hydrogen peroxide processes. It has been figured out that the optimization of reactor design, catalyst particle properties, and efficiency of UV lamp source will ultimately reduce the energy consumption. Moreover, different approaches presented dye pollution degradation using photocatalyst such as titanium dioxide (TiO_2). Some advances of this kind of oxidation have been performed for degradation of organic pollutants in wastewater (Teh, Mohamed 2011).

Chen *et al.* (2011) studied water pollution treatment for reusing water with TiO_2 photocatalyst method. They evaluated various parameters such as light source and retention time and mentioned that the proposed system can decrease the operating and maintenance costs of wastewater treatment for agricultural and domestic use.

2. The model of research

Table 1 illustrates the references of parameters as the model of research that has been done for adjusting the environmental, economical and chemical evaluation which was reported by authors. Table 2 demonstrates all information

Table 1. The model of research

Parameters	Reference
Mineralization	Lizama *et al.* 2002; Esplugas *et al.* 2002; Chang *et al.* 2008; Canizares *et al.* 2009a, 2009b; Tizaoui *et al.* 2011; Gómez-Pacheco *et al.* 2011; Krichevskaya *et al.* 2011; Autin *et al.* 2013; Prieto-Rodrıguez *et al.* 2013; Carra *et al.* 2013; Cabral da Silva *et al.* 2013
Volume of wastewater treated	Lizama *et al.* 2002; Esplugas *et al.* 2002; Chang *et al.* 2008; Canizares *et al.* 2009a, 2009b; Tizaoui *et al.* 2011; Gómez-Pacheco *et al.* 2011; Krichevskaya *et al.* 2011; Autin *et al.* 2013; Prieto-Rodrıguez *et al.* 2013; Carra *et al.* 2013; Cabral da Silva *et al.* 2013
Capital costs	Canizares *et al.* 2009b; Carra *et al.* 2013
Operating costs	Canizares *et al.* 2009a; Krichevskaya *et al.* 2011; Carra *et al.* 2013; Prieto-Rodrıguez *et al.* 2013; Cabral da Silva *et al.* 2013:
Energy consumption	Lizama *et al.* 2002; Canizares *et al.* 2009b; Krichevskaya *et al.* 2011; Cabral da Silva *et al.* 2013; Autin *et al.* 2013
pH levels	Lizama *et al.* 2002; Esplugas *et al.* 2002; Gómez-Pacheco *et al.* 2011; Chang *et al.* 2008; Canizares *et al.* 2009a; Tizaoui *et al.* 2011; Krichevskaya *et al.* 2011; Autin *et al.* 2013; Prieto-Rodrıguez *et al.* 2013; Carra *et al.* 2013; Cabral da Silva *et al.* 2013
Total time for decreasing pollution	Lizama *et al.* 2002; Esplugas *et al.* 2002; Chang *et al.* 2008; Canizares *et al.* 2009a, 2009b; Tizaoui *et al.* 2011; Gómez-Pacheco *et al.* 2011; Autin *et al.* 2013; Prieto-Rodrıguez *et al.* 2013; Carra *et al.* 2013; Krichevskaya *et al.* 2011; Cabral da Silva *et al.* 2013
Amount of chemical material used	Lizama *et al.* 2002; Chang *et al.* 2008; Canizares *et al.* 2009a; Krichevskaya *et al.* 2011; Tizaoui *et al.* 2011; Autin *et al.* 2013; Prieto-Rodrıguez *et al.* 2013
Pollution concentration	Krichevskaya *et al.* 2011; Cabral da Silva *et al.* 2013; Prieto-Rodrıguez *et al.* 2013; Carra *et al.* 2013; Autin *et al.* 2013

about criteria and sub-criteria that are used in different aspects of decision making. These criteria are organized based on environmental, economic and chemical factors for five AOPs: Ozonation, Fenton, UV/Photo-catalysis, Electrochemical Oxidation and UV/H_2O_2. Environmental factors represent the capacity of each technology for elimination of organic pollution and their capability for mineralization organic materials to mineral materials (Saeed, Sun 2012). Economic study is prepared to compare low cost and high cost technologies related the pollutant treatment. These indicators include capital costs, operating costs and energy consumption. Capital costs involve expenses incurred for fixed equipment, installation and organic matter mineralization for an industrial-scale plant. Operating costs consist of treatment plant and all machinery that are required to operate a process. Furthermore chemical process of waste water treatment is a process that consumes electrical energy. This can be an important factor to determine operation costs. The volumetric electrical energy consumption per order (EE/O) and the electrical energy consumption cost are determined according to rated power (kW) of the AOPs system and radiation time (h) per volume (m^3) of waste water that is treated (Kim *et al.* 2012). The third section of Table 2 is related to important chemical indicators such as total time for decreasing pollution, pH levels, and average amount of chemical materials that were used for treatment. The amounts of pollutions were studied in their basic amount in each treatment method. In this study the importance of pH level that has to be adjusted for different AOPs treatment has been considered. The pH level is important factor that can cause an increase in amount of pollution. Finally, the data in each column for each treatment method have been calculated in average such as the years of practice with the information extracted from references as mentioned in the Table 2.

Water treatment based AOPs acts in a way to reduce costs and increase process efficiency. There are multiple data that are reported in different articles and build a challenging situation for selecting suitable parameters. In this way the best parameters have been found within various articles.

Table 2. Environmental, economical and chemical factors for five AOPs

				Ozonation	Fenton	Electro-chemical Oxidation	UV Photo-catalysis	UV H_2O_2
Unit/ Parameters/Ref		The treatment method		Gómez-Pacheco *et al.* 2011; Chang *et al.* 2008; Canizares *et al.* 2009a; Tizaoui *et al.* 2011; Krichevskaya *et al.* 2011	Canizares et al.2009; Prieto-Rodrıguez *et al.* 2013; Carra *et al.* 2013; Krichevskaya *et al.* 2011; Canizares *et al.* 2009	Canizares *et al.* 2009a; Cabral da Silva *et al.* 2013; Canizares *et al.* 2009b	Tizaoui *et al.* 2011; Lizama *et al.* 2002; Autin *et al.* 2013; Prieto-Rodrıguez *et al.* 2013	Tizaoui *et al.* 2011; Autin *et al.* 2013; Esplugas *et al.* 2002
Environ-mental (C_1)	%	Mineralization (C_{1-1})	Max	52.85	64.25	88.90	85	87.50
	Litter	Volume of wastewater treated (C_{1-2})	Max	1.40	17.20	1.05	12.3	5.10
Economic (C_2)	€.m^{-2}	Capital costs (C_{2-1})	Min	88639	8568.50	50210	70400	10800
	€.m^{-3}	Operating costs (C_{2-2})	Min	546.59	6.20	12.19	27.30	50.50
	KW h m^{-3}	Energy consumption (C_{2-3})	Min	2070	1	219.28	31.75	5.4
Chemical (C_3)	–	pH levels (C_{3-1})	–	7	2.9	7.40	7.40	6.30
	min	Total time for decreasing pollution (C_{3-2})	Min	82.93	50	150	153.75	33.75
	mg/L	Amount of chemical material used (C_{3-3})	Min	54.41	1017.21	5025	434	696
	Mol/Lit	Pollution concentration (C_{3-4})	Max	1.23×10^{-3}	5.2×10^{-4}.	4.79×10^{-3}	2.87×10^{-3}	5.67×10^{-5}
Average Year of practice (C_4)	Year	–	Max	2010	2011	2010	2010	2009

3. Methodology

The general framework of this research is organized based on MCDM concept. The methodology of this research is established based on two different MCDM methods as a hybrid method. SWARA and WASPAS have been used together first time in a research in 2013. Hashemkhani Zolfani *et al.* (2013a) applied this hybrid method in a research for decision making in the business issue. Moreover in this short period of time some other researches are organized based on this new hybrid methodology. All of other researches based on SWARA-WASPAS are listed below:

Vafaeipour *et al.* (2014) in assessment of regions priority for implementation of solar plants; Bitarafan *et al.* (2014) in evaluation of real-time intelligent sensors for structural health monitoring of bridges.

3.1. Step-wise Weight Assessment Ratio Analysis (SWARA)

This method is focused on criteria and not alternatives. SWARA method is presented by Keršulienė *et al.* (2010) for evaluating criteria with a new perspective. There are some other methods for evaluating criteria like AHP, ANP and FARE. All these methods are organized based on pairwise comparison and related to direct relations of criteria. During these years after 2010, applications of SWARA method has widened. Also the main logic of this method has developed in some researches like: (Hashemkhani Zolfani, Saparauskas 2013) and (Hashemkhani Zolfani, Bahrami 2014) in general dimensions of method. Both previous researches believed that SWARA method is so suitable for policy making and decision making in top level. In fact each issue should be solved based on special circumstances eventually this topic should be considered. The experts are the main source of competency in SWARA methodology.

In comparison to other methods like AHP and ANP; SWARA has totally different characteristics. At the first step experts should make decision about priority of each issue. The priority should be based on needs and situation. SWARA can be deployed easily in abnormal or complicated situation. Experts are freer to make decision. At the next step experts should express their opinion about values of criteria. S_j is a symbol for this issue and each expert should express his/her opinion as a percentage. Percentage's units in all previous researches were based on 5% intervals.

Application of this method has increased in recent years. This method was applied in different methodological applications. In some researches SWARA is deployed singly but also this method developed in some new hybrid MCDM methodologies like SWARA-VIKOR, SWARA-COPRAS, SWARA-COPRAS-G and SWARA-WASPAS. All developments of decision making models based on SWARA method up to now are listed in Table 3.

Table 3. Methodological development of SWARA method

	Topic	Methodology	Reference
1	Design of products	SWARA	Hashemkhani Zolfani *et al.* (2013a)
2	Investigating on the success factors of online games based on explorer	SWARA	Hashemkhani Zolfani *et al.* (2013b)
3	Prioritizing Sustainability Assessment Indicators of Energy System	SWARA	Hashemkhani Zolfani, Saparauskas (2013)
4	Supplier selection in agile environment	SWARA-VIKOR	Alimardani *et al.* (2013)
5	Investment prioritizing investment in high tech industries	SWARA-COPRAS	Hashemkhani Zolfani, Bahrami (2014)
6	Decision making on business issues with foresight perspective	SWARA-WASPAS	Hashemkhani Zolfani *et al.* (2013c)
7	Evaluation of real-time intelligent sensors for structural health monitoring of bridges	SWARA-WASPAS	Bitarafan *et al.* (2014)
8	Assessment of regions priority for implementation of solar plants	SWARA-WASPAS	Vafaeipour *et al.* (2014)
9	Personnel selection	Game Theory-SWARA	Hashemkhani Zolfani, Seyed Agha Banihashemi (2014)
10	External wall insulation	SWARA-TODIM	Ruzgys *et al.* (2014)
11	Selection of a Packaging Design	SWARA	Stanujkic *et al.* (2015)
12	Glasshouse locating	SWARA-COPRAS	Haghnazar Kouchaksaraei *et al.* (2015)
13	Technology Foresight About R&D Projects Selection	SWARA	Hashemkhani Zolfani *et al.* (2015)
14	Planning the priority of high tech industries	SWARA-WASPAS	Ghorshi Nezhad *et al.* (2015)

The original procedure to the criteria weights determination using SWARA can be expressed as follows:

Step 1 – All criteria should be sorted based on expert ideas (Zavadskas, Vilutienė 2006).

Step 2 – From the second criterion, the comparative importance of the average value s_j should be determined as follows: the relative importance of the criterion j in relation to the previous $(j-1)$ criterion (Stanujkic *et al.* 2015).

Step 3 – Determine the coefficient k_j:

$$k_j = \begin{cases} 1 & j=1 \\ s_j + 1 & j>1. \end{cases} \quad (1)$$

Step 4 – Determine the recalculated weight w_j:

$$w_j = \begin{cases} 1 & j=1 \\ \dfrac{x_{j-1}}{k_j} & j>1. \end{cases} \quad (2)$$

Step 5 – The final step in the calculation of criteria weights:

$$q_j = \frac{w_j}{\sum_{k=1}^{n} w_j}, \quad (3)$$

where q_j denotes the relative weight of the criterion *j*. The calculating part of SWARA method is from determination of criteria weights till the end.

3.2. Weighted Aggregates Sum Product Assessment (WASPAS)

WASPAS method is a new MCDM method based on Weighted Sum Model (WSM) and Weighted Product Model (WPM). Zavadskas *et al.* (2012) illustrated that this aggregation makes WASPAS better in comparison with WSM and WPM. WASPAS calculation is based on these steps:

3.2.1. Normalized decision making matrix based on:

$$\bar{x}_{ij} = \frac{x_{ij}}{\underset{i}{opt}\, x_{ij}}, \text{ where } i=\overline{1,m};\ j=\overline{1,n}. \quad (4)$$

If opt value is max

$$\bar{x}_{ij} = \frac{\underset{i}{opt}\, x_{ij}}{x_{ij}}, \text{ where } i=\overline{1,m};\ j=\overline{1,n}. \quad (5)$$

If opt value is min.

3.2.2. Calculating WASPAS weighted and normalized decision making matrix for summarizing part:

$$\bar{\bar{x}}_{ij,sum} = \bar{x}_{ij} q_j, \text{ where } i=\overline{1,m};\ j=\overline{1,n}. \quad (6)$$

3.2.3. Calculating WASPAS weighted and normalized decision making matrix for multiplication part:

$$\bar{\bar{x}}_{ij,mult} = \bar{x}_{ij}^{\,q_j}, \text{ where } i=\overline{1,m};\ j=\overline{1,n}. \quad (7)$$

3.2.4. Final calculating for evaluating and prioritizing alternatives based on:

$$WPS_i = 0.5\sum_{j=1}^{n} \bar{\bar{x}}_{ij} + 0.5\prod_{j=1}^{n} \bar{\bar{x}}_{ij}, \text{ where } i=\overline{1,m};\ j=\overline{1,n}. \quad (8)$$

All the researches based on the WASPAS method up to now are described in several references as below:

Zavadskas *et al.* (2012) in developing WASPAS as a new methodology; Hashemkhani Zolfani *et al.* (2013c) in decision making on business issues with foresight perspective; Dejus and Antucheviciene (2013) in assessment of health and safety solutions at a construction site; Šiožinytė and Antuchevicienė (2013) in solving the problems of daylighting and tradition continuity in a reconstructed vernacular building, Hashemkhani Zolfani *et al.* (2016a) in presenting Prospective Multiple Attribute Decision Making (PMADM) and Hashemkhani Zolfani *et al.* (2016b) in presenting Multiple Attribute Decision Making (MADM) based scenarios.

4. Experts' information

In the process of this research, eight experts were participating. They were from different field of expertise such as: Chemistry, Environmental engineering, Chemical engineering, Electrical engineering, Industrial engineering, Mechanical engineering and Civil engineering. All the general information about them is shown in Table 4.

5. Findings

As mentioned before, this research is organized based on two MCDM methods as a hybrid method. This section is includes two sections. The first part is results of SWARA method and the second part is related to the results of WASPAS method. These two sections are prepared based on experts' evaluation.

5.1. SWARA's result

In this section priority of criteria and sub-criteria is shown. All information about priority and weights of criteria and sub-criteria is illustrated in Tables 5–8. S_js are calculated based on average of experts' evaluations. Priority of all criteria is defined again based on experts' evaluation similar to SWARA's methodology.

The weightage of main criteria and their priority are shown in Table 5. Environmental criteria were the most important in general.

Table 4. Background information of experts

	1	2	3	4	5	6	7	8
Gender	female	male	male	male	female	female	male	male
Field of Study	Chemistry	Environmental engineering	Chemical engineering	Chemistry	Electrical engineering	Industrial Engineering	Mechanical engineering	Civil engineering
Education Background	PhD	Master	PhD	PhD	Master	PhD	Bachelor	Bachelor
Work experience	15	18	20	10	10	12	16	14

Table 5. Final results of SWARA method in weighting criteria

Criterion	The comparative importance of average value s_j	Coefficient $k_j = s_j + 1$	Recalculated weight $w_j = \frac{x_{j-1}}{k_j}$	Weight $q_j = \frac{w_j}{\sum w_j}$
C_1		1	1.0000	0.3305
C_2	0.2188	1.2188	0.8205	0.2712
C_3	0.2313	1.2313	0.6664	0.2203
C_4	0.2375	1.2375	0.5385	0.1780

Table 6. Final results of SWARA method in weighting sub-criteria of environmental

Criterion	The comparative importance of average value s_j	Coefficient $k_j = s_j + 1$	Recalculated weight $w_j = \frac{x_{j-1}}{k_j}$	Weight $q_j = \frac{w_j}{\sum w_j}$	Final weights
C_{1-1}		1	1	0.5416	0.1790
C_{1-2}	0.1813	1.1813	0.8465	0.4584	0.1515

Table 7. Final results of SWARA method in weighting sub-criteria of economic

Criterion	The comparative importance of average value s_j	Coefficient $k_j = s_j + 1$	Recalculated weight $w_j = \frac{x_{j-1}}{k_j}$	Weight $q_j = \frac{w_j}{\sum w_j}$	Final weights
C_{2-1}		1.0000	1.0000	0.4001	0.1085
C_{2-2}	0.2313	1.2313	0.8121	0.3249	0.0881
C_{2-3}	0.1813	1.1813	0.6875	0.2750	0.0746

Table 8. Final results of SWARA method in weighting sub-criteria of chemical

Criterion	The comparative importance of average value s_j	Coefficient $k_j = s_j + 1$	Recalculated weight $w_j = \frac{x_{j-1}}{k_j}$	Weight $q_j = \frac{w_j}{\sum w_j}$	Final weights
C_{3-2}		1	1.0000	0.3167	0.0698
C_{3-3}	0.1938	1.1938	0.8377	0.2653	0.0584
C_{3-4}	0.1750	1.175	0.7129	0.2258	0.0497
C_{3-1}	0.1750	1.175	0.6067	0.1922	0.0423

The final weightage of each sub-criteria is calculated based on its criteria. Results of SWARA method are applied in next step.

5.2. WASPAS's results

Evaluating of four different alternatives is organized in this section based on WASPAS methodology. All alternatives evaluated based on criteria and sub-criteria. All the procedures of this methodology are shown in Tables 9–13.

Different sort of AOPs were ranked base on SWARA-WASPAS results. Five potential advance oxidation processes in this paper have been introduced as alternatives for wastewater purification plant. Priority of alternatives is shown in Table 13 based on calculating formula 4 and

Table 9. Decision making matrix

	C_1		C_2			C_3				C_4
	C_{1-1}	C_{1-2}	C_{2-1}	C_{2-2}	C_{2-3}	C_{3-1}	C_{3-2}	C_{3-3}	C_{3-4}	C_4
Weight	0.1790	0.1515	0.1085	0.0881	0.0746	0.0423	0.0698	0.0584	0.0497	0.1780
	Max	Max	Min	Min	Min	7 (Opt)	Min	Min	Max	Max
Ozonation	52.85	1.40	88639	546.59	2070	7	82.93	54.41	1.23×10^{-3}	2010
Fenton	64.25	17.20	8568.50	6.20	1	2.9	50	1017.21	5.2×10^{-4}	2011
Electrochemical Oxidation	88.90	1.05	50 210	12.19	219.28	7.40	150	5025	4.79×10^{-3}	2010
UV/ Photo-catalysis	85	12.3	70 400	27.30	31.75	7.40	153.75	434	2.87×10^{-3}	2010
UV/H_2O_2	87.50	5.10	10 800	50.50	5.4	6.30	33.75	696	5.67×10^{-5}	2009

* All the criteria are real data. Also there is an exception and that is C_{3-1}. The optimum rate for this criterion is 7 and max or min isn't suitable for this criteria.

Table 10. WASPAS normalized decision making matrix

	C_1		C_2			C_3				C_4
	C_{1-1}	C_{1-2}	C_{2-1}	C_{2-2}	C_{2-3}	C_{3-1}	C_{3-2}	C_{3-3}	C_{3-4}	C_4
Weight	0.1790	0.1515	0.1085	0.0881	0.0746	0.0423	0.0698	0.0584	0.0497	0.1780
	Max	Max	Min	Min	Min	7 (Opt)	Min	Min	Max	Max
Ozonation	0.5945	0.0814	0.0967	0.0113	0.0005	1.0000	0.4070	1.0000	0.2568	0.9995
Fenton	0.7227	1.0000	1.0000	1.0000	1.0000	2.4138	0.6750	0.0535	0.1086	1.0000
Electrochemical Oxidation	1.0000	0.0610	0.1707	0.5086	0.0046	0.9459	0.2250	0.0108	1.0000	0.9995
UV/ Photocatalysis	0.9561	0.7151	0.1217	0.2271	0.0315	0.9459	0.2195	0.1254	0.5992	0.9995
UV/H_2O_2	0.9843	0.2965	0.7934	0.1228	0.1852	1.1111	1.0000	0.0782	0.0118	0.9990

This table is calculated based on formula 1.

Table 11. WASPAS weighted and normalized decision making matrix for summarizing part

	C_1		C_2			C_3				C_4
	C_{1-1}	C_{1-2}	C_{2-1}	C_{2-2}	C_{2-3}	C_{3-1}	C_{3-2}	C_{3-3}	C_{3-4}	C_4
Ozonation	0.1064	0.0123	0.0105	0.0010	0.0000	0.0423	0.0284	0.0584	0.0128	0.1779
Fenton	0.1294	0.1515	0.1085	0.0881	0.0746	0.1021	0.0471	0.0031	0.0054	0.1780
Electrochemical Oxidation	0.1790	0.0092	0.0185	0.0448	0.0003	0.0400	0.0157	0.0006	0.0497	0.1779
UV/ Photocatalysis	0.1711	0.1083	0.0132	0.0200	0.0023	0.0400	0.0153	0.0073	0.0298	0.1779
UV/H_2O_2	0.1762	0.0449	0.0861	0.0108	0.0138	0.0470	0.0698	0.0046	0.0006	0.1778

This table is calculated based on formula 2.

Table 12. WASPAS weighted and normalized decision making matrix for multiplication part

	C_1		C_2			C_3				C_4
	$C_{1\text{-}1}$	$C_{1\text{-}2}$	$C_{2\text{-}1}$	$C_{2\text{-}2}$	$C_{2\text{-}3}$	$C_{3\text{-}1}$	$C_{3\text{-}2}$	$C_{3\text{-}3}$	$C_{3\text{-}4}$	C_4
Ozonation	0.9111	0.6838	0.7761	0.6739	0.5658	1.0000	0.9392	1.0000	0.9347	0.9999
Fenton	0.9435	1.0000	1.0000	1.0000	1.0000	1.0380	0.9729	0.8428	0.8955	1.0000
Electrochemical Oxidation	1.0000	0.6547	0.8254	0.9422	0.6689	0.9977	0.9011	0.7677	1.0000	0.9999
UV/ Photo-catalysis	0.9920	0.9505	0.7957	0.8776	0.7726	0.9977	0.8996	0.8858	0.9749	0.9999
UV/H_2O_2	0.9972	0.8318	0.9752	0.8313	0.8818	1.0045	1.0000	0.8617	0.8021	0.9998

This table is calculated based on formula 3.

Table 13. The results of WASPAS

	$0.5\sum_{j=1}^{n=} Xij, sum$	$0.5\prod_{j=1}^{n=} Xij, mult$	WSP_i	Ranking
Ozonation	0.2250	0.0809	0.3059	5
Fenton	0.4439	0.3596	0.8035	1
Electrochemical Oxidation	0.2679	0.1175	0.3855	4
UV/ Photocatalysis	0.2927	0.1971	0.4898	3
UV/H_2O_2	0.3158	0.2058	0.5216	2

in final column shows ranking. After creating decision matrix and other steps for receiving the final ranking, according to Table 13 which shows the ultimate results of WASPAS methodology, Alternative 2 (Fenton) is the best reagent for a chemical treated section in industrial wastewater plant.

Conclusions

In the current era many techniques are proposed to treat industrial wastewater especially chemical treatment which has been developed to help the wastewater treatment systems. Selecting an appropriate chemical treatment method which can rely on less expenses and more environmental impacts can be interesting for decision makers in managing water pollution crises. Since Advance Oxidation Processes system (AOPs) is one of the leading chemical treatments method in degradation and mineralization of pollutants, this research has been accomplished based on it. This research followed five important sorts of AOPs: Ozonation, Fenton, electrochemical Oxidation, UV/Photo-catalysis, and UV/H_2O_2 which are compared by MCDM methods. SWARA is applied for evaluating and weighting the criteria and WASPAS method is deployed for evaluating distinctive alternatives of the AOPs.

Different case studies have been chosen to check the applicability of influential factors such as ability of each method to mineralize the pollution, capital and operating costs, pH level as a significant factor of industry outlet and the amount of chemicals that are used in each technology. Eight experts from distinct scientific fields participated in this research and integrated SWARA-WASPAS decision model showed that the Fenton process is the best option in AOPs area for removing many hazardous and organic pollutants regarding environmental, economical and chemical factors. The results of this investigation can be used for decision making in real cases of future researches. Also, governments can use this methodology for basic wastewater treatment process.

References

Alimardani, M.; Hashemkhani Zolfani, S.; Aghdaie, M.; Tamosaitiene, J. 2013. A novel hybrid SWARA and VIKOR methodology for Supplier selection in an agile environment, *Technological and Economic Development of Economy* 19(3): 533–548. https://doi.org/10.3846/20294913.2013.814606

Autin, O.; Hartb, J.; Jarvis, P.; MacAdam, J.; Parsonsa, S.; Jefferson, B. 2013. Comparison of UV/TiO_2 and UV/H_2O_2 processes in an annular photoreactor for removal of micro-pollutants: influence of water parameters on metaldehyde removal, quantum yields and energy consumption, *Applied Catalysis B: Environmental* 138–139: 268–275. https://doi.org/10.1016/j.apcatb.2013.02.045

Bitarafan, M.; Hashemkhani Zolfani, S.; Lale Arefi, S.; Zavadskas, E. K.; Mahmoudzadeh, A. 2014. Evaluation of real-time intelligent sensors for structural health monitoring of bridges based on SWARA-WASPAS; A case in Iran, *Baltic Journal of Road and Bridge Engineering* 9(4): 333–340. https://doi.org/10.3846/bjrbe.2014.40

Cabral da Silva, A. J.; Santos, E. V.; Carla de Oliveira Morais, C.; Martinez-Huitle, C. A.; Leal Castro, S. S. 2013. Electrochemical treatment of fresh, brine and saline produced water generated by petrochemical industry using Ti/IrO2–Ta2O5 and BDD in flow reactor, *Chemical Engineering* 233: 47–55. https://doi.org/10.1016/j.cej.2013.08.023

Canizares, P.; Hernandez-Ortegab, M.; Rodrigoa, M. A.; Barrera-Diazb, C. E.; Roa-Morales, G.; Saeza, C. 2009b. A comparison between Conductive-Diamond Electrochemical Oxidation and other Advanced Oxidation Processes for the treatment of synthetic melanoidins, *Hazardous Materials* 164: 120–125. https://doi.org/10.1016/j.jhazmat.2008.07.134

Canizares, P.; Paz, R.; Saez, C.; Rodrigo, M. 2009a. Costs of the electrochemical oxidation of wastewaters: a comparison with ozonation and Fenton oxidation processes, *Environmental Management* 90: 410–420. https://doi.org/10.1016/j.jenvman.2007.10.010

Carra, I.; Ortega-Gómez, E.; Santos-Juanes, L.; Casas López, J. L.; Sánchez Pérez, J. A. 2013. Cost analysis of different hydrogen peroxide supply strategies in the solar photo-Fenton process, *Chemical Engineering* 224: 75–81. https://doi.org/10.1016/j.cej.2012.09.067

Chang, E. E.; Hsing, H. J.; Chiang, P. C.; Chenc, M. Y.; Shyng, J. Y. 2008. The chemical and biological characteristics of coke-oven wastewater by ozonation, *Hazardous Materials* 156: 560–567. https://doi.org/10.1016/j.jhazmat.2007.12.106

Chen, Y.; Lo, S.; Kuo, J. 2011. Effects of titanate nanotubes synthesized by a microwave hydrothermal method on photocatalytic decomposition of perfluorooctanoic acid, *Water Research* 45: 4131–4140. https://doi.org/10.1016/j.watres.2011.05.020

Dejus, T.; Antucheviciene, J. 2013. Assessment of health and safety solutions at a construction site, *Journal of Civil Engineering and Management* 19(5): 728–737. https://doi.org/10.3846/13923730.2013.812578

Esplugas, S.; Gimenez, J.; Contreras, S.; Pascual, E.; Rodrıguez, M. 2002. Comparison of different advanced oxidation processes for phenol degradation, *Water Research* 36(4): 1034–1042. https://doi.org/10.1016/S0043-1354(01)00301-3

Ghorshi Nezhad, M. R.; Hashemkhani Zolfani, S.; Moztarzadeh, F.; Zavadskas, E. K.; Bahrami, M. 2015. Planning the priority of high tech industries based on SWARA-WASPAS methodology: the case of the nanotechnology industry in Iran, *Economic Research-Ekonomska Istraživanja* 28(1): 1111–1137. https://doi.org/10.1080/1331677X.2015.1102404

Gómez-Pacheco, C. V.; Sánchez-Polo, M.; Rivera-Utrilla, J.; López-Penalver, J. 2011. Tetracycline removal from waters by integrated technologies based on ozonation and biodegradation, *Chemical Engineering* 178: 115–121. https://doi.org/10.1016/j.cej.2011.10.023

Haghnazar Kouchaksaraei, R.; Hashemkhani Zolfani, S.; Golabchi, M. 2015. Glasshouse locating based on SWARA-COPRAS approach, *International Journal of Strategic Property Management* 19(2): 111–122. https://doi.org/10.3846/1648715X.2015.1004565

Hashemkhani Zolfani, S.; Aghdaie, M. H.; Derakhti, A.; Zavadskas, E. K.; Morshed Varzandeh, M. H. 2013c. Decision making on business issues with foresight perspective; an application of new hybrid MCDM model in shopping mall locating, *Expert Systems with Applications* 40: 7111–7121. https://doi.org/10.1016/j.eswa.2013.06.040

Hashemkhani Zolfani, S.; Bahrami, M. 2014. Investment prioritizing in high tech industries based on SWARA-COPRAS approach, *Technological and Economic Development of Economy* 20(3): 534–553. https://doi.org/10.3846/20294913.2014.881435

Hashemkhani Zolfani, S.; Farrokhzad, M.; Turskis, Z. 2013b. Investigating on successful factors of online games based on explorer, *E & M: Ekonomie a Management* 16(2): 161–169.

Hashemkhani Zolfani, S.; Maknoon, R.; Zavadskas, E. K. 2016a. An introduction to Prospective Multiple Attributes Decision Making (PMADM), *Technological and Economic Development of Economy* 22(2): 309–326. https://doi.org/10.3846/20294913.2016.1150363

Hashemkhani Zolfani, S.; Maknoon, R.; Zavadskas, E. K. 2016b. Multiple Attribute Decision Making (MADM) based scenarios, *International Journal of Strategic Property Management* 20(1): 101–111. https://doi.org/10.3846/1648715X.2015.1132487

Hashemkhani Zolfani, S.; Salimi, J.; Maknoon, R.; Kildiene, S. 2015. Technology foresight about R&D projects selection; Application of SWARA method at the policy making level, *Inzinerine Ekonomika-Engineering Economics* 26(5): 571–580. https://doi.org/10.5755/j01.ee.26.5.9571

Hashemkhani Zolfani, S.; Saparauskas, J. 2013. New Application of SWARA method in prioritizing sustainability assessment indicators of energy system, *Inzinerine Ekonomika-Engineering Economics* 24(5): 408–414.

Hashemkhani Zolfani, S.; Seyed Agha Banihashemi, S. 2014. Personnel selection based on a novel model of game theory and MCDM approaches, in *8th International Scientific Conference "Business and Management 2014"*, 191–198.

Hashemkhani Zolfani, S.; Zavadskas, E. K.; Turskis, Z. 2013a. Design of products with both International and Local perspectives based on Yin-Yang balance theory and SWARA method, *Economic Research-Ekonomska Istraživanja* 26(2): 153–166. https://doi.org/10.1080/1331677X.2013.11517613

Keršulienė, V.; Zavadskas, E. K.; Turskis, Z. 2010. Selection of rational dispute resolution method by applying new step-wise weight assessment ratio analysis (SWARA), *Journal of Business Economics and Management* 11(2): 243–258. https://doi.org/10.3846/jbem.2010.12

Khan, D.; Samadder, S. R. 2015. A simplified multi-criteria evaluation model for landfill site ranking and selection based on AHP and GIS, *Journal of Environmental Engineering and Landscape Management* 23(4): 267–278. https://doi.org/10.3846/16486897.2015.1056741

Kim, Tae-Hun.; Kim, S. D.; Kim, H. Y.; Lim, S. J.; Lee, M.; Yu, S. 2012. Degradation and toxicity assessment of sulfamethoxazole and chlortetracycline using electron beam, ozone and UV, *Hazardous Materials* 227–228: 237–242. https://doi.org/10.1016/j.jhazmat.2012.05.038

Krichevskaya, M.; Klauson, D.; Portjanskaja, E.; Preis, S. 2011. The Cost Evaluation of advanced Oxidation Processes in Laboratory and Pilot-Scale Experiments, *International Ozone Association* 33: 211–223. https://doi.org/10.1080/01919512.2011.554141

Kucas, A. 2010. Location prioritization by means of multicriteria spatial decision-support systems: a case study of forest fragmentation-based ranking of forest administrative areas, *Journal of Environmental Engineering and Landscape Management* 18(4): 312–320. https://doi.org/10.3846/jeelm.2010.36

Leyva-Díaz, J. C.; López-López, C.; Martín-Pascual, J.; Muñío, M. M.; Poyatos, J. M. 2015. Kinetic study of the combined processes of a membrane bioreactor and a hybrid moving bed biofilm reactor-membrane bioreactor with advanced oxidation processes as a post-treatment stage for wastewater treatment, *Chemical Engineering and Processing* 91: 57–66. https://doi.org/10.1016/j.cep.2015.03.017

Liu, P.; He, L.; Yu, X. 2016. Generalized hybrid aggregation operators based on the 2-dimension uncertain linguistic information for multiple attribute group decision making, *Group Decision and Negotiation* 25(1): 103–126. https://doi.org/10.1007/s10726-015-9434-x

Liu, P.; Jin, F. 2012. A multi-attribute group decision-making method based on weighted geometric aggregation operators of interval-valued trapezoidal fuzzy numbers, *Applied Mathematical Modelling* 36(6): 2498–2509. https://doi.org/10.1016/j.apm.2011.09.006

Liu, P.; Jin, F., Zhang, X.; Su, Y.; Wang, M. 2011. Research on the multi-attribute decision-making under risk with interval probability based on prospect theory and the uncertain linguistic variables, *Knowledge-Based Systems* 24(4): 554–561. https://doi.org/10.1016/j.knosys.2011.01.010

Liu, P.; Shi, L. 2015. The generalized hybrid weighted average operator based on interval neutrosophic hesitant set and its application to multiple attribute decision making, *Neural Computing and Applications* 26(2): 457–471. https://doi.org/10.1007/s00521-014-1736-4

Liu, P.; Zhang, X.; Jin, F. 2012. A multi-attribute group decision-making method based on interval-valued trapezoidal fuzzy numbers hybrid harmonic averaging operators, *Journal of Intelligent & Fuzzy Systems* 23(5): 159–168.

Lizama, C.; Freer, J.; Baeza, J.; Mansilla, H. 2002. Optimized photodegradation of Reactive Blue 19 on TiO_2 and ZnO suspensions, *Catalysis Today* 76: 235–246. https://doi.org/10.1016/S0920-5861(02)00222-5

Mantzavinos, D.; Psillakis, E. 2004. Review Enhancement of biodegradability of industrial wastewaters by chemical oxidation pre-treatment, *Journal of Chemical Technology and Biotechnology* 79: 431–454. https://doi.org/10.1002/jctb.1020

Oppenländer, T. 2003. *Photochemical purification of water and air*, WILEY-VCH.

Prieto-Rodrıguez, L.; Oller, I.; Klamerth, N.; Aguera, A.; Rodriguez, E. M.; Malato S. 2013. Application of solar AOP and ozonation for elimination of micropollutants in municipal wastewater treatment plant effluents, *Water research* 47: 1521–1528. https://doi.org/10.1016/j.watres.2012.11.002

Ruzgys, A.; Volvačiovas, R.; Ignatavičius, Č.; Turskis, Z. 2014. Integrated evaluation of external wall insulation in residential buildings using SWARA-TODIM MCDM method, *Journal of Civil Engineering and Management* 20(1): 103–110. https://doi.org/10.3846/13923730.2013.843585

Saeed, T.; Sun, G. 2012. A review on nitrogen and organics removal mechanisms in subsurface flow constructed wetlands: dependency on environmental parameters, operating conditions and supporting media, *Environmental Management* 112: 429–448. https://doi.org/10.1016/j.jenvman.2012.08.011

Sharma, V. K.; Triantis, T. M.; Antoniou, M. G.; He, X.; Pelaez, M.; Han, C.; Song, W.; O'Shea, K. E.; de la Cruz, A. A.; Kaloudis, T.; Hiskia, A.; Dionysiou, D. D. 2012. Destruction of microcystins by conventional and advanced oxidation processes: a review, *Separation and Purification Technology* 91: 3–17. https://doi.org/10.1016/j.seppur.2012.02.018

Sievers, M. 2011. Advanced oxidation processes. Earth Systems and Environmental Sciences, *Treatise on Water Science* 4: 377–408. https://doi.org/10.1016/B978-0-444-53199-5.00093-2

Šiožinytė, E.; Antuchevičienė, J. 2013. Solving the problems of daylighting and tradition continuity in a reconstructed vernacular building, *Journal of Civil Engineering and Management* 19(6): 873–882. https://doi.org/10.3846/13923730.2013.851113

Stanujkic, D.; Karabasevic, D.; Zavadskas, E. K. 2015. A Framework for the Selection of a Packaging Design Based on the SWARA Method, *Inzinerine Ekonomika – Engineering Economics* 26(2): 181–187.

Teh, C. M.; Mohamed, A. R. 2011. Roles of titanium dioxide and ion-doped titanium dioxide on photocatalytic degradation of organic pollutants (phenolic compounds and dyes) in aqueous solutions: a review, *Journal of Alloys and Compounds* 509(5): 1648–1660. https://doi.org/10.1016/j.jallcom.2010.10.181

Tizaoui, C.; Mezughi, K.; Bickley, R. 2011. Heterogeneous photocatalytic removal of the herbicide clopyralid and its comparison with UV/H_2O_2 and ozone oxidation techniques, *Desalination* 273: 197–204. https://doi.org/10.1016/j.desal.2010.11.036

Turskis, Z.; Lazauskas, M.; Zavadskas, E. K. 2012. Fuzzy multiple criteria assessment of construction site alternatives for non-hazardous waste incineration plant in Vilnius City, Applying ARAS-F and AHP Methods, *Journal of Environmental Engineering and Landscape Management* 20(2): 111–120. https://doi.org/10.3846/16486897.2011.645827

Vafaeipour, M.; Hashemkhani Zolfani, S.; Morshed Varzandeh, M. H.; Derakhti, A.; Keshavarz Eshkalag, M. 2014. Assessment of regions priority for implementation of solar plants in Iran: new application of a hybrid multi-criteria decision making approach, *Energy Conversion and Management* 86: 653–663. https://doi.org/10.1016/j.enconman.2014.05.083

Zavadskas, E. K.; Turskis, Z.; Antucheviciene, J.; Zakarevicius, A. 2012. Optimization of weighted aggregated sum product assessment, *Elektronika ir Elektrotechnika* 122(6): 3–6. https://doi.org/10.5755/j01.eee.122.6.1810

Zavadskas, E. K.; Vilutiene, T. 2006. A multiple criteria evaluation of multi-family apartment block's maintenance contracts: I-Model for maintenance of contractor evaluation and determination of its selection criteria, *Building and Environment* 41: 621–632. https://doi.org/10.1016/j.buildenv.2005.02.019

Mohamad Reza KHODADADI received his B.S degree in Chemical Engineering in 2012 and a Master degree in Nanochemistry in 2014. He is currently pursuing a PhD in Faculty of chemistry, Picardie Jules Verne University, Amiens, France. Mohamad background and focus on chemistry and Nano technology has led him to tackle the various challenges in engineering, environment and sustainability.

Sarfaraz HASHEMKHANI ZOLFANI is PhD candidate of Technology Foresight in Amirkabir University of Technology (Tehran Polytechnic). He is in cooperating with Futures Studies Research Institute of Amirkabir University of Technology (Tehran Polytechnic) and Sustainability office of Amirkabir University of Technology (Tehran Polytechnic). He is part time research associate in council for futures studies, Iranian Academy of Sciences and adviser to the president, Iranian Society of Future Studies. He is Council member of International Association of Grey System and Uncertain Analysis (GSUA) and member of EURO Working Group *OR in Sustainable Development and Civil Engineering*. He is reviewer in journals like: *Technological and Economic Development of Economy, Technological Forecasting and Social Change, Evaluation and Program Planning, Journal of Business Economics and Management, International Journal of Strategic Property Management*, and etc. He is author of more than 50 scientific papers that presented, published or reviewed at/for International Conferences and Journals (including ISI-cited publications). He has published in journals such as: *Expert Systems with Applications, Energy Conversion and Management, Journal of Cleaner Production, Technological and Economic Development of Economy, Journal of Business Economics and Management, International Journal of Strategic Property Management, Archives of Civil and Mechanical Engineering*, and etc. His research interests include: Performance Evaluation, Strategic Management, Decision-making Theory, Supply Chain Management, (Dynamic/Prospective) Multi Criteria Decision Making, Futures Studies, Sustainable Development and Game Theory.

Morteza YAZDANI is a PhD Scholar in Business Economic at Universidad Europea de Madrid, Spain. His main research interests are application of multi criteria decision making in different fields as material science, supply chain management, construction management and also strategic planning. He is member of editorial board at International Journal of Business and Systems research (INDERSCIENCE) and reviewer of journal "Sustainability". He has published in top international journals such as Journal of Civil Engineering and Management, Technological and Economic Development of Economy, Expert Systems with Applications, Materials and Design and International Journal of Logistics Research and applications.

Edmundas Kazimieras ZAVADSKAS. PhD, DSc, h.c.multi. Prof., the Head of the Department of Construction Technology and Management at Vilnius Gediminas Technical University, Lithuania. Senior Research Fellow at the Research Institute of Smart Building Technologies. PhD in Building Structures (1973). DrSc in Building Technology and Management (1987). A member of Lithuanian and several foreign Academies of Sciences. Doctore Honoris Causa from Poznan, Saint Petersburg and Kiev universities. Honorary International Chair Professor of the National Taipei University of Technology. A member of international organizations; a member of steering and programme committees at many international conferences; a member of the editorial boards of several research journals; the author and co-author of more than 400 papers and a number of monographs in Lithuanian, English, German and Russian. The editor-in-chief of journals *Technological and Economic Development of Economy* and *Journal of Civil Engineering and Management*. Research interests: building technology and management, decision-making theory, automation in design and decision support systems.

5

AN INNOVATIVE CONTINUOUS FLOW BNR-IC PROCESS FOR NUTRIENTS REMOVAL AND PHOSPHORUS RECOVERY FROM SYNTHETIC AND REAL DOMESTIC WASTEWATER

Haiming ZOU[a,b], Xiwu LU[b]

[a] *School of Resource and Environment, Anhui Science and Technology University, Donghua Road, Fengyang 233100, PR China*

[b] *School of Energy and Environment, Southeast University, Sipailou Road, Nanjing 210096, PR China*

Abstract. An innovative continuous flow process linking biological nutrients removal (BNR) with induced crystallization (IC) was used to remove nutrients and recover phosphorus (P) from synthetic and real domestic wastewater. The results showed that a good nutrients removal performance was found regardless of feeding solutions. P recovery efficiency from synthetic wastewater was 70.2% slightly less than that from real domestic sewage (74.2%). Importantly, P recovery can effectively enhance the subsequent biological P removal. Polymerase chain reaction-denaturing gradient gel electrophoresis (PCR-DGGE) analysis displayed an obvious shift in microbial community structure when switching feeding synthetic solution to real wastewater. A total of 13 bands were detected in sludge samples using synthetic and real domestic sewage, affiliated with 8 phyla or classes domain Bacteria (*Alphaproteobacteria, Betaproteobacteria, Gammaproteobacteria, Flavobacteria, Actinobacteria, Sphingobacteria, Epsilonproteobacteria and Chlorobia*). The results obtained here suggest that the continuous flow BNR-IC process is feasible for nutrients removal and P recovery from domestic sewage and is a promising technology for wastewater treatment combined with recycling of P elements.

Keywords: nutrients removal, phosphorus recovery, induced crystallization, denitrifying polyphosphate accumulating organisms, denaturing gradient gel electrophoresis.

Introduction

Phosphorus (P) is one of critical elements required for biomass growth. According to the Liebig law of the minimum, the biomass (experience formula: $C_{106}H_{263}O_{110}N_{16}P$) (Moheimani *et al.* 2013) growth productivity is mainly determined by the supply of P in water bodies. Nowadays, eutrophication or algae blooming of water bodies is an increasing pollution problem worldwide due to the excessive discharge of P loading into lakes, rivers or reservoirs. Except Oceania and Africa, about 50% of all reservoirs and lakes are subject to eutrophication in all continents (Sengupta, Pandit 2011). In China, lakes of Dianchi, Chaohu and Taihu are eutrophic. Currently, eutrophication is of particular concern in water bodies, since it results in the water environmental degradation, threatens the survival of aquatic species and affects the safety of drinking water. More than 0.02 mg/L of P in lakes can cause eutrophication (Seviour *et al.* 2006). In domestic sewage, the fluctuation of P concentration ranges from 3 mg/L to 10 mg/L, often requiring mandatory removal before discharge into water bodies. Consequently, a strict P limit in effluents from municipal wastewater treatment plants (WWTP) was set in legislation by governments in the world. For instance, there is a compulsory requirement of TP concentration in effluents from WWTP to be less than 0.1 mg/L in North America and 0.5 mg/L in China (Zou *et al.* 2014a). Methods for P removal from domestic wastewater are commonly biological, physical or chemical. Compared with chemical or physical methods, recent advances in enhance biological phosphorus removal (EBPR) have demonstrated that it was regards as an effective and economic technology for P removal from domestic wastewater (Ong *et al.* 2014; Yang *et al.* 2013; Zhang *et al.* 2013).

Today, another a notable aspect of phosphorus problem is that P is a non-renewable mineral resource, mostly obtained from phosphate rock. If current exploitation rate of phosphate rock is continued it is predicted to run out in next 50 years (Gilbert 2009). Thus, it is very urgent to reuse and recycle P from human activities, i.e. phosphorus

Corresponding author: Haiming Zou
E-mail: hmzou@126.com

recovery, which has drawn considerable attention in recent years (Acelas *et al.* 2014; Hutnik *et al.* 2013; Qiu, Ting 2014). It is noteworthy that about 3~10 mg/L of P in domestic sewage may provide a valuable potential market for P recovery. Current technology for P recovery from domestic wastewater mainly adopted chemical precipitation and the commonly used processes were the two types of struvite (NH_4MgPO_4, MAP) (Cusick, Logan 2012; Ichihashi, Hirooka 2012) and hydroxylapatite ($Ca_{10}(PO_4)_6(OH)_2$, HAP) (Zou *et al.* 2014b), which were regarded as a slow release fertilizer (Yetilmezsoy, Sapci-Zengin 2009). In addition to, Al and Fe ions also have a great affinity for P binding (Westholm 2006). Since phosphate fertilizer accounted for about 80% of used phosphate rock in the world (Guney *et al.* 2008), MAP or HAP (similar to fertilizer component) precipitation should be preferred to recover P element from solution.

The discussion above shows that both natural phosphate rock diminution and pollution due to P excessive discharge should be solved quickly. This may lead to a new research on P removal coupled with P recovery from domestic sewage to meet the effluent standards and recycle phosphate rock. Recently, several attempts have been conducted to develop new combination process that remove nutrients and recover P (Ichihashi, Hirooka 2012; Song *et al.* 2014; Yuan *et al.* 2012), in particular to investigate the P recovery from excess sludge. However, most of those reported on the combination process about biological nutrients removal coupled with chemical precipitation. Of them, a well-known wastewater treatment process linking nutrients removal with P recovery is the Phostrip (Yuan *et al.* 2012), adopting biological nutrients removal combined with chemical precipitation, where the P in supernatant generated from excess sludge digestion was recovered by MAP or HAP precipitation. The disadvantages of the process are obvious: (1) increasing cost in excess sludge digestion system; (2) lower P recovery efficiency due to recover P only from excess sludge; (3) chemical precipitation product (such as MAP or HAP) is not directly used to industrial or agricultural activities owe to contain high moisture or impurity content, where the induced crystallization technology may be a preferable alternative to recovery P from solution.

More recently, our research team developed an innovative process of biological nutrients removal (BNR) coupled with induced crystallization (IC), referred herein to as "BNR-IC", responsible for P removal with P recovery (Shi *et al.* 2012; Zou *et al.* 2014a). The process overcame disadvantages of conventional process mentioned above. Although good nutrients removal and P recovery performances were found in our previous results via serial batch experiments (Zou *et al.* 2014c), the behavior of BNR-IC operation in continuous flow is not still unclear, especially for treatment real domestic wastewater rather than synthetic wastewater. Most studies on domestic wastewater treatment were nowadays performed by feeding the artificial wastewater. However, the component of real domestic sewage is more extensive and complex than that of synthetic wastewater, thus resulting in the infeasibility of application of results obtained from synthetic wastewater treatment to practice. Consistently, there is a large discrepancy in results obtained from between batch tests and continuous flow experiments. For this, in the innovative BNR-IC process, real domestic sewage treatment operated by continuous flow has to be performed in the further study.

Therefore, the aim of this study was to investigate whether stable and high nutrients removal and P recovery from synthetic and real domestic wastewater can be obtained in the continuous flow BNR-IC process. The carbon, nitrogen and phosphorus content in the influent and effluent were monitored. Furthermore, the differences of microbial community structure between feeding synthetic wastewater and feeding real domestic sewage were also compared by microbial analysis. The results obtained from here may serve as a good suggestion for application of the novel BNR-IC process to practice.

1. Materials and methods

1.1. Experimental set-up and operation

The BNR-IC process (Fig. 1, left) proposed here consisted of biological nutrients removal (BNR) system and induced crystallization (IC) P recovery column (Fig. 1, right). The BNR system is an anaerobic/anoxic/aerobic process with double-sludge, in which denitrifying polyphosphate accumulating organisms (DPAO) performed denitrifying simultaneous nitrogen and phosphorus removal. The IC column was comprised of crystallization, buffer and settling zones, which was proposed and described by our

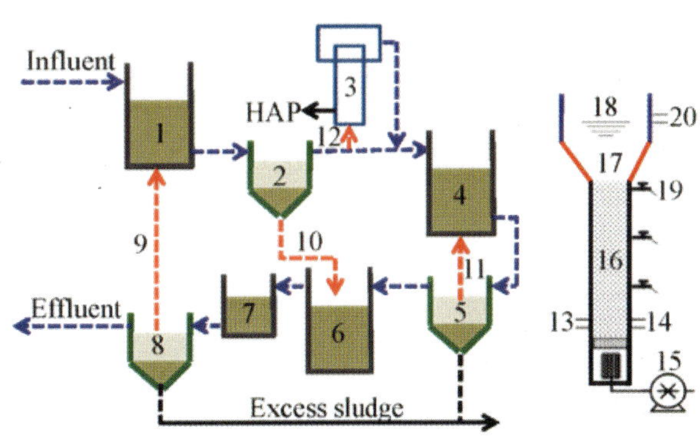

Fig. 1. Schematic diagram of the BNR-IC process.1– anaerobic tank; 2– settling tank; 3– IC column; 4– nitrification tank; 5– settling tank; 6– anoxic tank; 7– post-aeration tank; 8– settling tank; 9– sludge return; 10– sludge bypass return; 11– sludge return; 12– later flow; 13– influent of P-rich solution; 14– feeding CaCl2 solution; 15– aeration; 16– IC reaction zone; 17– buffer zone; 18– settling zone; 19– sample point; 20– effluent

research group (Zou *et al.* 2014b). In the BNR-IC process, the P-rich supernatant in settling tank 2 was partly introduced into IC column to recover P and the sludge from settling tank 2 was led into anoxic tank 6 for denitrifying simultaneous nitrogen and phosphorus removal. The post-aeration tank enhanced the final effluent quality.

Synthetic wastewater and real domestic sewage were used and compared in this study. The influent flow was 18 L/d, and the bypass sludge flow (10), return sludge flow (5, 8) were all 7.2 L/d. The later flow ratio (12) was 6.3 L/d, i.e. 35% of supernatant from settling tank 2 introduced into the IC column, where 4.5 mM $CaCl_2$ solution was led into the IC column with the flow rate of 0.63 L/d. The aeration flow was 90 L/h both in aerobic tank and post-aeration tank, resulting in the dissolve oxygen (DO) high than 2 mg/L and that was 270 L/h in IC column, for CO_2 stripping and fluidization of seed crystals (calcite used here).

1.2. Synthetic and real domestic wastewater

The synthetic domestic wastewater contained carbon, nitrogen and phosphorus in the typical species of real domestic sewage, specific composition detailed in Table 1. COD, NH_4^+-N, TN and TP concentrations in the influent were 239.2~259.5 mg/L, 38.2~41.8 mg/L, 39.6~43.8 mg/L and 8.72~11.40 mg/L, respectively. Real domestic wastewater was collected from the Southeast University, Wuxi, China and its main characteristics were described in Table 2.

Table 1. Composition of synthetic wastewater used in this study.

Composition of feeds	Concentration, g/L	Composition of nutrient solution	Concentration, g/L
CH_3COONa	0.322	$FeCl_3·6H_2O$	1.50
KH_2PO_4	0.044	H_3BO_3	0.15
$(NH_4)_2SO_4$	0.047	$CuSO_4·5H_2O$	0.03
$CaCl_2$	0.005	KI	0.18
$MgSO_4·7H_2O$	0.05	$MnCl_2·4H_2O$	0.12
Nutrient solution	0.30 mL/L	$Na_2MoO_4·2H_2O$	0.06
		$ZnSO_4·7H_2O$	0.12
		$CoCl_2·6H_2O$	0.15
		EDTA	10.00

Table 2. Characteristics of real domestic sewage used in this study.

Contents	Range	Average
COD (mg/L)	153.71-223.63	188.94 ± 16.26
NH_4^+-N (mg/L)	38.50-55.72	48.26 ± 3.85
TN (mg/L)	50.35-67.28	60.01 ± 3.85
TP (mg/L)	3.76-6.43	5.03 ± 0.67
pH	7.08-7.69	7.36 ± 0.19

1.3. Experimental procedure

Before the continuous flow experiments, the two types of activated sludge responsible for phosphorus removal and nitrification have been cultivated in batch tests (Haiming *et al.* 2014), i.e. the start-up of the BNR-IC system. And then, the operating in continuous flow was conducted and divided into two main phases described as follows: (1) from day 0 to day 30, the BNR-IC was operated by feeding synthetic wastewater; (2) at day 31, the real domestic sewage treatment was started for 60 days. During this period, the mixed liquid suspended solid (MLSS) were 3400~3600 mg/L both in anaerobic and anoxic tank and that was 3000~3200 mg/L in nitrification tank. The solids retention times (SRTs) of nitrification sludge and phosphorus removal sludge were 12 and 16 days, respectively.

1.4. Analytical methods

50 mL of samples collected (once a day) from influent and effluent of the BNR-IC were filtered through 0.45-micron membrane filters using a microfiltration apparatus before analysis. COD and MLSS were monitored in accordance with the standard methods (APHA, 2005). TP, NH_4^+-N and TN were performed by using segmented flow analysis (AutoAnalyzer3, SEAL, UK). DO was determined by a DO meter analyzer (YSI DO200, USA) and pH was assessed by a pH meter analyzer (YSI pH100, USA). Statistical treatment (arithmetic mean and standard deviations) of experimental data was done using ORIGIN software (Microcal Software, Northhampton, MA).

Sludge samples (100 mL) were collected at day 30 and 90 for Polymerase Chain Reaction-Denaturing Gradient Gel Electrophoresis (PCR-DGGE) analysis, described in our previous report (Zou *et al.* 2014c), mainly including sludge sample pretreatment, total DNA extraction, PCR amplification, DGGE analysis and sequencing of PCR-amplified 16S ribosomal (r) DNA.

2. Results and discussion

2.1. Removal of nutrients from synthetic and real domestic wastewater

Figures 2 and 3 and Table 3 show the carbon, nitrogen and phosphorus concentrations of influent and effluent in the BNR-IC process, using synthetic and real domestic wastewater. In the final periods of operation, a similar nutrients removal behavior was found in the system at day 30 and day 90, respectively. At the stable state, COD, NH_4^+-N, TN and TP concentrations of effluent remained at a low level and were on average 18.75±3.50 mg/L, 5.64±0.26 mg/L, 8.99±0.97 mg/L and 0.42±0.03 mg/L in synthetic wastewater and 25.35±9.72 mg/L, 4.35±0.52 mg/L, 11.89±1.54 mg/L and 0.46±0.02 mg/L in real domestic sewage, respectively, meeting the discharge standard of

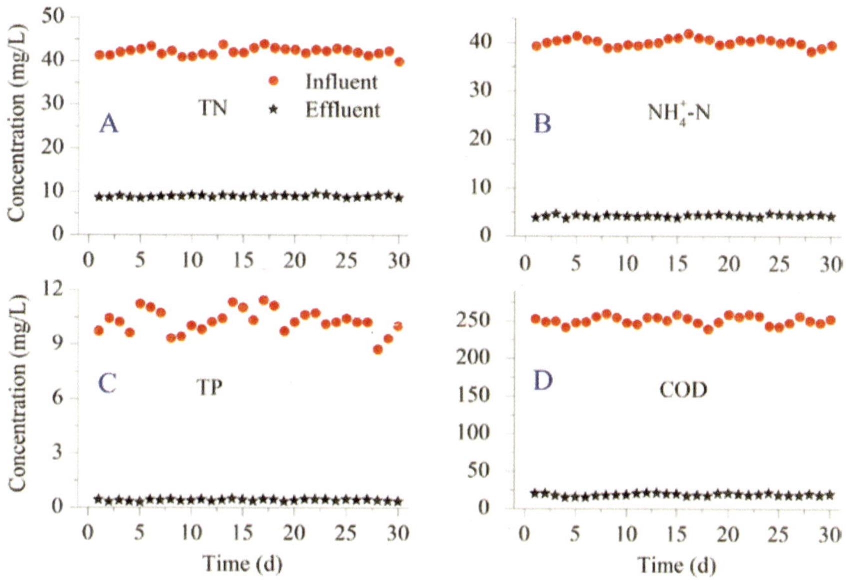

Fig. 2. Variations of COD, NH4-N, TN and TP in the influent and effluent and removal efficiency during synthetic wastewater treatment

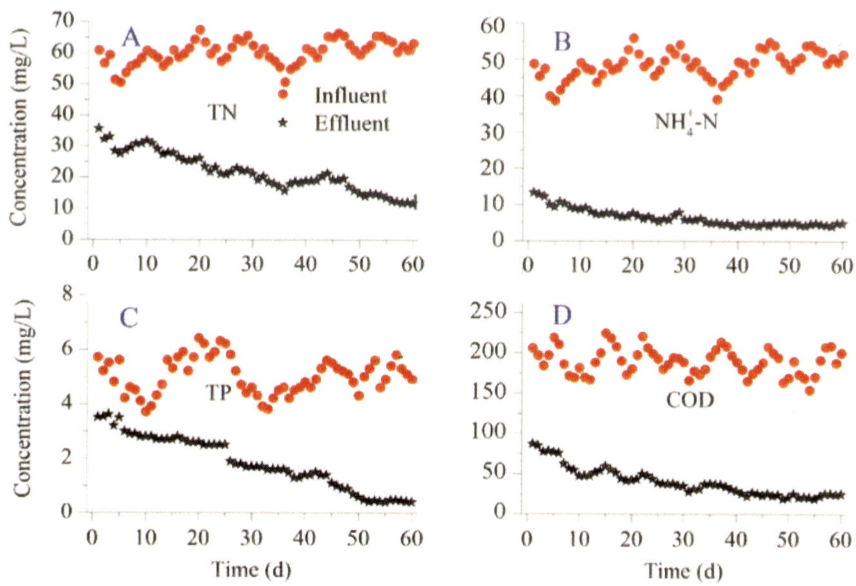

Fig. 3. Variations of COD, NH4-N, TN and TP in the influent and effluent and removal efficiency during real domestic sewage treatment

Table 3. Averaged concentrations of COD, NH_4-N, TN and TP in the influent and effluent and those averaged removal efficiencies when synthetic or real domestic wastewater

Contents (mg/L)	Synthetic domestic wastewater			Real domestic wastewater		
	Influent	Effluent	RE (%)	Influent	Effluent	RE (%)
COD	250.00 ± 2.30	18.75 ± 3.50	92.51 ± 1.41	188.94 ± 16.26	25.35 ± 9.72	87.33 ± 2.05
NH_4^+-N	40.00± 0.82	5.64 ± 0.26	85.93 ± 1.28	48.26 ± 3.85	4.35 ± 0.52	90.21 ± 1.08
TN	42.00 ± 1.21	8.99 ± 0.97	78.62 ± 1.36	60.01 ± 3.85	11.89 ± 1.54	79.66 ± 2.09
TP	10.00 ± 0.46	0.42 ± 0.03	95.20 ± 0.99	5.03 ± 0.67	0.46 ± 0.02	90.67 ± 1.18

municipal domestic sewage treatment (GB 18918-2012) formulated by Chinese government. The low P content in effluent was attributed to the joint action of induced crystallization P recovery and biological phosphorus removal by DPAO. In conventional wastewater treatment process, P content of effluent was hard to meet the stringent discharge standard due to the lack of carbon in real domestic sewage, i.e. low C/N. However, in the BNR-IC system, although the C/N ratios (5.9±0.8 in synthetic wastewater and 3.2±0.3 in real domestic sewage) were still low, P concentrations of effluent were both less than 0.5 mg/L, this suggesting that P recovery effectively enhance the subsequent biological P removal by DPAO due to the decrease at P loading in solution. Another reason may be that DPAO mainly performed the denitrifying simultaneous nitrogen and phosphorus removal in anoxic tank, considerably saving the consumption of carbon source (Lv *et al.* 2015).

During the treatment of synthetic wastewater (30 days), the COD, NH_4^+-N, TN and TP removal performances (Fig. 2) were still constant stable and more than 90% of C and P and around 80% of N were removed from synthetic solution, where those removal efficiencies were 92.51±1.41%, 85.93±1.28%, 78.62±1.36% and 95.20±0.99%, respectively. This is mainly due to the fact that two types of activated sludge responsible for P removal and nitrification were cultured by feeding synthetic wastewater.

For real domestic sewage treatment, as was expected, the nutrients removal efficiencies were extremely low and fluctuant during the beginning period (Fig. 3). With the operation in continuous flow, carbon, nitrogen and phosphorus concentrations in effluent decreased gradually and those removal efficiencies finally remained stable and high after approximately 55 days. The COD, NH_4^+-N, TN and TP removal efficiencies were 87.33±2.05%, 90.21±1.08%, 79.66±2.09% and 90.67±1.18%, respectively, in which, compared with synthetic wastewater treatment, both COD and P removal rates were slightly lower while the N removals including NH_4^+-N and TN were somewhat higher. The lower COD and P removal performances probably result from a decrease in DPAO when treating real domestic sewage. A high content of microorganisms responsible for only denitrifying rather than denitrifying simultaneous nitrogen and phosphorus in domestic wastewater may lead to the high nitrogen removal performance.

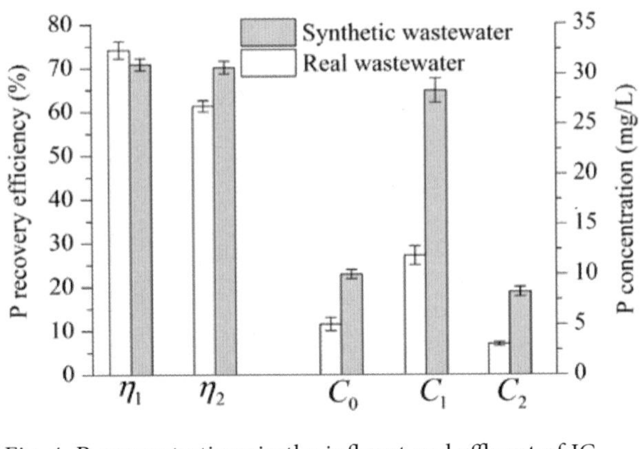

Fig. 4. P concentrations in the influent and effluent of IC column and its recovery efficiencies from synthetic and real wastewater

2.2. Recovery of phosphorus from synthetic and real domestic wastewater

This study introduces an innovative technology and approach to recover P from effluent of anaerobic phase by induced crystallization in the BNR-IC system. The P removal efficiency in IC column and recovery efficiency per liter of wastewater were calculated with the following equations (1) and (2), respectively.

$$\eta_1 = \frac{C_1 - C_2}{C_1} \times 100\%\,; \tag{1}$$

$$\eta_2 = \frac{0.35\,(C_1 - C_2)}{C_0} \times 100\%\ , \tag{2}$$

where: C_0 is the P concentration in synthetic or real domestic wastewater (mg/L); C_1 is the P concentration in the effluent of settling tank 2 (mg/L); C_2 is the P concentration of IC column effluent (mg/L); 0.35 represents the later flow ratio.

Figure 4 presents the influent and effluent characteristics including C_0, C_1 and C_2, P removal efficiency in IC column (η_1) and P recovery efficiency in BNR-IC (η_2), calculated according to Eqs. (1) and (2), during treatment of synthetic and real wastewater.

As can be seen from Figure 4, the thickener P-rich supernatant was obtained in anaerobic tank (C_1) and partly sent to the induced crystallization column for recovering P. During the crystallization process, P content in the solutions fed to the IC column gradually decreased ranging from 28.31 to 8.35 mg/L for synthetic wastewater and 11.85 to 3.06 mg/L for real domestic wastewater, displaying a good P removal performance during induced crystallization process. A specific P concentration of 10 mg/L in the synthetic wastewater was assumed, with a resulting 70.2% of P recovery efficiency. Nevertheless, when P concentration was 5.03 mg/L in real domestic sewage, 74.2% of P recovery efficiency was found. This suggests that lower P concentration in wastewater may lead to a higher P recovery efficiency.

2.3. Change of microbial community structure with feeding synthetic and real domestic wastewater

A total of 13 bands with strong intensity were observed in the two DGGE band patterns, as given in Figure 5. Although there was no significant difference on nutrients

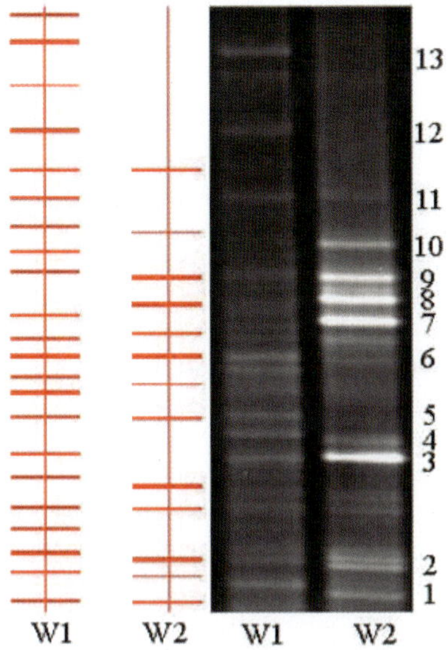

Fig. 5. DGGE fingerprint of 16Sr DNA fragments (right) and lanes comparison (left). W1-real domestic wastewater; W2-synthetic domestic wastewater

Table 4. NCBI BLAST search results of sequences from DGGE bands.

Band	Closest match	Similarity %	Phylogenetic affiliation
1	*Acinetobacter spp.*	98	*Gammaproteobacteria*
2	*Rhodocyclus spp.*	96	*Betaproteobacteria*
3	*Flavobacterium spp.*	100	*Flavobacteria*
4	*Brachymonas spp.*	98	*Betaproteobacteria*
5	*Tetrasphaera spp.*	99	*Actinobacteria*
6	*Dechlorosoma spp.*	95	*Betaproteobacteria*
7	*Pseudomonas spp.*	100	*Gammaproteobacteria*
8	*Terrimonas spp.*	100	*Sphingobacteria*
9	*Thermomonas spp.*	100	*Gammaproteobacteria*
10	*Simplicispira spp.*	100	*Betaproteobacteria*
11	*Arcobacter spp.*	100	*Epsilonproteobacteria*
12	*Chlorobaculum spp.*	87	*Chlorobia*
13	*Methylocystis spp.*	98	*Alphaproteobacteria*

removal efficiencies between using synthetic and real domestic wastewater, sludge sample with real domestic sewage displayed a larger diversity of microbial community structure than that with synthetic wastewater. This suggests that influent composition appears to have an influence on the microbial populations, which was consistent with the report (Boelee *et al.* 2011). This may be due to more complex compositions in real domestic sewage than that in synthetic wastewater, probably supporting other bacterial growth. Moreover, real domestic wastewater may be introduced in various microorganisms when passing through sewage conduits (Thai *et al.* 2014). Here, Dice coefficient, comparing the similarities of DGGE fingerprints from different feeding, was 0.42, also suggesting a large shift of microbial community structure when switching feeding synthetic wastewater to real domestic wastewater.

Of these 13 bands, 6 (band 1, 2, 3, 4, 6 and 11) bands were common to both the synthetic wastewater sludge and real domestic sewage sludge. The rest, 3 (band 5, 12 and 13) bands were only found in real domestic sewage sludge and another 4 (7, 8, 9 and 10) only presented in synthetic wastewater sludge. As can be seen in Table 4 and Figure 5, 13 distinct DGGE bands revealed that microbial species were affiliated with 8 phyla or classes domain Bacteria: *Alphaproteobacteria*, *Betaproteobacteria*, *Gammaproteobacteria*, *Flavobacteria*, *Actinobacteria*, *Sphingobacteria*, *Epsilonproteobacteria* and *Chlorobia*.

16S rRNA fragments of bands 3, 7, 8, 9, 10 and 11 belong to genus *Flavobacterium* spp., *Pseudomonas* spp., *Terrimonas* spp., *Thermomonas spp.*, *Simplicispira* spp. and *Arcobacter spp.* Of them, *Pseudomonas* spp., *Terrimonas* spp. and *Thermomonas spp.* with high content in the sludge were only found as feeding synthetic wastewater, probably responsible for P removal. *Tetrasphaera spp.* (band 5), *Chlorobaculum spp.* (band 12) and *Methylocystis spp.* (band 13) were only detected within the DGGE patterns of feeding real domestic sewage. *Tetrasphaera spp.* was regarded as a PAO in wastewater treatment system (Kristiansen *et al.* 2013). It is probably that the corresponding organism of band 12, *Chlorobaculum* spp., is responsible for a sulfate reduction and was often found in secondary sedimentation tanks (Zhang *et al.* 2012). It was introduced into the BNR-IC system possibly from real domestic wastewater. In addition to, the band 2 corresponding to *Rhodocyclus spp.* is a typical DPAO capable of simultaneous nitrogen and phosphorus removal and was often found in biological enhanced phosphorus removal system (Huang *et al.* 2015; Lv *et al.* 2015; Zengin *et al.* 2011).

Conclusions

This study showed that effective nutrients removal and phosphorus recovery were achieved in the novel BNR-IC process regardless of feeding types (synthetic or real domestic wastewater). The major phosphorus recovery mechanism was induced crystallization of HAP. When the BNR-IC system achieved the stable state, the COD, NH_4^+-N, TN and TP concentrations of effluent remained at a low level both for treating real domestic sewage and synthetic wastewater, meeting the discharge standard of municipal domestic sewage treatment (GB 18918-2012) formulated by Chinese government. Phosphorus recovery efficiency from synthetic wastewater was 70.2% slightly less than that from real domestic sewage (74.2%). Importantly, phosphorus recovery effectively enhanced the subsequent biological phosphorus removal. It was found that an obvious shift in microbial community structure was observed by PCR-DGGE, and diversity of sludge using real domestic sewage was higher than that with synthetic wastewater. The results of this study suggest that the BNR-IC operated in continuous flow is feasible for nutrients removal and phosphorus recovery from domestic sewage.

Acknowledgements

This research is supported by grant 1508085ME90 from Natural Science Foundation of Anhui Province, China and grant KJ2015A173 from Natural Science Foundation of the Ministry of Education of Anhui province, China.

References

Acelas, N. Y.; López, D. P.; Brilman, D. W.; Kersten, S. R.; Kootstra, A. M. J. 2014. Supercritical water gasification of sewage sludge: gas production and phosphorus recovery, *Bioresource Technology* 174: 167–175. http://dx.doi.org/10.1016/j.biortech.2014.10.003

Boelee, N. C.; Temmink, H.; Janssen, M.; Buisman, C.; Wijffels, R. H. 2011. Nitrogen and phosphorus removal from municipal wastewater effluent using microalgal biofilms, *Water Research* 45(18): 5925–5933. http://dx.doi.org/10.1016/j.watres.2011.08.044

Cusick, R. D.; Logan, B. E. 2012. Phosphate recovery as struvite within a single chamber microbial electrolysis cell, *Bioresource Technology* 107: 110–115. http://dx.doi.org/10.1016/j.biortech.2011.12.038

GB 18918-2012. *Discharge Standard of Pollutants for Municipal Wastewater Treatment Plant.*

Gilbert, N. 2009. Environment: the disappearing nutrient, *Nature News* 461(7265): 716–718. http://dx.doi.org/10.1038/461716a

Guney, K.; Weidelener, A.; Krampe, J. 2008. Phosphorus recovery from digested sewage sludge as MAP by the help of metal ion separation, *Water Research* 42(18): 4692–4698. http://dx.doi.org/10.1016/j.watres.2008.08.016

Haiming, Z.; Xiwu, L.; Abualhail, S.; Jing, S.; Qian, G. 2014. Enrichment of PAO and DPAO responsible for phosphorus removal at low temperature, *Environment Protection Engineering* 40(1): 67–83.

Huang, W.; Li, B.; Zhang, C.; Zhang, Z.; Lei, Z.; Lu, B.; Zhou, B. 2015. Effect of algae growth on aerobic granulation and nutrients removal from synthetic wastewater by using sequencing batch reactors, *Bioresource Technology* 179: 187–192. http://dx.doi.org/10.1016/j.biortech.2014.12.024

Hutnik, N.; Kozik, A.; Mazienczuk, A.; Piotrowski, K.; Wierzbowska, B.; Matynia, A. 2013. Phosphates (V) recovery from phosphorus mineral fertilizers industry wastewater by continuous struvite reaction crystallization process, *Water Research* 47(11): 3635–3643. http://dx.doi.org/10.1016/j.watres.2013.04.026

Ichihashi, O.; Hirooka, K. 2012. Removal and recovery of phosphorus as struvite from swine wastewater using microbial fuel cell, *Bioresource Technology* 114: 303–307. http://dx.doi.org/10.1016/j.biortech.2012.02.124

Kristiansen, R.; Nguyen, H. T. T.; Saunders, A. M.; Nielsen, J. L.; Wimmer, R.; Le, V. Q.; McIlroy, S. J.; Petrovski, S.; Seviour, R. J.; Calteau, A. 2013. A metabolic model for members of the genus Tetrasphaera involved in enhanced biological phosphorus removal, *The ISME Journal* 7(3): 543–554. http://dx.doi.org/10.1038/ismej.2012.136

Lv, X.; Shao, M.; Li, J.; Li, C. 2015. Metagenomic analysis of the sludge microbial community in a lab-scale denitrifying phosphorus removal reactor, *Applied Biochemistry and Biotechnology*: 175(7): 3258–3270. http://dx.doi.org/10.1007/s12010-015-1491-8

Moheimani, N. R.; Cord-Ruwisch, R.; Raes, E.; Borowitzka, M. A. 2013. Non-destructive oil extraction from Botryococcus braunii (Chlorophyta), *Journal of Applied Phycology* 25(6): 1653–1661. http://dx.doi.org/10.1007/s10811-013-0012-9

Ong, Y. H.; Chua, A.; Fukushima, T.; Ngoh, G. C.; Shoji, T.; Michinaka, A. 2014. High-temperature EBPR process: The performance, analysis of PAOs and GAOs and the fine-scale population study of Candidatus "Accumulibacter phosphatis", *Water Research* 64: 102–112. http://dx.doi.org/10.1016/j.watres.2014.06.038

Qiu, G.; Ting, Y. 2014. Direct phosphorus recovery from municipal wastewater via osmotic membrane bioreactor (OMBR) for wastewater treatment, *Bioresource Technology* 170: 221–229. http://dx.doi.org/10.1016/j.biortech.2014.07.103

Sengupta, S.; Pandit, A. 2011. Selective removal of phosphorus from wastewater combined with its recovery as a solid-phase fertilizer, *Water Research* 45(11): 3318–3330. http://dx.doi.org/10.1016/j.watres.2011.03.044

Seviour, R. J.; Mino, T.; Onuki, M. 2006. The microbiology of biological phosphorus removal in activated sludge systems, *FEMS Microbiology Reviews* 27(1): 99–127. http://dx.doi.org/10.1016/S0168-6445(03)00021-4

Shi, J.; Lu, X.; Yu, R.; Zhu, W. 2012. Nutrient removal and phosphorus recovery performances of a novel anaerobic-anoxic/nitrifying/induced crystallization process, *Bioresource Technology* 121: 183–189. http://dx.doi.org/10.1016/j.biortech.2012.06.064

Song, Y.; Dai, Y.; Hu, Q.; Yu, X.; Qian, F. 2014. Effects of three kinds of organic acids on phosphorus recovery by magnesium ammonium phosphate (MAP) crystallization from synthetic swine wastewater, *Chemosphere* 101: 41–48. http://dx.doi.org/10.1016/j.chemosphere.2013.11.019

Thai, P. K.; O'Brien, J.; Jiang, G.; Gernjak, W.; Yuan, Z.; Eaglesham, G.; Mueller, J. F. 2014. Degradability of creatinine under sewer conditions affects its potential to be used as biomarker in sewage epidemiology, *Water Research* 55: 272–279. http://dx.doi.org/10.1016/j.watres.2014.02.035

Westholm, L. J. 2006. Substrates for phosphorus removal—Potential benefits for on-site wastewater treatment?, *Water Research* 40(1): 23–36. http://dx.doi.org/10.1016/j.watres.2005.11.006

Yang, X.; Peng, Y.; Ren, N.; Guo, J.; Tang, X.; Song, J. 2013. Nutrient removal performance and microbial community structure in an EBPR system under the limited filamentous bulking state, *Bioresource Technology* 144: 86–93. http://dx.doi.org/10.1016/j.biortech.2013.06.103

Yetilmezsoy, K.; Sapci-Zengin, Z. 2009. Recovery of ammonium nitrogen from the effluent of UASB treating poultry manure wastewater by MAP precipitation as a slow release fertilizer, *Journal of Hazardous Materials* 166(1): 260–269. http://dx.doi.org/10.1016/j.jhazmat.2008.11.025

Yuan, Z.; Pratt, S.; Batstone, D. J. 2012. Phosphorus recovery from wastewater through microbial processes, *Current Opinion in Biotechnology* 23(6): 878–883. http://dx.doi.org/10.1016/j.copbio.2012.08.001

Zengin, G. E.; Artan, N.; Orhon, D.; Satoh, H.; Mino, T. 2011. Effect of aspartate and glutamate on the fate of enhanced biological phosphorus removal process and microbial community structure, *Bioresource Technology* 102(2): 894–903. http://dx.doi.org/10.1016/j.biortech.2010.09.023

Zhang, G.; Zhao, Q.; Jiao, Y.; Wang, K.; Lee, D.; Ren, N. 2012. Efficient electricity generation from sewage sludge using-biocathode microbial fuel cell, *Water Research* 46(1): 43–52. http://dx.doi.org/10.1016/j.watres.2011.10.036

Zhang, H.; Fang, W.; Wang, Y.; Sheng, G.; Xia, C.; Zeng, R. J.; Yu, H. 2013. Species of phosphorus in the extracellular polymeric substances of EBPR sludge, *Bioresource Technology* 142: 714–718. http://dx.doi.org/10.1016/j.biortech.2013.05.068

Zou, H.; Lu, X.; Li, T. 2014a. A novel process for nutrients removal and phosphorus recovery from domestic wastewater by combining BNR with induced crystallization, *Journal of Environmental Engineering and Landscape Management* 22(4): 274–283. http://dx.doi.org/10.3846/16486897.2014.931284

Zou, H.; Lu, X.; Saad, A. 2014b. Characterization of microorganisms responsible for phosphorus removal linking operation performance with microbial community structure at low temperature, *Environmental Technology* 35(9): 1101–1109. http://dx.doi.org/10.1080/09593330.2013.861858

Zou, H. M.; Lu, X. W.; Li, T. 2014c. Effect of side-stream phosphorus recovery on biological phosphorus removal performance investigated by chemical and microbial analyses in a novel BNR-IC process, *Water Science & Technology* 70(9): 1441–1447. http://dx.doi.org/10.2166/wst.2014.351

Haiming ZOU. Dr, Associate Professor, Department of Resource and Environment, Anhui Science and Technology University; School of Energy and Environment, Southeast University. Research interests: with nitrogen and phosphorus removal from domestic wastewater.

Xiwu LU. Dr, Professor, School of Energy and Environment, Southeast University. Research interests: research on nitrogen and phosphorus removal from wastewater.

6

ANALYSIS AND ASSESSMENT OF HEAVY METALS CONCENTRATIONS IN NEMUNAS RIVER BOTTOM SEDIMENTS AT ALYTUS CITY TERRITORY

Vaidotas VALSKYS, Roberta VALSKIENĖ, Gytautas IGNATAVIČIUS

Ecology and Environmental Center, Vilnius University, Vilnius, Lithuania

Abstract. Concentrations and spatial distribution of heavy metals on the left and right banks of the river Nemunas bottom sediments are analyzed in this article. The research methodology of X-ray fluorescence spectrometry for bottom sediments and operating principles of XL2 spectrometer used for analysis are overviewed. The results of analysis are presented and compared with LAND 20-2005 requirements as well as studies that were carried out previously. The influence of Alytus city for Nemunas river sediments quality is assessed. Dischargers formed additional samples which were taken and included to the list of ordinary samples. The trend of heavy metals (Cd, Cu, Cr) concentrations showed the growth of pollution downstream the urban area. The estimated Zd (total pollution) values clearly indicated higher contamination by heavy metals on the left bank of Nemunas River. Extensive surveys of river sediments allow assessing the extent of anthropogenic impact, which can be harmful to the river ecosystem and human health.

Keywords: x-ray fluorescence spectrometry, heavy metals, sediments, water pollution.

Introduction

The present day river bottom sediments are complex mechanical, mineral and chemical composition heterogeneous polydispersed system which is sensitive to the change of surrounding environment physical and chemical conditions. Typically river sediments are formed from the mineral part of the river banks surrounding landscape, various chemical and mechanical maturity subsoil and rock soil constituents (Jankauskaitė *et al.* 2008). In addition, dead and unevenly decomposed biomass, manganese, iron and other elements hydroxides participates in sediments formation. In recent decades, the particularly important factors that determined sediments macro- and microelement composition has become of human productive and economic activity (Chalmers *et al.* 2006).

Heavy metals (HM) transport, access to river and accumulation in sediments are determined by natural and anthropogenic factors as a result of the self-purification of river water (Taraškevičius, Zinkutė 2011). HM which have low solubility and biodegradation characteristics are more accumulated in sediments. Usually in unaffected environments, the concentration of most of the metals in rivers is very low and is mostly derived from weathering of rock and soil (Reza, Singh 2010). In addition to natural factors concentrations of heavy metals in river sediments depend on the type of anthropogenic effect, intensity and duration. Urban environment is a complicated system, because 70–90% of all the soil lies under the streets, pedestrian paths, buildings and therefore, its natural environmental recovery takes a lot of time, a natural energy and material circulation are destroyed (Vasarevičius *et al.* 2010). The highest accumulation of metals occurs in low flow period and the lowest accumulation of metals occurs during the high flow period, suggesting that the concentration of heavy metals in bed sediments is brought about by changes of water flow (Bartoli *et al.* 2011; Li *et al.* 2012).

Contaminated sediments have negative impact to the bottom and bottom-dwelling organisms. Such environment leads to the loss or extinction of molluscs, fish and other hydrobionts (Farkas *et al.* 2007; Lourino-Cabana *et al.* 2011). In addition, the sediments that are heavily contaminated with toxic substances cause not only diseases or extinct of aquatic wildlife but are also dangerous to animals and humans that are consuming fish. Even if trace metals are initially emitted in their elemental forms, usually considered as non-bioavailable, their ultimate fate

Corresponding author: Vaidotas Valskys
E-mail: vaidas.valskys@gmail.com

remains unknown. Therefore, urban runoffs not only degrade the environmental quality of aquatic ecosystems, but also lead to increasing contaminants in aquatic organisms through bioaccumulation and biomagnification, potentially causing elevated trace metal concentration in the food chain (Yujun *et al.* 2011; Sen, Varol 2011). Thus, the primary purposes of sediment quality guidelines are to protect the aquatic biota from the deleterious effects associated with sediment-bound contaminants, to rank and/or prioritize contaminated areas or chemicals of concern for further investigation (Honglei *et al.* 2007; Fu *et al.* 2014).

The analysis of HM concentrations and determination of allowable exceedance limits oblige to take measures to improve the quality of human life and the environment. Environmental pollution by HM is a global problem. This phenomenon is very important for river ecosystems which accumulate HM that get from natural and anthropogenic sources (Peng *et al.* 2008). It has long been accepted that HM are very important at ecotoxicological point of view as these elements are very persistent and can be toxic. In order to assess the ecological status of surface sediment s and their effects on the biota and human health it is important to measure the concentrations of HM (Davutluoglu *et al.* 2011).

Mainly due to the lack of methodology and equipment analyses of heavy metals basically have not been carried out until 1990 in Lithuania. First environmental monitoring system was developed in 1991–1992 in Lithuania. Later in 1997 first Lithuanian Environmental Monitoring Program which provided research of HM (Zn, Cu, Cr, Pb, Ni, Hg, Cd) in surface water and sediments has been prepared and approved. Despite the investigations carried out there is still not enough information about the distribution of HM in Lithuanian river sediments (Lithuanian Environmental Protection Agency 2003).

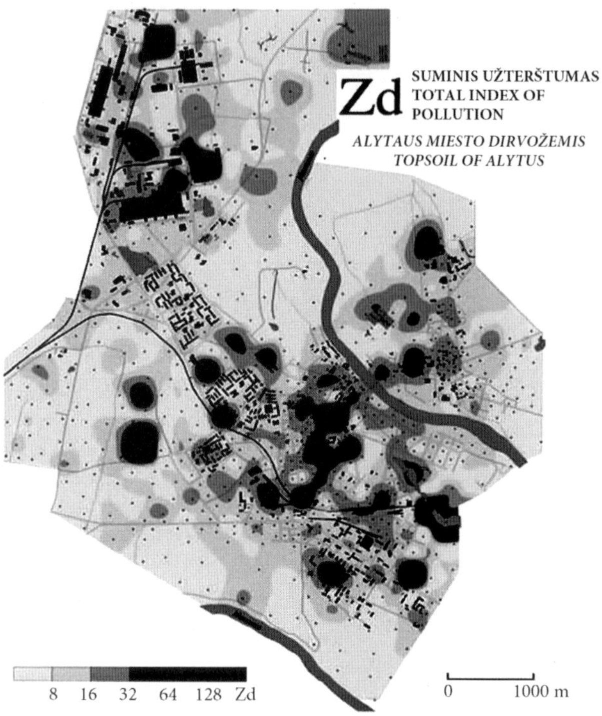

Fig. 1. Soil contamination with heavy metals in Alytus town according to the index Zd (Gregorauskienė 2006)

The objective of this work was to investigate and assess the contamination by heavy metals of Nemunas river bottom sediments at Alytus city territory using X-ray fluorescence spectrometry. This territory was chosen due to the lack of this type of research in this area. Also industries are well-developed in this city thus it can lead to an increased concentrations of heavy metals. Such detailed research which is important in determining the pollution sources shows the real distribution of heavy metals concentrations.

Results of researches that were carried out in Alytus city territory clearly shows industry and arbitrary household waste areas that are contaminated with zinc, copper, silver and other elements. Anomalies of heavy metals concentrations were determined near "Snaige" refrigerator factory, textile factory, railway, agricultural machinery factory, garages, lake "Dailide" and other territories (Fig. 1). All industrial and other polluting objects are located in the western part (left bank of the river Nemunas) of the city (Gregorauskienė 2006).

1. Methodology

1.1. General description of the study area

Sampling locations are set out on the left and right banks of the Nemunas River. 4.5 km sampling length was selected and it is actually bounded by two bridges in the area. Sampling started above Alytus city from the ruins of the railway bridge (destroyed during the World War I). The end of sampling was below Alytus city near the new bridge of the city ("Lithuania Millennium Bridge").

4.5 km sampling length was divided into equal parts at 300 m (Fig. 2, grey round points). Coordinates of sampling points were set with Garmin GPS Navigator "Nuvi 40". Dischargers and tributaries formed additional samples which were taken and included to the list of ordinary samples (Fig. 2, white square points).

1.2. Sampling and preparation

Samples were taken from the surface (0–0.1 m) sediment layer 1 m away from the shore. Samples were taken using stainless steel unpainted 7 liter bucket. Sample collection was held in summer which is low runoff period (summer period: 15–18% of total runoff). Samples were taken 1 m away from the shore thus granulometric composition of samples were sand and sandy loam (85–95% of physical sand). Samples were placed in separate polyethylene containers to avoid the cross-contamination. Containers were registered indicating the river side and the sample sequence number. Sediment samples were taken to the laboratory and placed in petri dishes. Samples were dried to constant weight in drying oven at (105±10) °C temperature. Dried samples were grinded in porcelain mortar until homogenous mass. The obtained mass of the sample was filtered through 2.00 mm, 250 μm and 125 μm mesh filter. Filtrated mass were placed in special caps which were inserted into spectrometer (Shu-hai *et al.* 2006; Suthar *et al.* 2009). Samples were tested in Thermo Scientific Niton® XL2 series X-ray fluorescence spectrometer.

Measurement time varies from 30 to 600 seconds depending on the intended quality of the results. The optimal 240-second analysis time was chosen to analyze bottom sediments of the Nemunas River (US EPA 2006).

28 elements were analyzed in bottom sediments (As, Hg, Cd, Ba, Sb, Sn, Ag, Pd, Zr, Sr, Rb, Pb, Se, Au, Zn, W, Cu, Ni, Co, Fe, Mn, Cr, V, Ti, Sc, Ca, K, S). Further analysis involved those metals that are anthropogenic and potentially dangerous.

1.3. Research device

Research was carried out using X-ray fluorescence spectrometer – Niton XL2. Spectroscopy is an analytical technique in which dry river bottom sediment sample is exposed to X-rays. The X-rays from the source have the appropriate excitation energy that causes elements in the sample to emit characteristic X-rays. A qualitative elemental analysis is possible from the characteristic energy, or wavelength, of the fluorescent X-rays emitted (Fig. 3).

A quantitative elemental analysis is possible by counting the number (intensity) of X-rays at a given wavelength (Wrobel, Czyzycki 2013). The energy of this X-ray radiation is unique for each element (US EPA 2006).

Each sample is analysed with 3 spectra using 4 different energy lines. Device automatically counts each spectra and each energy line values giving counted errors, element concentration values and detection limit. Detection limit with 240 seconds scanning time gives 2–4 ppm detection limit depending on each element.

1.4. The reliability of the X-ray fluorescence (XRF) spectrometry results

The reliability of XRF analysis method depends on sample preparation (drying, grinding, sieving). Unreliable results are recorded when there are no compliance with recommendations for sampling and preparation (Guoren *et al.* 2013). The most important requirements for the preparation of samples are grinding and sieving. The particle size of 250 μm (preferably 125 μm) must be achieved. Continuous dry matter should be 3 to 5 grams in the sample cup.

An experiment was carried out analyzing lead levels in Massachusetts (USA). Fully prepared XRF samples showed excellent correlation with laboratory Atomic Absorption Spectrometry (AAS) for material split after the final grinding, sieving, and homogenization (Fig. 4). A set of 20 fully prepared XRF samples (oven dried, ground to 0.125 mm), including 11 bridge site samples, 6 residential lead samples, and 3 NIST SRM soils, gave a linear regression slope of 1.004 and an R^2 of 0.995. For the 17 samples with lead concentrations above 100 ppm, the mean recovery of the XRF relative to AAS was 0.952 and the standard deviation of the recovery was 0.068, for relative standard deviation (RSD) of 7.1 percent.

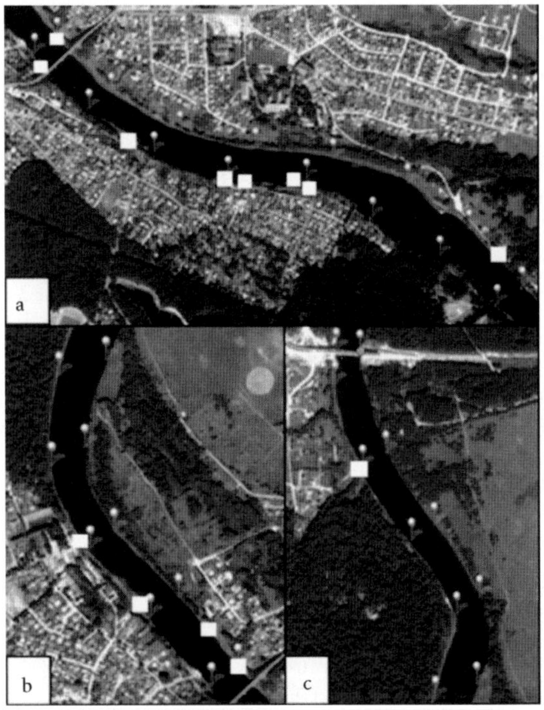

Fig. 2. Sampling locations downstream – Nemunas river above (a), middle (b) and below (c) Alytus city

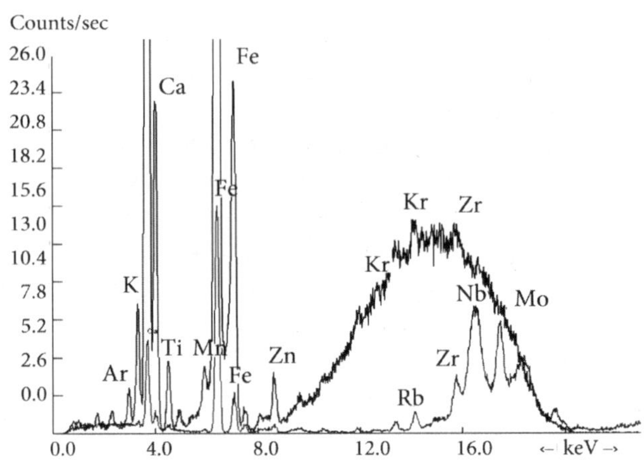

Fig. 3. Niton XL2 window showing wavelength and intensity measurement

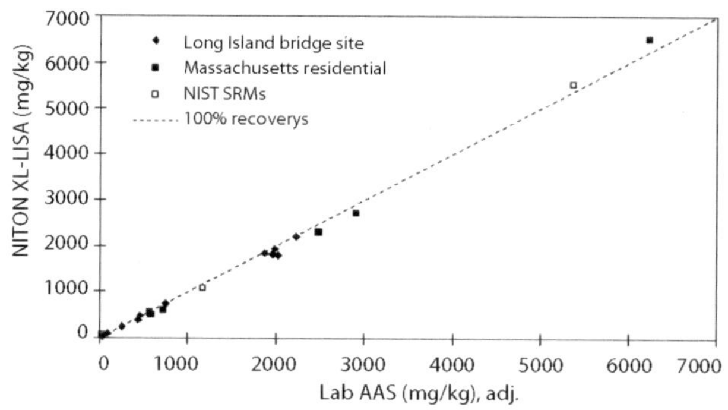

Fig. 4. Comparison of fully prepared XRF and laboratory AAS results

The subset of 11 bridge site samples gave a linear regression slope of 0.958 and correlation coefficient (R^2) of 0.994. The subset of 6 Massachusetts residential samples gave a linear regression slope of 1.010 and an R^2 of 0.994 (Shefsky 1997).

The XRF is much cheaper than the AAS analysis method thus it provides an opportunity to analyze a greater number of samples and obtain a more detailed distribution of HM concentrations in the territory.

1.5. Total contamination index (Zd)

Zd of HM (Cd, Ni, Cr, Cu, Zn, As, Mn) in bottom sediments was calculated for the left and right bank of the Nemunas River. These HM were chosen due to their potential anthropogenic origin.

Geochemical background values for these elements were taken from Lithuanian Hygiene Standard 60: 2004. There are values given for loam and clay that were chosen as a background values. This decision was made after the comparison of these concentrations with concentrations given in comparative characteristics of change of trace elements background values in Lithuanian surface sediments (Kadūnas, Radzevičius 2003). These values were closest.

Total contamination index is calculated by the formula:

$$Z_d = \sum K_K - (n-1). \qquad (1)$$

Values of coefficient of element concentration (K_K) are summed up only if K_K values are greater than 1.

n – number of chemical elements (pollutants);

K_K – coefficient of each element concentrations which is calculated for each element separately by the formula:

$$K_K = C / C_f, \qquad (2)$$

C – determined concentration of chemical element in the sample (mg/kg);
C_f – background concentration of chemical element in the sample (mg/kg).
Z_d index is used in cases when the subject that is analysed is contaminated by several substances or chemical elements (Lithuanian Hygiene Standard 60: 2004).

2. Results

42 samples were analyzed using XRF method and concentrations of heavy metals were determined on the right and left banks of Nemunas River at Alytus city territory.

Nemunas river sediments are not as polluted as some western European countries rivers (Table 1) such as Ireland (Glendasan). This contamination is caused by extraction of metals (mines), heavy industry companies and other polluting objects. There are no such polluting objects in Lithuania thus it leads to the less contaminated soils and river sediments. Heavy metal concentrations in Nemunas river sediments are similar to neighboring countries (Latvia, Poland, etc.).

Table 1. Comparison of HM concentrations in Nemunas river and some rivers in other Europe countries

River	Heavy metals (mg/kg)					
	Cu	Pb	Cr	Ni	Cd	Zn
Glendasan (Ireland)	291	10.22	–	–	56	17.5
Mala Welna (Poland)	13.8	13.4	3.9	8.3	0.85	43.4
Bogdanas (Greece)	44.35	10.98	–	5.23	0.12	130.2
Nemunas (left bank)	26.88	–	60.06	61.1	14.67	27.19
Nemunas (right bank)	21.07	–	39.42	–	13.23	21.6

Although concentrations are much lower in Nemunas river than in rivers that are in western Europe countries as industry develops there is a need to investigate the geochemical changes in Lithuania.

Concentrations of 21 metals were determined which were higher than the detection limit. Lithuanian environmental normative document 20-2005 (LAND 20-2005) requirements for sludge categorization by heavy metals were used to compare and evaluate the quality of the sediments.

There is no normative document that regulates concentrations of heavy metals in river bottom sediments in Lithuania. According to the measured concentrations of heavy metals mean values are calculated separately for the left and right banks of the Nemunas River. Those mean values are used to evaluate the general condition of the sludge and compare with the requirements of LAND 20-2005. Most of HM (Cr, Cu, Zn) concentrations in the surface (0–0.1 m deep) layer sediments meets I sludge category.

Higher pollution by heavy metals (Cd and Ni) is observed on the left bank of the Nemunas River. Those metals concentrations correspond to the II sludge category. Only Cd concentration on the right bank of the Nemunas River exceeds the LAND 20-2005 requirements (Table 2).

Table 2. Sludge categorization by HM concentrations (LAND 20-2005)

Sludge category	Concentrations of heavy metals, mg/kg (*LOD-Level of Detection)						
	Pb	Cd	Cr	Cu	Ni	Zn	Hg
I	<140	<1.5	<140	<75	<50		<1.0
II	140–750	1.5–20	140–400	75–1000	50–300	300–2500	1.0–8.0
III	>750	>20	>400	>1000	>300	>2500	>8.0
Right bank of the Nemunas River	<LOD*	13.23	39.42	21.07	<LOD*	21.6	<LOD*
Left bank of the Nemunas River	<LOD*	14.67	60.06	26.88	61.1	27.19	<LOD*

Such distribution of HM concentrations shows higher contamination load on the left river bank which is possibly caused by industries located in the urban area on the left bank of the river. There are no industrial facilities or other concentrated contamination hotspots on the right river bank.

Cadmium concentrations 8.8 and 9.8 times are greater than required for the I sludge category while nickel concentration exceeds 1.2 times only on the left bank of the Nemunas River. In terms of the worse Indicators River sediments on both banks are assigned to the II sludge category (LAND 20-2005).

Most of the heavy metals concentrations in bottom sediments of the Nemunas River have steadily increased with the exception of Pb and Zn. Cr and Ni concentrations are highest in River sediments (Fig. 5). New industrial companies in 2009–2010 were established in the western part of Alytus city instead of old auto repair, motor transport companies. These industrial companies are producing plastic, rubber products and electronics, thus it could have caused the change of heavy metals concentrations. The increase of Cr, Ni and Cd could probably be caused by the change of industrial activity.

Sediments of the Nemunas River were quite polluted in compare with clean (Skroblus) river: it dozens of times exceeded background values with the exception of Pb which was below the detection limit.

Cu concentration was similar in Kulpė and Nemunas Rivers bottom sediments (Fig. 6) while Cd concentration was 16 times higher in Kulpė river sediments (Lithuanian Environmental Protection Agency 2003).

2.1. The trend of HM concentrations in the sediments of the Nemunas River going downstream

Sample numbers are given on the x axis in Figures 7–12 (dis. – discharges and trib. – tributaries). The change of heavy metals (Cr, Cu, Cd) concentrations on the right and left banks of the Nemunas River was calculated using the trend feature. It is noticeable that Cd and Cu concentrations on the left and right banks increase downstream (Figs 7–12). Concentrations of Cr on the right bank of

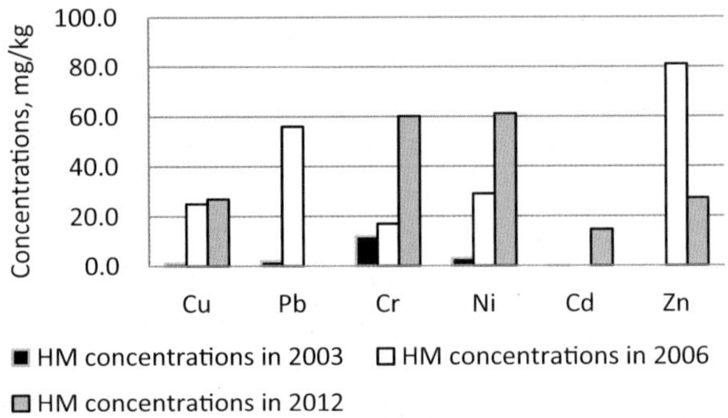

Fig. 5. HM concentrations in Nemunas River bottom sediments

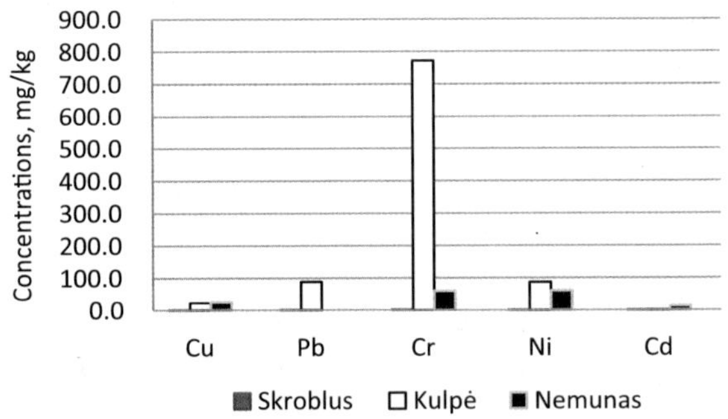

Fig. 6. HM concentrations in bottom sediments of the most contaminated and the cleanest Lithuanian rivers

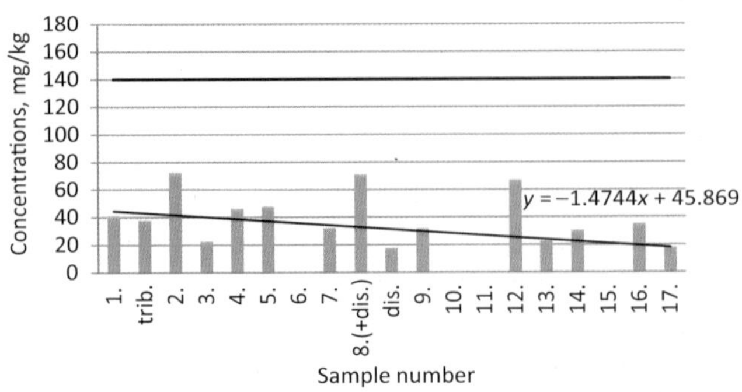

Fig. 7. Cr concentrations on the right bank, downstream trend (*y*), the concentration of the I sludge category (bold line)

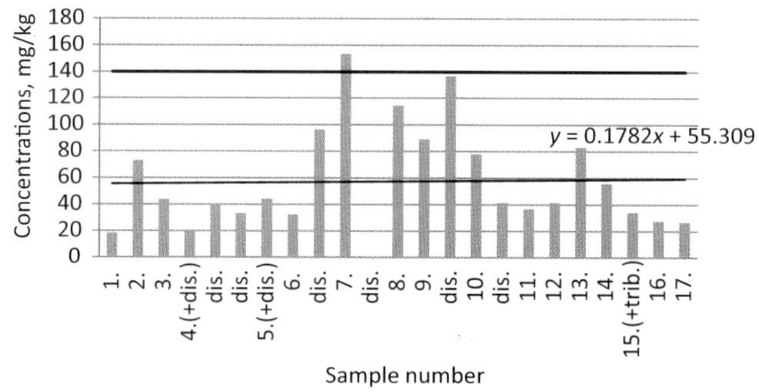

Fig. 8. Cr concentrations on the left bank, downstream trend (*y*), the concentration of the I sludge category (bold line)

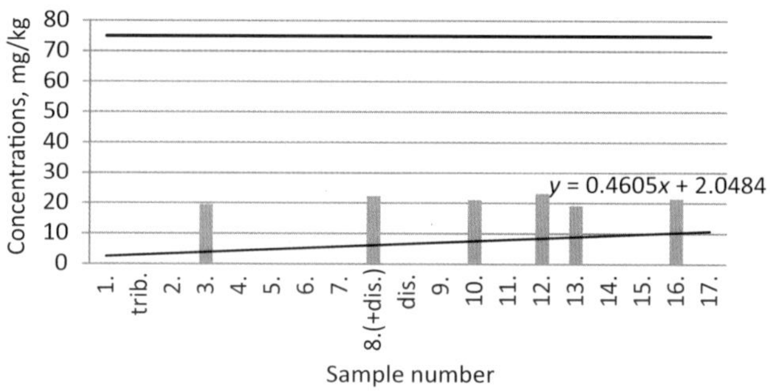

Fig. 9. Cu concentrations on the right bank, downstream trend (*y*), the concentration of the I sludge category (bold line)

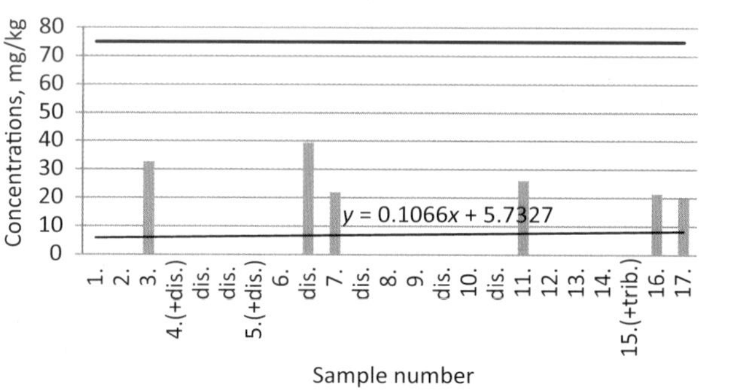

Fig. 10. Cu concentrations on the left bank, downstream trend (*y*), the concentration of the I sludge category (bold line)

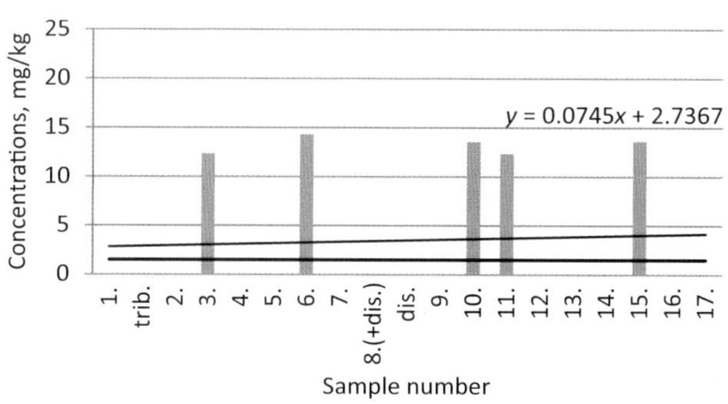

Fig 11. Cd concentrations on the right bank, downstream trend (*y*), the concentration of the I sludge category (bold line)

the Nemunas River bottom sediments decreases downstream while on the left bank increases (Figs 7 and 8). Also concentrations of heavy metals on the left bank are much higher than on the right bank. This obviously shows higher pollution by heavy metals on the left bank of Nemunas river where are much more dischargers. Those dischargers bring more pollutants from Alytus city and its contamination sources.

All concentrations of heavy metals (Cr, Cu, Cd) are noticeably higher on the left bank of the Nemunas River (Figs 7–12). This is due to the fact that industry is concentrated in this part of Alytus city. Bottom sediments near some of discharges were highly contaminated by heavy metals, especially on the left river bank. Most of the concentrations of HM increased downstream.

The Total pollution index Zd showed that the left bank of the Nemunas River bottom sediments pollution by HM is greater than the right bank. Zd of the right bank is 68.3 while Zd of the left bank is 78.6. These Zd values are attributed to hazardous contamination category ($32 < Z_d < 128$) according to Lithuanian Hygiene Standard 60: 2004. It also shows the higher contamination by HM on the left bank of Nemunas River.

Conclusions

1. The analysis of heavy metals of the Nemunas River bottom sediments showed increased concentrations of 5 elements (Cd, Ni, Cr, Cu, Zn) in 2012. These heavy metals are considered to be the most dangerous.

2. The highest concentrations that exceed the permissible environmental standards were set to Cd and Ni. The major exceedances are specific to Cd from 8.8 to 9.8 times above the permissible limits respectively on the right and left banks of the Nemunas River.

3. Sediments were assigned to the II sludge category according to LAND 20-2005 in terms of the highest concentrations of Cd and Ni in bottom sediments on both banks of the Nemunas River.

4. Heavy metals concentrations in bottom sediments dozens of times exceed the cleanest river (Skroblus) values

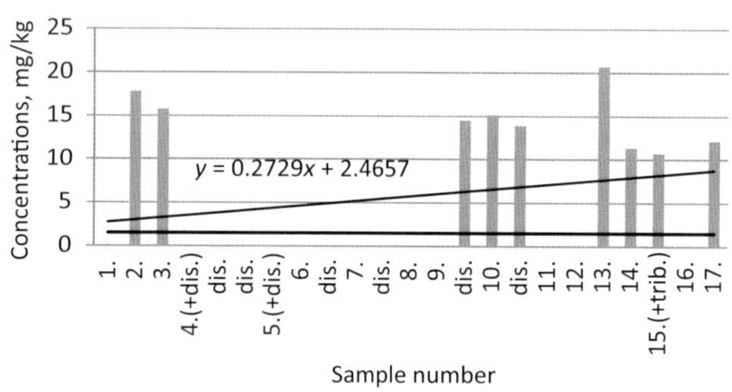

Fig 12. Cd concentrations on the left bank, downstream trend (*y*), the concentration of the I sludge category (bold line)

with the exception of Pb which was below the detection limit. Most of the heavy metals concentrations do not exceed the most contaminated river (Kulpė) values.

5. Zd of the right bank is 68.3 while Zd of the left bank is 78.6. According to Zd index – 13.1% higher contamination by heavy metals is observed on the left bank of the Nemunas River. Such distribution of contamination is due to the anthropogenic load and industrial enterprises located on the left part of the Nemunas River bank.

6. Trend equations showed the rising concentrations of heavy metals (Cr (except right bank), Cu, Cd) going downstream the urban area. This shows the anthropogenic impact of the Alytus city on the Nemunas River ecosystem.

References

Bartoli, G.; Papa, S.; Sagnella, E.; Fioretto, A. 2011. Heavy metal content in sediments along the Calore river: relationships with physical-chemical characteristics, *Journal of Environmental Management* 95: 9–14. http://dx.doi.org/10.1016/j.jenvman.2011.02.013

Chalmers, A. T.; Van Metre, P. C.; Callender, E. 2006. The chemical response of particle-associated contaminants in aquatic sediments to urbanization in New England, U.S.A., *Journal of Contaminant Hydrology* 91: 4–25. http://dx.doi.org/10.1016/j.jconhyd.2006.08.007

Davutluoglu, I. O.; Seckin, G.; Ersu, B. C.; Yilmaz, T.; Sari, B. 2011. Heavy metal content and distribution in surface sediments of the Seyhan River Turkey, *Journal of Environmental Management* 92: 2250–2259. http://dx.doi.org/10.1016/j.jenvman.2011.04.013

Farkas, A.; Erratico, C.; Vigano, L. 2007. Assessment of the environmental significance of heavy metal pollution in surficial sediments of the River Po, *Chemosphere* 68: 761–768. http://dx.doi.org/10.1016/j.chemosphere.2006.12.099

Fu, J.; Zhao, C.; Luo, Y.; Liu, C.; Kyzas, Z. G.; Luo, Y.; Zhao, D.; An, S.; Zhu, H. 2014. Heavy metals in surface sediments of the Jialu River, China: their relations to environmental factors, *Journal of Hazardous Materials* 270: 102–109. http://dx.doi.org/10.1016/j.jhazmat.2014.01.044

Gregorauskienė, V. 2006. Mapping of geochemical contamination in urban areas of Lithuania, *Journal of Environmental Engineering and Landscape Management* 14(1): 52–57.

Guoren, X.; Mingwei, L.; Guibai, L. 2013. Stabilization of heavy metals in lightweight aggregate made from sewage sludge and river sediment, *Journal of Hazardous Materials* 260: 74–81. http://dx.doi.org/10.1016/j.jhazmat.2013.04.006

Honglei, L.; Liqing, L.; Chengqing, Y.; Baoqing, S. 2007. Fraction distribution and risk assessment of heavy metals in sediments of Moshui Lake, *Journal of Environmental Sciences* 20: 390–397. ISSN 1001-0742.

Jankauskaitė, M.; Taraškevičius, R.; Zinkutė, R.; Veteikis, D. 2008. Relationship between landscape self-regulation potential and topsoil additive contamination by trace elements in Vilnius city, *Journal of Environmental Engineering and Landscape Management* 16(1): 5–14. http://dx.doi.org/10.3846/1648-6897.2008.16.5-14

Kadūnas, V.; Radzevičius, A. 2003. Comparative characteristics of change of trace element background values and associations in Lithuanian surface sediments, *Litosfera* 7: 80–87. ISSN 1392-334X

LAND 20-2005. Requirements for the use of sewage sludge for fertilization and recultivation. Lithuanian Environmental Normative Document. Ministry of Environment of Lithuania.

Li, F.; Zhang, H.; Meng, X.; Chen, L.; Yin, D. 2012. Contamination by persistent toxic substances in surface sediment of urban rivers in Chaohu City, China, *Journal of Environmental Sciences* 24: 1934–1941. http://dx.doi.org/10.1016/S1001-0742(11)61033-4

Lithuanian Environmental Protection Agency. 2003. *Accumulation of heavy metals in fish and sediments.* Final project report of Lithuanian Environmental Protection Agency.

Lithuanian Hygiene Standard 60: 2004. Maximum allowable concentrations of hazardous substances in soil. Ministry of Health of the Republic of Lithuania.

Lourino-Cabana, B.; Lesven, L.; Charriau, A.; Billon, G.; Ouddane, B.; Boughriet, A. 2011. Potential risks of metal toxicity in contaminated sediments of Deule river in Northern France, *Journal of Hazardous Materials* 186: 2129–2137. http://dx.doi.org/10.1016/j.jhazmat.2010.12.124

Peng, J.; Song, Y.; Yuan, P.; Cui, X.; Qiu, G. 2008. The remediation of heavy metals contaminated sediment, *Journal of Hazardous Materials* 161: 633–640. http://dx.doi.org/10.1016/j.jhazmat.2008.04.061

Reza, R.; Singh, G. 2010. Heavy metal contamination and its indexing approach for river water, *International Journal of Environmental Science and Technology* 7(4): 785–792. ISSN 1735-1472. http://dx.doi.org/10.1007/BF03326187

Sen, B.; Varol, M. 2011. Assessment of nutrient and heavy metal contamination in surface water and sediments of the upper Tigris River, Turkey, *Catena* 92: 1–10. http://dx.doi.org/10.1016/j.catena.2011.11.011

Shefsky, S. 1997. Comparing field portable x-ray fluorescence (XRF) to laboratory analysis of heavy metals in soils, *International Symposium of Field Screening Methods for Hazardous Wastes and Toxic Chemicals*, 29–31 January 1997, Las Vegas, Nevada, USA.

Shu-Hai, G.; Xiao-Li, W.; Yu, L.; Jie-Jiang, C.; Jun-Cheng, Y. 2006. Investigation on Fe, Mn, Zn, Cu, Pb, and Cd fractions in the natural surface coating samples and surficial sediments in the Songhua River, China, *Journal of Environmental Sciences* 18: 1193–1198. ISSN 1001-0742. http://dx.doi.org/10.1016/S1001-0742(06)60061-2

Suthar, S.; Nema, K. A.; Chabukdhara, M.; Gupta, K. S. 2009. Assessment of metals in water and sediments of Hindon River, India: impact of industrial and urban discharges, *Journal of Hazardous Materials* 171: 1088–1095. http://dx.doi.org/10.1016/j.jhazmat.2009.06.109

Taraškevičius, R.; Zinkutė, R. 2011. Urban geochemical anomalies of Lithuania and their spread, *Baltica* 24: 163–168. ISSN 0067-3064.

US EPA. 2006. *XRF technologies for measuring trace elements in soil and sediment.* US EPA Innovative Technology Verification Report, 4–8.

Vasarevičius, S.; Mineikaitė, A.; Vaitiekūnas, P. 2010. Investigation into heavy metals in storm wastewater from Vilnius Žirmunai district and pollutants spread model in Neris river, *Journal of Environmental Engineering and Landscape Management* 18: 242–249. ISSN 1648-6897. http://dx.doi.org/10.3846/jeelm.2010.28

Wrobel, P.; Czyzycki, M. 2013. Direct deconvolution approach for depth profiling of element concentrations in multi-layered materials by confocal micro-beam X-ray fluorescence spectrometry, *Talanta* 113: 62–67. http://dx.doi.org/10.1016/j.talanta.2013.03.087

Yujun, Y.; Zhifeng, Y.; Shanghong, Z. 2011. Ecological risk assessment of heavy metals in sediment and human health risk assessment of heavy metals in fishes in the middle and lower reaches of the Yangtze River basin, *Environmental Pollution* 159: 2575–2585.

Vaidotas VALSKYS. PhD in Ecology and Environmental Studies at the Ecology and Environmental Research Centre, Faculty of Natural Sciences, Vilnius University. Publications: author/co-author of 3 scientific papers, 1 scientific presentation in international conference and 2 scientific presentations in Lithuanian scientific conferences. Research interests: hydrology, water quality, limnology, environmental chemistry.

Roberta VALSKIENĖ. M.S. in Environmental Studies and Environment Management, at the Ecology and Environmental Research Centre, Faculty of Natural Sciences, Vilnius University. Now works at the laboratory of Genotoxicology, Institute of Ecology, Nature Research Center as a biologist. Publications: author of 1 scientific paper. Research interests: ecotoxicology, effects on the biota, water quality.

Gytautas IGNATAVIČIUS. Dr, Associate Professor of Environmental Engineering at the Ecology and Environmental Research Centre, Faculty of Natural Sciences, Vilnius University. He gained his Environmental Engineering PhD in 2002 at Vilnius Gediminas Technical University. Publications: author/co-author of over 50 scientific papers and over 30 scientific presentations in Lithuanian and international scientific conferences. Research interests: environmental chemistry, filtration process, soil chemistry.

7

THE CORRELATION OF WATER WITH SETTLEMENT AND TRANSPORTATION NETWORK

Ömer ATABEYOĞLU

Department of Landscape Architecture, Faculty of Agrigulture, Ordu University in Ordu, Turkey

Abstract. Cities, with their ancient history, are living formations that develop and change constantly. Although there are several economical, safety and transportation reasons for their foundation purposes and for the locations where cities founded, geographical conditions also significantly affect these reasons. One of the most important elements determining the area where the cities have been founded is the water. This article examined how the geographical water resources affect the humankind settlements and the transportation network. The relationships, similarities and differences between transport and water networks were revealed using fractal analysis method. The effect of water on the location of living areas was also determined by using the bufferzone method. Swot analysis was made for evaluated to affects of the roads on ecosystem and landscape. The study was carried out on the base of 81 provinces in Turkey. The roads in the country that was netted with dense transportation networks were passed through considering the geographic elements. Rivers, lakes and seas are important factors in determining the transit routes of the roads. The water elements use the routes for which geographic conditions are the most suitable and this serves as a guide for the transport network. Additionally, provincial and district centers have also been established to be particularly close to the water element due to its landscape and economic factors, the appropriate geographic and climatic conditions.

Keywords: landscape management, environmental sustainability, fractal analysis, river network, road network.

Introduction

Water is one of the crucial determinant and formative element in all aspects of life. Humans, as they cannot survive without water, the very foundation of their life, have always inhabited the places that are close to water and water sources and easily accessible. They have chosen and designed the settlement places particularly in accordance with this. With the contribution of other geological and natural factors, they have provided with the formation of our today's world, cities, residences and lifestyles.

Together with other geological and natural factors, water factor is the most effective on determining itineraries, forming routes and the arteries that connects our habitats. The transportation network was mostly followed the natural pathway of the closest sea, surrounded the lakes and relied on the guideline of the rivers.

"It was determined that most of ancient people lived on the beaches and coasts of the dried and drying lakes, alluvial plains and cones of deposit which provided easy access to their water need. The tumulus, as their surroundings have fertile soils, were very convenient areas to meet the basic needs of the first age people. They made their own living by hunting wild animals in swampy areas and in lands near lakes, they also occupied with a small-scale agriculture on those fertile plains" (Topur 2009).

Roads and rivers are the important corridors for urban expansion too. Before the invention of automobiles, cities developed along river sand coastlines because waterborne transportation was so important. Rivers are important natural corridors (Tian, Wu 2015). Old roads are curvy and follow the landforms, whereas the modern roads often ignore landforms and cross the county in straight lines (Seppala 1999). Interaction between stream and transportation networks is more complex, with the roads paralleling the streams in some locations. In small valleys the transportation network follows the trunk stream more closely (Blanton, Marcus 2009).

According to Liu *et al.* (2014), transportation infrastructures built in line with a city's expansion, particularly road networks, cause both direct and indirect losses of habitat, which, in turn, influence the abundance and distribution of plant and animal species (Geneletti 2003; Eigenbrod *et al.* 2007).

Corresponding author: Ömer Atabeyoğlu
E-mail: atabey6@hotmail.com

Recent large-scale studies have often explored the relationship between road networks and landscape pattern by applying road density and road effect as the parameters of road networks' impact (Hawbaker, Radeloff 2004; Liu *et al.* 2008; Fu *et al.* 2010; Patarasuk 2013; Jenelius, Mattsson 2015).

The rivers that compatible with topography are the most compatible spaces. The ways circulate circle of lake, continue down the parallel of river or sea. All time, water is evidence of civilization and is accepted as guide and route. Waterfronts have scenery, optimum climate, fertile soil as agriculture and source as needs. The aim of this study to reveal in the cities of Turkey to the physical effect of the rivers, lakes and seas on determined of the route and plan for road, and to similarity between road network and hydrology network, and to relation between locations of current settlements and water elements.

Aim of the study sea, lake and rivers put forth their direct and indirect effects on transportation function. Running and still waters effect on environment at both urban and suburban. The effect of this element on the one of the human made functions puts forth to relationship between nature and human. The study emphasizes important of natural factors to designs and plans in our urban and suburban areas. Another aim expresses that the works made according to the natural factors important for succeed in terms of function and aesthetic. At the same time it indicates that the human follows nature knowingly and unknowingly in the developments.

1. Methodology

1.1. General description of the study area

The study material includes stagnant and running water network, transportation network, provincial and district centers found in Turkey. Turkey do have an area, including lakes, 783.577 km^2. The length of Turkey's land border is 2.949 km, with its coastal length is 7.816 kilometers (DSİ 2016).

1.2. Preparation and analyzing

The method of the study constitutes the evaluation of the maps prepared based on Photoshop 9.0 and ArcGIS 9.1 programs. All maps are based on the physical and political maps obtained from website of the General Command of Mapping (HGK 2013a; HGK 2013b). All flowing water routes, stagnant waters' shorelines, the centers of province and district were obtained from those maps. The main reasons for the preference of those maps are that they are reliable and scaled military maps with high standards. Harfa5.3 program was used to make fractal analyze of all maps. To evaluate the fractal analysis method, Kaya (2003), Atabeyoğlu *et al.* (2012) is followed.

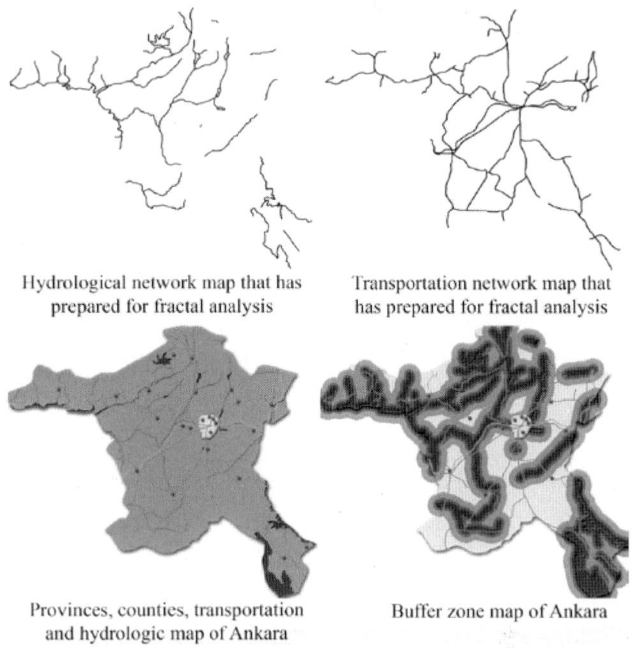

Fig. 1. The map of hydrology, road, province and county centers, Ankara city case

All 81 provinces were taken under evaluation in this study. For each province separately, black-and-white maps showing the transport network and hydrological network were prepared (Fig. 1). Fractal analysis of all maps by using Harfa program was made and fractal analysis values obtained for each map were used to compare two maps of the same province.

All 81 provinces in Turkey have been mapped on the basis of their hydrological transport network (rivers, lakes and seas) and transport network. Fractal analysis of 162 maps of 81 cities have been made with the use of Harfa software and their fractal values have been determined. The higher fractal value reveals the complexity, homogenous dispersion and the density of transport or hydrologic network. On the other hand, its low value expresses simplicity, heterogeneity, and low density. Similarity of the fractal values of both maps of the same city also indicates that they have similarities in terms of their characteristics. As the values get closer to each other, the relationship between the maps increases.

After the determination of fractal values, the fractal values obtained from the maps related with both water elements and transport were grouped in order to compare with each other and to reveal the similarities and the differences. For this grouping, the difference between the second highest and lowest fractal values was divided by 5. The obtained value was added to the second lowest value so that 5 groups were formed and the fractal values were classified under these 5 groups (Table 1). Thus, 5 similarity groups were formed and then the similarities of the maps containing transport network and water network of the same province was evaluated. The evaluated method was adapted than Atabeyoğlu and Bulut (2007) and Atabeyoğlu (2004).

Table 1. Fractal value groups

Hydrology + Road	
Fractal value	Explanation
≤ 1.169	1
1.170–1.292	2
1.293–1.415	3
1.416–1.538	4
1.539 ≤	5

After grouping the fractal values, water and transport fractal values of the same province were compared with each other and the difference between fractal groups was determined. Based upon the series of fractal values, fractal values of the cities were grouped into 5 and compared with each other after that the differences between the maps were evaluated, resulting in the differences ranging from 0 to 4. The value of "0" demonstrates that there is no relative difference between the 2 maps of the same city and "4" is the value demonstrating that there is no relative similarity between the two maps. It can be said that there is a parallelism between the hydrological network and the transport network of the cities defined with 0, the transport network follows the water network with a strong correlation between these two and the water elements are dominant and determinative on transportation route. On the other hand, the case for the cities defined with "4" refers exactly the opposite. Furthermore, percent similarity ratios related to these difference were expressed as percent value of the two values and added to the table.

Buffer Zone Method was used to determine the relationship between the places where the cities and counties were established and the water elements. Using the maps obtained from the site of General Command of Mapping, based on the centers of cities and counties, 2,5, 5 and 10 m zones were constituted in the circle of water elements and the number of the cities and counties placed in and out the zones were determined. Thus, in choosing the places for city and county centers, the effect of closeness to water was tried to put forward. The relationships were visualized by using the maps related to the values of fractal analysis and buffer zone analysis.

1.3. Statistical analysis

Statistical analysis was conducted to determine the relationship of water with road and settlement. Statistical analysis was carried out to strengthen the evaluation of relationships of water network with the road network and with city and district settlements. Accordingly, coastal and non-coastal provinces were classified and the relationships and the interactions between road and water networks in both provincial and country bases were determined. The data of fractal values were displayed as mean±standard error of the mean (SEM), standard deviation. The assumptions of data normality, which are prerequisite for correlation analysis, were tested with the Kolmogorov-Smirnov normality test. The Pearson correlation coefficient was calculated to assess the concordance fractal values between water and road. Pearson chi-square test were performed for the locations of the counties (have/haven't coast to sea and lake) to assess significant differences among distance to water elements (2.5 km, 5 km, 10 km, 10 km<). Likelihood/Fisher's exact chi-square tests were used when cell sizes were less than 5. The alpha level was set at 5%. All calculations were performed with SPSS 23 statistical software.

Finally, SWOT analysis was made. SWOT analysis is a good method for evaluated positive, negative, true and wrong sides. It was evaluated strengths, weaknesses, opportunities and threats on rivers, lakes and sea of roads follow shores with SWOT.

2. Results

Turkey is a geographic region with high density of water elements. Several large and small rivers, lakes, dams and the sea surround the territory of Turkey like a network (Fig. 2). These natural elements are often found in both urban and rural areas in every province. Turkey also has a very rugged geography. The roads in the country that was netted with dense transportation networks were passed through considering the geographic elements (Fig. 3). In this context, residential and transport network should be related with geographic elements and flowing and stagnant waters. Fractal analysis was used to reveal and objectify these relations.

Information concerning the maps of hydrological and transport networks belonging to the provinces was converted to numerical expressions and compared. For this reason, the relationships among fractal values with their similarities and differences were considered. The fractal similarity ratio represents the similarity relation of the provincial fractal values as percentage. A 100% resemblance ratio indicates that the classification groups of fractal values are the same. Among the cities, 12 of them are located on the coastal area, but the other 12 in the inland. The number of cities with seashore and a resemblance ratio over 50% is 25, while the number of non-coastal is 42 (Table 2, Fig. 4).

Water is a crucial factor for all settlements in terms of strategic, economic, visual, and climatic aspects. The relation of hydrology network with settlement is important

Fig. 2. Map of hydrology network

Fig. 3. Map of road network

Fig. 4. Range of the cities according to the groups of fractal value

Table 2. Fractal values, values groups and similarity of the cities

Provinces	Fractal value (water-road)	Fractal value group (water-road)	Fractal similarity rate (%)	Provinces	Fractal value (water-road)	Fractal value group (water-road)	Fractal similarity rate (%)
The cities haven't coast to sea and lake							
Adiyaman	1,338–1,311	3–3	100	Bayburt	1,080–1,256	1–2	50
Afyon	1,179–1.501	2–4	50	Karaman	1,112–1,274	1–2	50
Ağri	1,233–1,152	2–1	50	Kirikkale	1,098–1,255	1–2	50
Amasya	1,141–1,276	1–2	50	Batman	1,154–1,292	1–2	50
Ankara	1,526–1,485	4–4	100	Şirnak	1,105–1,211	1–2	50
Bilecik	1,093–1,326	1–3	33	Ardahan	1,148–1,197	1–2	50
Bingöl	1,476–1,303	4–3	75	Iğdir	1,301–1,123	3–1	33
Bolu	1,122–1,312	1–3	33	Karabük	1,059–1,282	1–2	50
Burdur	1,240–1,213	1–2	50	Kilis	1.025–1,248	1–2	50
Çankiri	1,080–1,262	1–2	50	Osmaniye	1,061–1,340	1–3	33
Çorum	1,135–1,000	1–1	100	The cities have coast to sea and lake			
Denizli	1,166–1,251	1–2	50	Adana	1,613–1,482	5–4	80
Diyarbakir	1,221–1,263	2–2	100	Antalya	1,090–1,418	1–4	25
Elaziğ	1,341–1,403	3–3	100	Artvin	1,156–1,271	1–2	50
Erzincan	1,207–1,260	1–2	50	Aydin	1,273–1,332	1–3	50
Erzurum	1,789–1,625	5–5	100	Balikesir	1,408–1,510	3–4	75
Eskişehir	1,149–1.277	1–2	50	Bitlis	1,173–1,254	2–2	100
Gaziantep	1,507–1,372	4–3	75	Bursa	1,534–1,580	4–5	80
Gümüşhane	1,000–1,180	1–1	100	Çanakkale	1,000–1,591	1–5	20
Hakkari	1.103–1.176	1–2	50	Edirne	1,303–1,547	3–5	60
Isparta	1,218–1,294	2–3	67	Giresun	1,258–1,237	2–2	100
Kars	1,235–1,307	2–3	67	Hatay	1.191–1,307	2–3	67
Kayseri	1,371–1,282	3–2	67	Içel	1,354–1.472	3–4	75
Kirşehir	1,238–1,352	2–3	67	Istanbul	1,517–1,475	4–4	100
Konya	1,399–1,661	3–5	60	Izmir	1,414–1,000	3–1	33
Kütahya	1,000–1,585	1–5	20	Kastamonu	1,359–1,653	3–5	60
Malatya	1,509–1.509	4–4	100	Kirklareli	1,485–1,423	4–4	100
Manisa	1,532–1,521	4–4	100	Kocaeli	1,306–1,415	3–3	100
K.maraş	1,149–1,291	1–2	50	Muğla	1,702–1,253	5–2	40
Mardin	1,047–1,330	1–3	33	Ordu	1,234–1,469	2–4	50
Muş	1,150–1,294	1–3	33	Rize	1,217–1,288	2–2	100
Nevşehir	1,107–1,315	1–3	33	Sakarya	1,216–1,299	2–3	67
Niğde	1,165–1,268	1–2	50	Samsun	1,310–1,368	3–3	100
Siirt	1,137–1,222	1–2	50	Sinop	1,331–1,322	3–3	100
Sivas	1,165–1,000	1–1	100	Tekirdağ	1,317–1,513	3–4	75
Tokat	1,344–1,394	3–3	100	Trabzon	1,281–1,259	2–2	100
Tunceli	1,212–1,328	2–3	67	Van	1,409–1,000	3–1	33
Şanliurfa	1,292–1,000	2–1	50	Zonguldak	1,240–1,337	2–3	67
Uşak	1,179–1,347	2–3	67	Bartin	1,182–1,273	2–2	100
Yozgat	1,247–1,645	2–5	40	Yalova	1,232–1,243	2–2	100
Aksaray	1,182–1,223	2–2	100	Düzce	1,209–1,235	2–2	100

as well as with the transport network. The relationship between the hydrological network with the settlements was evaluated on the basis of provincial and district locations (Fig. 5). The evaluation was made on the basis of 970 districts of Turkey's 81 cities (E-Interior 2016). The coastal and non-coastal cities to the large water elements and their districts are evaluated separately. According to this evaluation, there are 28 coastal cities with their 411 districts, and 53 non-coastal cities with their 559 districts. In general, a total of 70 provinces (42 coastal and 28 non-coastal) was located within the area that is 10 km away from existing water elements (Table 2, Fig. 5).

According to fractal values, in terms of water elements, 32 cities have very low fractal value of 40%, and 3 cities have a very high value of 4%. When evaluated in terms of road network, 8 cities have very low value of 10%, and 8 cities have very high fractal value of 10%. The distribution of fractal values concerning water and transport networks indicates that, though it seems the of opposite the distribution related to the intervals of the highest and lowest values, there is a parallelism between them when low and high values included (Table 3).

Table 3. The rate of the groups of fractal similarity

	Water network		Road network	
	Frequency	Percent	Frequency	Percent
Very low	32	39,5	8	9,9
Low	22	27,2	30	37,0
Middle	16	19,8	23	28,4
High	8	9,9	12	14,8
Very high	3	3,7	8	9,9
Total	81	100,0	81	100,0

The number of provinces and counties was determined from the bufferzone maps prepared for determining provinces and counties settlements within the distances of 2.5, 5 and 10 km from the water elements. In total, 79% of the counties is located in an area of 10 km (Table 4).

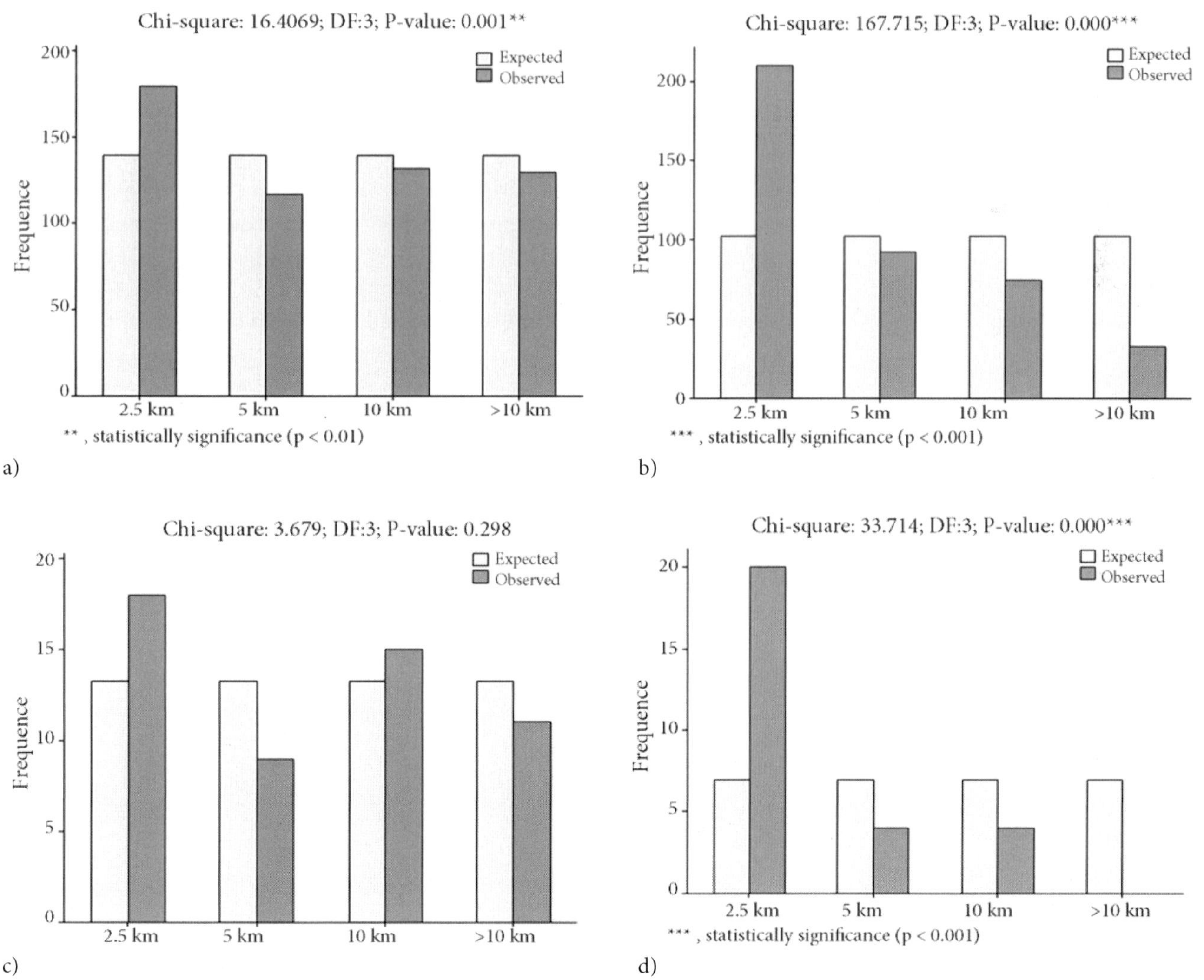

Fig. 5. Number of provinces and counties according to distances: a) number of counties with no coast to big water masses to distances; b) number of counties with coast to big water masses to distances; c) number of provinces with no coast to big water masses to distances; d) number of provinces with coast to big water masses to distances

Table 4. Distribution of counties and provinces according to distances

		2.5 km	5 km	10 km	10 km<
Counties	The coastal cities to the large water elements	% 19	% 9.9	% 7.6	% 3.5
	The non-coastal cities to the large water elements	% 17.5	% 11.6	% 13.1	% 13
	Total	% 36.6	% 21.5	% 20.8	% 16.6
Provinces	The non-coastal cities to the large water elements	% 22.2	% 11.1	% 18.5	% 13.5
	The coastal cities to the large water elements	% 24.6	% 4.9	% 4.9	% 0
	Total	% 46.8	% 16	% 23.4	% 13.5

Table 5. Pearson correlation coefficients of fractal value between water and road

Counties	Region	*n*	Mean±SEM	Standard deviation	Correlation coefficient	P-Value
The cities have coast to sea and lake	Road	30	1.361±0.028	0.155	0.055	0.771 [NS]
	Water	30	1.311±0.028	0.152		
The cities haven't coast to sea and lake	Road	51	1.306±0.020	0.145	0.447	0.001**
	Water	51	1.217±0.022	0.159		
All of the cities	Road	81	1.326±0.017	0.150	0.334	0.002**
	Water	81	1.252±0.018	0.162		

* SEM, Standard error of mean; ** statistically significant ($p < 0.01$); *** statistically significant ($p < 0.001$); NS, statistically not significant ($p > 0.05$)

In the assessment of urban settlements, among the non-coastal provinces to large water masses 18 cities are located in 2,5 km, 9 cities in 5 km, 15 cities in 10 km and 11 cities in +10 km away from existing water sources. As for the cities having shores to the large water masses, 20 city centers are located in 2,5 km, 4 in 5 km, and 4 in 10 km distance to water element.

Based on fractal analysis comparisons, substantial similarities between hydrological and transportation networks of the provinces have been identified. Between the two maps of the cities, considering the percentage values, high similarity rate is 75.4% with 61 cities, little similarity rate is 18.5% with 15 cities and very little or dissimilarity rate is 6.1% with 5 cities. The average percentage of similarity is 74% in reference with the fractal values of coastal cities, while it is 62% in the non-coastal ones. The fractal similarity percentage across the country is 66%. According to this result, there is a high parallelism between hydrological and transport networks in Turkey.

Though similarity includes the majority of the provinces, when it is compared with information regarding the geographical location of the provinces showing high rate of similarity, it corresponds to varying topography and highlands. Even though they are mostly in high altitudes, it is observed that the provinces showing low or no similarities, have been placed calm topographic, and lower mobility areas. The transport network mostly comply with existing valleys, and the water routes in areas with high geographical restrictions, and the diverse topographic changes, while in a relatively quiet topography there is no similar compliances due to better geographical conditions (Fig. 6).

According to Pearson correlation tests (Table 5). The cities identified as the same" or "very similar" are approximately 75% of the country's whole provinces. The total percentage of the cities within the groups of little similarity, no similarity or very low similarity cities percentages remained at 25 % (Table 6).

For the cities have coast to water; mean of road network is 1.361±0.028, mean of water network is 1.311±0.028. The coefficient of correlation between water and road networks is 0.055, but it is not statistically significant ($p > 0.05$). For the cities haven't coast to water; mean of road network is 1.306± 0.020, mean of water

Table 6. Distribution of provinces to groups of fractal values

Groups of fractal values	Number of provinces	Percentage
0 – same	24	% 29,6
1 – very similar	37	% 45,8
2 – little similar	15	% 18,5
3 – very little similar	3	% 3,7
4 – no similar	2	% 2,4

network is 1.217±0.022. The coefficient of correlation between water and road networks is 0.447, and it is statistically significant ($p < 0.01$). If the fractal value of road network increases, the fractal value of water network increases at the rate of 44.7%. For the all of the cities; mean of road network is 1.326±0.017, mean of water network is 1.252±0.018. The coefficient of correlation between water and road networks is 0.334, and it is statistically significant ($p < 0.01$).

Together with obtained results, SWOT analysis was made for evaluated to affects of the roads that follow river, lake and sea.

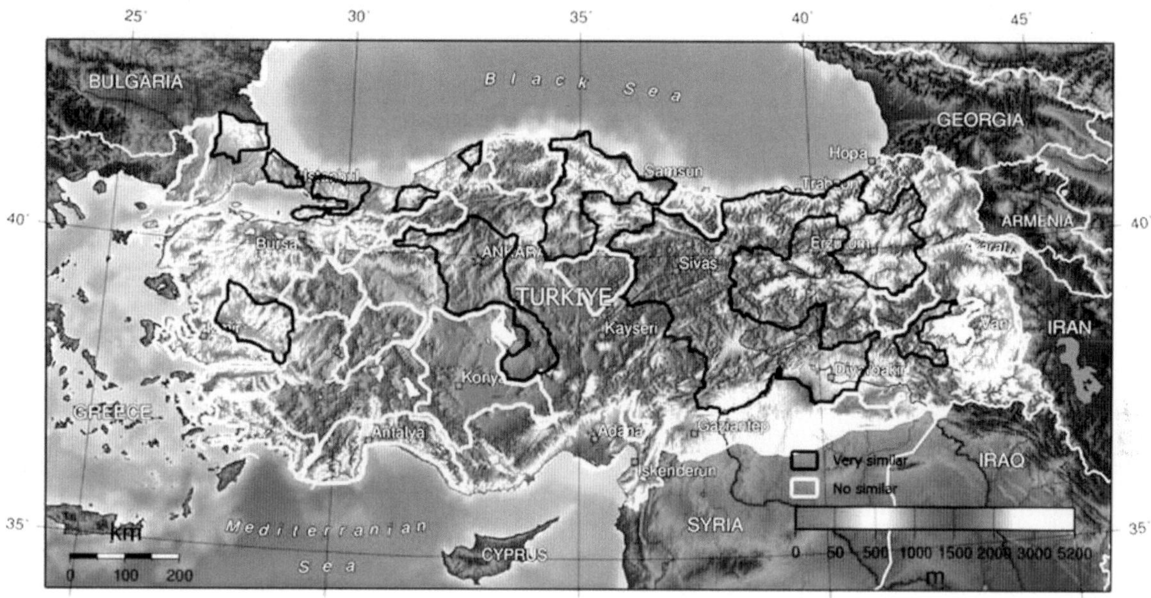

Fig. 6. Topographic map for the similarity of hydrology and road networks

Strengths	Weaknesses
- Aesthetic value and scenery of shores and river lines are high. - Travel enjoyment is high. - These roads require minimum amount of excavation and filling, because these roads, e. g. water pass form in optimal land conditions. - The roads following the water are economic, as this road doesn't need excavation–filling and tunnels. - Transportation to city near the water is easy. - These roads have connection with water transportation. - Vegetable material is rich, intensive and various. - These roads have transportation safety. - It is the shortest route to transportation. - It is in harmony with nature. - It is one of the most harmless methods to nature, as engineering activities are almost not needed. - Extreme weather does not influence the roads because they are in low altitude - It is economic, easy to transport	- The roads weaken aesthetic value of water shores and divide scenery. - The roads have too much of meanders. - River corridors have intensive air movements. - It is necessary to bridge in order to pass rivers. - Agricultural estates and fertile lands are divided by roads. - It is impossible to follow sudden water changes. River network may not be available everywhere suitable for all destinations.

Opportunities	Threats
- It provides opportunity to improve waterway transportation. - It provides new recreation areas. - It provides increased recreational activities based on water.	- There are threats of flooding and overflow. - There are threats of landslide and dent. - The roads cause destructions on natural areas. - It is a threat to the migration routes of the birds. The fauna in shores and environment of water are impressed. - The exhaust gas, fuel and oil pollute the waters. - Transportation noise impress the fauna. - Noise, dust, gas, and density prevent to use the water shores for recreational aims. - It affects on existing recreational areas in water shores - It divides and impresses agricultural estate and fertile soils.

In result of evaluation, there are positive and negative sides of the roads that follow water shores. The positive side consists of mostly economic reasons and enjoy of travelling. And the negative side is mostly ecologic reasons like negative effect on the nature and intensive increasing of recreation areas.

Conclusions

Settlements set near waterfront or underground water source because of agriculture and farming (Tunçdilek 1985; Tapur 2009; Demir 2003; Selim 2011; Keleş *et al.* 2012). Important cities have water elements near them. At the same time river, lake or sea are as the hunting-ground for settlers (Tunçdilek 1988). Water elements are natural barriers because of difficulty to pass them. Therefore they offer thesecurity in settlements (Şahinalp 2006). Stream, lake, and sea are important parts of landscape. They have effects on the aspects of environment, e. g. geology, history, culture, and biology (Yıldırım *et al.* 2013). They have an effect on transportation, too. The water elements are used both to transportation and as a guide to roads (Değerliyurt 2014). Turkey has some important settlements of the world in point of the water. Turkey has 26 basins and they give form to settlements in Turkey (Selim 2011).

According to the results; rivers, lakes, and seas are important factors in determining the transit routes of the roads. The water elements use the routes for which geographic conditions are the most suitable and this serves as a guide for the transport network. Because these itineraries are relatively flatter, more convenient and also micro climatic areas with more views due to the proximity to the water element. The very same fact became the reason of the road transit root as well, considering the geographical conditions, using the shortest possible transitions and straight lines with mostly parallel and close to the water element. When the results are evaluated, it is seen that the road networks generally follow the water networks. The road networks knowingly or unknowingly built in parallel to water's edge. This situation increases scenery, transportation comfort and travel pleasure. In the same time, it also derives a profit in economic. Additionally, provincial and district centers have also been established to be particularly close to the water elements due to its landscape and economic factors, the appropriate geographic and climatic conditions.

Acknowledgements

I thank Prof. Dr. Şevket Metin Kara for the language editing and I thank Assist. Prof. Dr. Yeliz Kaşko Arıcı for the statistical analysis. I value the constructive comments of reviewers and editors as they significantly improved the quality of the final manuscript.

References

Atabeyoğlu, Ö. 2004. *The evaluation of the gardens of the institution and public houses in the respect of landscape architecture:* Master Thesis. Atatürk University Graduate School of Natural and Applied Science, Erzurum.

Atabeyoğlu, Ö.; Bulut, Y. 2007. Evaluation of outdoor quality efficiency of public institutions with scoring method, *The Journal of Süleyman Demirel University Faculty of Forestry* A(1): 92–106.

Atabeyoğlu, Ö.; Turgut, H.; Yeşil, P.; Yılmaz, H. 2012. Evaluation of alterations on historical urban structure of Erzurum with fractal analysis method, *The Journal of Süleyman Demirel University Faculty of Forestry* 13: 162–171.

Blanton, P.; Marcus, W. A. 2009. Railroads, roads and lateral disconnection in the river landscapes of the continental United States, *Geomorphology* 112: 212–227. http://dx.doi.org/10.1016/j.geomorph.2009.06.008

Değerliyurt, M. 2014. *Factors affecting use of space in cities. Urban Studies I.* Ankara: Detay Publication.

Demir, E. 2003. The contribution of the Southeastern Anatolian Project to the domestic economy and its effect on the settlements areas, *Journal of Gazi Education Faculty* 23(3): 189–205.

DSİ. 2016. *Soil, water resource. General directorate of state hydraulic works* [online], [cited 03 April 2016]. Available from Internet: http://www.dsi.gov.tr/toprak-ve-su-kaynaklari

Eigenbrod, F.; Hecnar, S. J.; Fahrig, L. 2007. Accessible habitat: an improved measure of the effects of habitat loss and roads on wildlife populations, *Landscape Ecology* 23: 159–168. http://dx.doi.org/10.1007/s10980-007-9174-7

E-Interior. 2016. *The Counties of Turkey. Ministry of Interior* [online], [cited 18 December 2012]. Available from Internet: https://www.e-icisleri.gov.tr/Anasayfa/MulkiIdariBolumleri.aspx

Fu, W.; Liu, S.; Dong, S. 2010. Landscape pattern changes under the disturbance of road networks, *Procedia Environmental Sciences* 2: 859–867. http://dx.doi.org/10.1016/j.proenv.2010.10.097

Geneletti, D. 2003. Biodiversity impact assessment of roads: an approach based one ecosystem rarity, *Environmental Impact Assessment Review* 23: 343–365. http://dx.doi.org/10.1016/S0195-9255(02)00099-9

Hawbaker, T. J.; Radeloff, V. C. 2004. Roads and landscape pattern in Northern Wisconsin based on a comparison of four road data sources, *Conservation Biology* 18(5): 1233–1244. http://dx.doi.org/10.1111/j.1523-1739.2004.00231.x

HGK. 2013a. *Administrative map of Turkey, General Command of Mapping* [online], [cited 20 January 2013]. Available from Internet: http://www.hgk.msb.gov.tr/tematik-haritalar

HGK. 2013b. *Physical Map of Turkey. General Command of Mapping* [online], [cited 20 January 2013]. Available from Internet: http://www.hgk.msb.gov.tr/tematik-haritalar

Jenelius, E.; Mattsson, L. G. 2015. Road network vulnerability analysis: Conceptualization, implementation and application, *Computers, Environment and Urban Systems* 49: 136–147. http://dx.doi.org/10.1016/j.compenvurbsys.2014.02.003

Kaya, H. S. 2003. *Evaluating richness of urban space by using chaos theory and fraktal geometry:* Master's Thesis. Istanbul Technique University, Graduate School Of Natural And Applied Sciences, Istanbul, Turkey.

Keleş, R.; Hamamcı, C.; Çoban, A. 2012. *Environmental policy.* Ankara: İmge Bookstore.

Liu, S.; Dong, Y.; Deng, L.; Liu, Q.; Zhao, H.; Dong, S. 2014. Forest fragmentation and landscape connectivity change associated with road network extension and city expansion: a case study in the Lancang River Valley, *Ecological Indicators* 36: 160–168. http://dx.doi.org/10.1016/j.ecolind.2013.07.018

Liu, S. L.; Cui, B. S.; Dong, S. K.; Yang, Z. F.; Yang, M.; Holt, K. 2008. Evaluating the influence of road networks on landscape and regional ecological risk – a case study in Lancang River Valley of Southwest China, *Ecological Engineering* 34(2): 91–99. http://dx.doi.org/10.1016/j.ecoleng.2008.07.006

Patarasuk, R. 2013. Road network connectivity and land-cover dynamics in Lop Buri province, Thailand, *Journal of Transport Geography* 28: 111–123. http://dx.doi.org/10.1016/j.jtrangeo.2012.11.011

Selim, S. 2011. The potential effects of ecosystem integrity on human source activities on the river valley: a case of Çağlayan River, *Journal of Natural and Applied Sciences* 15(2): 94–101.

Seppala, M. 1999. Geomorphological aspects of road construction in a cold environment, Finland, *Geomorphology* 31: 65–91. http://dx.doi.org/10.1016/S0169-555X(99)00073-2

Şahinalp, M. S. 2006. The factors effecting on establishment of Şanlıurfa City, *Journal of Geographical Sciences* 4(1): 105–127.

Topur, T. 2009. Effect on first settlements of ancient Konya Lake, *Blacksea Research* 6(23): 99–115.

Tian, G.; Wu, J. 2015. Comparing urbanization patterns in Guangzhou of China and Phoenix of the USA: The influences of roads and rivers, *Ecological Indicators* 52: 23–30. http://dx.doi.org/10.1016/j.ecolind.2014.11.024

Tunçdilek, N. 1985. *Relief forms and land uses in Turkey.* İstanbul: İstanbul University Publications, No. 3279.

Tunçdilek, N. 1988. *Population dynamics in the world.* İstanbul: İstanbul University Publications, No: 3501.

Yıldırım, E.; Yılmaz, T.; Benliay, A. 2013. Important of stream ecology on landscape planning, *Turkish Journal of Scientific Reviews* 6(1): 51–54.

Ömer ATABEYOĞLU. PhD, Assist. Prof. Dr in Department of Landscape Architecture, Faculty of Agriculture, Ordu University in Ordu/Turkey. Research interests: landscape design, urban design, urban planning, historical places. Published 41 scientific paper, made 19 scientific presentations in conferences.

8

FORECASTING SURFACE WATER LEVEL FLUCTUATIONS OF LAKE SERWY (NORTHEASTERN POLAND) BY ARTIFICIAL NEURAL NETWORKS AND MULTIPLE LINEAR REGRESSION

Adam PIASECKI[a], Jakub JURASZ[b], Rajmund SKOWRON[a]

[a]*Department of Hydrology and Water Management, Nicolaus Copernicus University, Lwowska 1, 87-100 Toruń, Poland*

[b]*Department of Engineering Management, Faculty of Management, AGH University of Science and Technology, ul. Gramatyka 10, 30-067, Kraków, Poland*

Abstract. The aim of this study is to assess the possibility of forecasting water level fluctuations in a relatively small (<100 km^2), post-glacial lake located in a temperate climate zone by means of artificial neural networks and multiple linear regression. The area of study was Lake Serwy, located in northeastern Poland.

Two artificial neural network (ANN) multilayer perceptron (MLP) and multiple linear regression (MLR) models were built. The following explanatory variables were considered: maximal and minimal temperature (Tmax, Tmin) wind speed (WS), vertical circulation (VC) and water level from previous periods (WL). Additionally, a binary variable describing the period of the year (winter, summer) has been considered in one of the two MLP and MLR models. The forecasting models have been assessed based on selected criteria: mean absolute percentage error (MAPE), root mean squared error (RMSE), coefficient of determination (R2) and mean biased error. Considering their values and absolute deviations from observed values it was concluded that the ANN model using an additional binary variable (MLP_B+) has the best forecasting performance. Absolute deviations from observed values were the determining factor which made this model the most efficient. In the case of the MLP_B+ model, those values were about 10% lower than in other models.

The conducted analyses indicated good performance of ANN networks as a forecasting tool for relatively small lakes located in temperate climate zones. It is acknowledged that they enable water level forecasting with greater precision and lower absolute deviations than the use of multiple linear regression models.

Keywords: ANN, multiple linear regression, lake Serwy, water level, post-glacial lake.

Introduction

Lakes are very sensitive parts of the environment, responding very quickly to changes in inflow water volume. Hence, systematic measurement of their water level allows knowledge of the range of seasonal and long-term variations in inflow and outflow to be gathered. They may also indirectly be a source of information on transformations in water balance, or regional climate changes (Hartmann 1990; Angel, Kunkel 2010).

Information regarding water level fluctuations is essential in coastal zone management, for the location both of water management facilities (*inter alia* potable water intakes) and of tourism/recreation (beaches, decks, educational trails). Water level changes determine the vegetation and occurrence of certain kinds of plant life that will flourish along a coastal zone (Nicol, Ganf 2000; Riis, Hawes 2002). The impact of water level regime on the development of different plant communities has been demonstrated experimentally (Weiher, Keddy 1995; Nielsen, Chick 1997).

The ability to accurately forecast water level changes in lakes enables conducting of a well thought-out water management within given area or catchment. Additionally it indirectly provides information on the volume of water resources which is of great importance in case of energy generation and irrigation. Precise forecasts makes evidencing and detection of variation in water ecosystem functioning much easier and enable faster adjustments to those changes.

Corresponding author: Jakub Jurasz
E-mail: jakubkamiljurasz@gmail.com

As part of the water ecosystem of a given area, lakes are influenced by a number of natural and anthropogenic factors. Human activities manifest themselves as interference in the volume and velocity of water outflow from the drainage basin. In the case of natural factors, the basic determinant shaping water conditions is climate. All attempts to model the lake water level must be preceded by an in-depth local analysis of the lake and its catchment in order to properly identify all explanatory variables. At the same time, it is important to note that it may be an arduous or even impossible task to take into consideration all factors influencing water level fluctuations. The best example of a reason for this is miscellaneous human interference in the water resources of a given area: these activities are often not registered, and are hard to specify and take into account in any model. Therefore, it is important to consider in the future application of any model created, that some explanatory variables have not been considered.

So far in papers focusing on water level fluctuations, authors have considered diverse sets of hydrological and hydro-meteorological parameters (Bengtsson, Malm 1997; Jones *et al.* 2001). The most commonly used explanatory variables include (Talebizadeh, Moridnejad 2011):

- inflow and outflow rates;
- historical time series of water levels;
- sum of precipitation over a given period;
- evaporation;
- wind speed;
- humidity;
- temperature.

In practice, several methods and statistical tools are used to forecast water level fluctuations. One of the most commonly used is Auto Regressive Moving Average (ARMA). It has been used by (Wiche, Vecchia 1995) in the case of Devils Lake, USA. A modified version of ARMA (ARIMA) has been successfully applied by Irvine and Eberhard (1992) for Lakes Erie and Ontario. Another approach has been implemented by Sen *et al.* (2000) to model the water level of Lake Van, situated in eastern Turkey – the authors of that study used Fourier transformations combined with second-order Markov chains. One of the newer approaches in the area of water level forecasting is Artificial Neural Networks (Ondimu, Murase 2007; Çimen, Kisi 2009; Coulibaly 2010). So far, research has focused on lakes of greater than several thousand square kilometres, or those located in warm climate zones. With their ability to forecast the variability of a given time series with determined accuracy, ANNs have been used multiple times to provide broad models of water resources. As an example, they were effectively used for modelling drainage of rainwater (Shamseldin 1997; Sajikumar, Thandaveswara 1999), predicting flow (Clair, Ehrman 1998; Campolo *et al.* 1999; Zealand *et al.* 1999), forecasting groundwater level (Chełmicki *et al.* 2003; Daliakopoulos *et al.* 2005; Nayak *et al.* 2006; Yang *et al.* 2009; Adamowski, Chan 2011), proglacial discharge simulation (Nowak, Sobota 2015) and also in issues related to water management in cities (Jain *et al.* 2001; Muleta, Nicklow 2005; Herrera *et al.* 2010; Campisi-Pinto *et al.* 2012).

The aim of this study is to compare the accuracy of water level forecasts based on artificial neural networks and multiple linear regression for a relatively small, post-glacial lake located in a temperate climate zone. The main limitations of this study is the fact that it concentrates on only one lake (due to the data availability) and some potentially influencing factors impacting water level – such as snow cover has been neglected (due to the data availability) but mentioned. What is more this study compares only two commonly used forecasting methods but the obtained results build foundation for further research.

1. Area of study

The object of this study was Lake Serwy in northeastern Poland, which, similarly to most of the lakes located in the Central European Lowlands, originated during the recent Baltic Glaciation (Fig. 1).

The selection of this lake is supported by following arguments. The meteorological and hydrological measurements (in a form of time series) are taken by a meteorological station which is located not far from Lake Serwy. What is also important, those time series encompass relatively long time period and do not exhibit any missing records. Additionally morphometric parameters of the Lake Serwy make this particular lake representative for this part of Europe. Finally there is no recorded significant impact of anthropogenic factors on water level fluctuations in this lake.

In terms of area and volume, the analyzed lake is relatively large with regard to other lakes in this area. The lake has three islands, whose surfaces amount to 0.12 km^2. The basic morphometric parameters are presented in Table 1.

Table 1. Morphometric parameters of Lake Serwy which enable its comparison with other lakes in this area

Parameter	Value	Unit
Area (A)	4.6	km^2
Volume (V)	$67.1 \cdot 10^6$	m^3
Maximal length (L_{max})	6900	m
Maximal width (W_{max})	1100	m
Extension ($\lambda = L_{max}/W_{max}$)	6.2	–
Mean width ($W_{mean} = A/L_{max}$)	667	m
Shoreline length (SL)	16100	m
Maximal depth (H_{max})	41.5	m
Mean depth ($H_{mean} = V/A$)	14.1	m

The lake is supplied with water from several small watercourses and ditches draining the surrounding marshes and wetlands. The remaining water supply sources are: direct precipitation, surface runoff from own catchment, and underground inflow. The last of these water supplies seems to be of particular significance in the case of the analyzed lake because of its considerable depth. The catchment area amounts to 53.9 km^2. Forests (mainly coniferous) cover 79% of its area, the remainder consists of agricultural land (farmlands and pastures). A small drainage channel is located in the southern part of the lake.

The climatic conditions of this area are typical of a temperate maritime/continental climate with significant influence of continentalism, which manifests itself in very low winter air temperatures, often exceeding –20 °C. Subzero mean monthly temperatures occur from December to February, but also frequently in March in many years. The annual mean temperature does not exceed 7 °C and the annual sum of precipitation is 586 mm (Fig. 2). Precipitation occurs mainly in summer (June–August).

The highest water level is usually observed during spring (April–May). In the subsequent months the water level tends to decline as a result of intensified evaporation. In September and October the water level reaches its nadir (Fig. 3). Over the analyzed period the annual water level amplitude reached 42 cm.

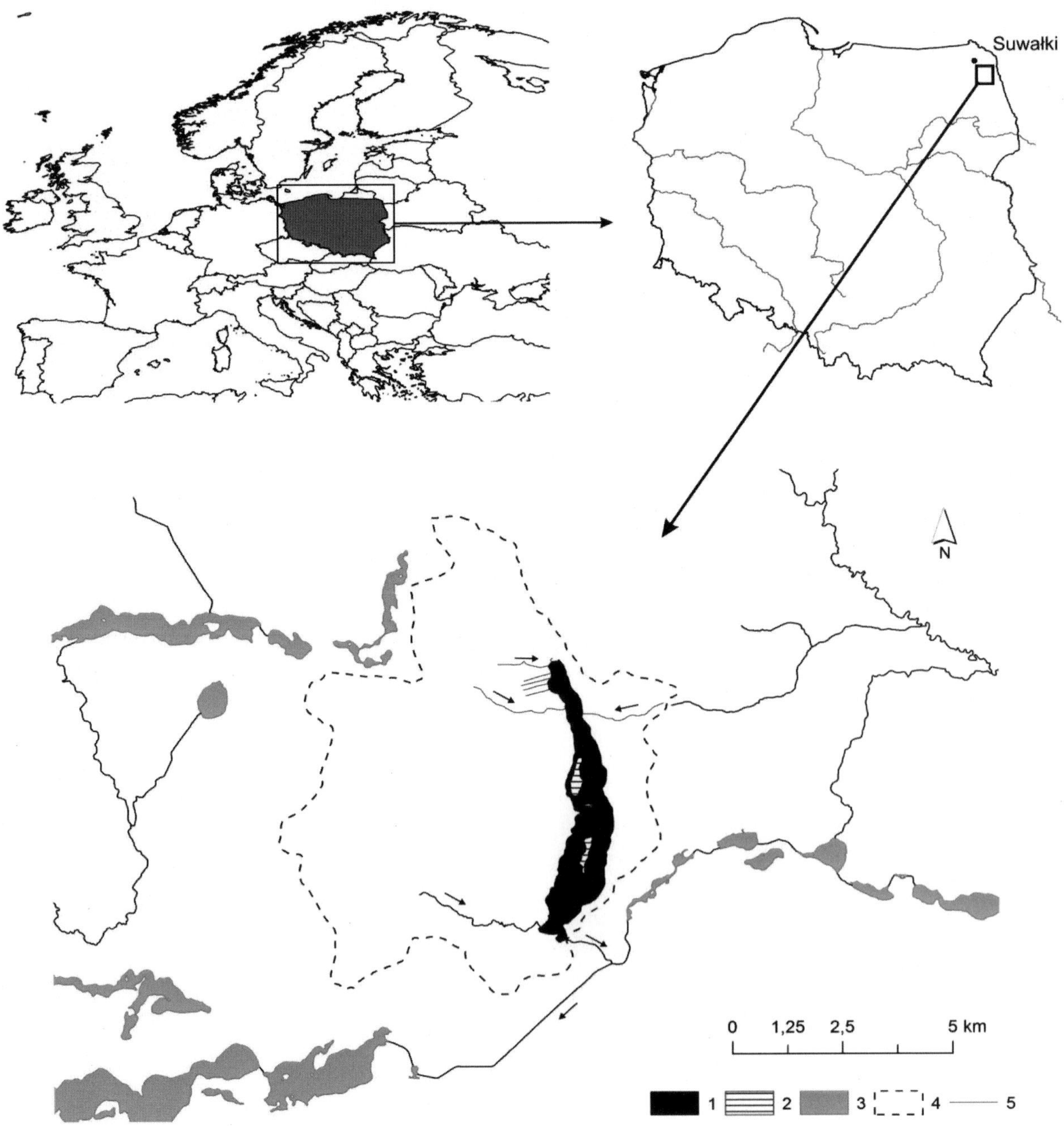

Fig. 1. Lake Serwy location from the perspective of the map of Europe and Poland. Key: 1 – Lake Serwy; 2 – islands; 3 – other lakes; 4 – Lake Serwy catchment; 5 – watercourses

Annual water level changes over a 32-year period are as shown in Figure 4.

2. Materials and methods

In this study, monthly values of water level (WL) in Lake Serwy over the years 1980–2012 have been used. They have been calculated based on daily observations conducted by the Institute of Meteorology and Water Management: State Research Institute (IMGW-PIB). Preliminarily, the following meteorological parameters have been considered as independent variables for forecasting models:

- mean air temperature (T);
- maximal air temperature (T_{max});
- minimal air temperature (T_{min});
- relative humidity (RH);
- sum of rainfall – precipitation (PPT);
- wind speed (WS);
- evaporation (Evap).

Besides the above, vertical circulation (the difference between evaporation and precipitation) has also been considered. All these parameters have been observed and recorded by the meteorological station in Suwałki, which is 25 kilometres to the north of the analyzed lake. Consequently, those monthly values correspond to the same time period as the water level values.

The explanatory variables initially described above have been tested by means of a correlation coefficient, in order to determine whether they have a significant impact on water level (Tables 2 and 3). Based on the calculations conducted a meaningful correlation between some of the variables has been observed. From further analyses, all those which exhibit mutual correlation have been excluded and only those presenting the highest correlation with water level were used as input variables. As a result, the following variables were used in the modeling process:

- maximal and minimal air temperature (T_{max}, T_{min});
- wind speed (WS);
- vertical circulation;
- water level from previous periods (WL).

It should be pointed out that, during the forecasting process, only explanatory variables of historical values referring to one (T-1) and two (T-2) previous months were used.

The structure of forecasting models (their architecture or formulas) is presented in Table 4. The quantitative

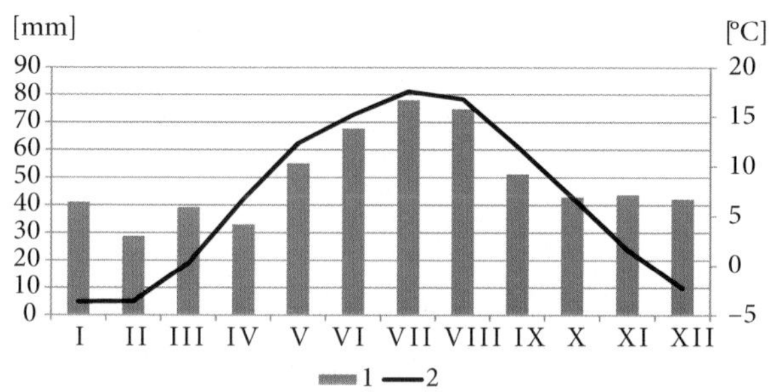

Fig. 2. Monthly sums of precipitation and monthly mean temperatures observed at the meteorological station in Suwałki over the period 1980–2012. Key: 1 – sum of precipitation; 2 – mean air temperature

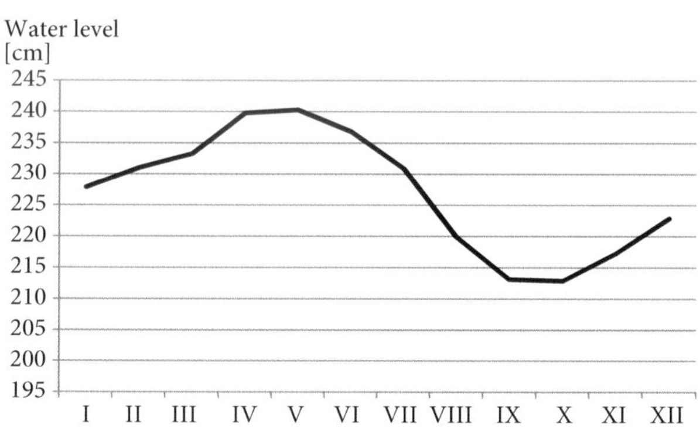

Fig. 3. Lake Serwy monthly mean water level, 1980–2012

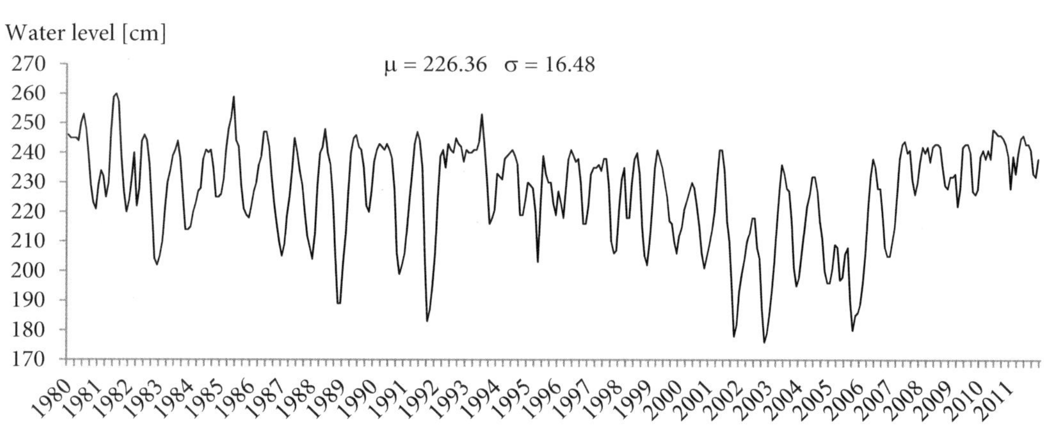

Fig. 4. Lake Serwy water level fluctuations in the years 1980–2012. Over this period the mean water level was equal to μ = 226.36 cm with a standard deviation of σ = 16.48 cm

Table 2. Coefficient of correlation matrix for one-month-lagged exogenous variables, bold font indicated correlation coefficients which exhibit statistical significance – P value <0.05. *PPT* stands for precipitation, whereas *Temp* for temperature

	Water level T-1	Evapo-ration	Temp	Wind speed	Temp Max	Temp Min	Relative humidity	PPT	Vertical circulation
Water level	**0.885**	**−0.132**	**−0.260**	**0.245**	**−0.252**	**−0.267**	0.028	−0.075	0.077
Water level T−1	1.000	**0.177**	0.006	0.061	0.021	−0.018	**−0.197**	−0.011	**−0.180**
Evaporation	**0.177**	1.000	**0.898**	**−0.488**	**0.921**	**0.837**	**−0.838**	**0.324**	**−0.757**
Temp	0.006	**0.898**	1.000	**−0.484**	**0.995**	**0.988**	**−0.661**	**0.409**	**−0.597**
Wind speed	0.061	**−0.488**	**−0.484**	1.000	**−0.506**	**−0.428**	**0.402**	−0.081	**0.423**
Temp Max	0.021	**0.921**	**0.995**	**−0.506**	1.000	**0.971**	**−0.707**	**0.382**	**−0.638**
Temp Min	−0.018	**0.837**	**0.988**	**−0.428**	**0.971**	1.000	**−0.557**	**0.444**	**−0.515**
Relative humidity	**−0.197**	**−0.838**	**−0.661**	**0.402**	**−0.707**	**−0.557**	1.000	−0.067	**0.775**
PPT	−0.011	**0.324**	**0.409**	−0.081	**0.382**	**0.444**	−0.067	1.000	**0.374**
Vertical circulation	**−0.180**	**−0.757**	**−0.597**	**0.423**	**−0.638**	**−0.515**	**0.775**	**0.374**	1.000

Table 3. Coefficient of correlation matrix for two-month-lagged exogenous variables, bold font indicated correlation coefficients which exhibit statistical significance. Presented abbreviations are coherent with those presented in Table 2

	Water level T-2	Evapo-ration	Temp	Wind speed	Temp Max	Temp Min	Relative humidity	PPT	Vertical circulation
Water level	**0.653**	**−0.406**	**−0.472**	**0.363**	**−0.473**	**−0.462**	**0.242**	**−0.145**	**0.298**
Water level T-2	1.000	**0.172**	0.003	0.073	0.017	−0.021	**−0.191**	−0.022	**−0.183**
Evaporation	**0.172**	1.000	**0.899**	**−0.491**	**0.922**	**0.839**	**−0.838**	**0.316**	**−0.761**
Temp	0.003	**0.899**	1.000	**−0.485**	**0.995**	**0.988**	**−0.663**	**0.405**	**−0.602**
Wind speed	0.073	**−0.491**	**−0.485**	1.000	**−0.508**	**−0.429**	**0.409**	−0.084	**0.422**
Temp Max	0.017	**0.922**	**0.995**	**−0.508**	1.000	**0.971**	**−0.709**	**0.378**	**−0.643**
Temp Min	−0.021	**0.839**	**0.988**	**−0.429**	**0.971**	1.000	**−0.559**	**0.439**	**−0.520**
Relative humidity	**−0.191**	**−0.838**	**−0.663**	**0.409**	**−0.709**	**−0.559**	1.000	−0.063	**0.775**
PPT	−0.022	**0.316**	**0.405**	−0.084	**0.378**	**0.439**	−0.063	1.000	**0.375**
Vertical circulation	**−0.183**	**−0.761**	**−0.602**	**0.422**	**−0.643**	**−0.520**	**0.775**	**0.375**	1.000

Table 4. Structure, parameters and equations for MLP and MLR models. *SOS* stands for sum of squared error, *BFGS* – is a teaching algorithm and its name is an abbreviation from its developers, namely Broyden–Fletcher–Goldfarb–Shanno

Name	Structure	Teaching quality	Testing quality	Validation quality	Teaching algorithm	Error function	Activation hidden layer	Activation output layer
MLP_Bin+	MLP 8-6-1	0.963	0.934	0.920	BFGS 108	SOS	Tanh	Tanh
MLP_Bin−	MLP 7-8-1	0.958	0.917	0.915	BFGS 60	SOS	Tanh	Logistic
MLR_Bin+	$WL(t)=1.1985\times WL(T-1)+1.0123\times WS(T-1)\ 0\ 0.2118\times T_{min}(T-1)-0.361\times WL(T-2)-0.2808\times WS(T-2)-0.06836\times T_{max}(T-2)+0.0091\times CV(T-2)-0.7577\times \mathrm{Bin}+35.8324$							
MLR_Bin−	$WL(t)=1.1978\times WL(T-1)+1.1271\times WS(T-1)-0.2647\times T_{min}(T-1)-0.3654\times WL(T-1)-0.3414\times WS(T-2)-0.0357\times T_{max}(T-2)+0.0185\times VC(T-2)+36.1983$							

Conventions: *WL* – water level, *WS* – wind speed, T_{min} – minimal temperature, T_{max} – maximal temperature, *VC* – vertical circulation, $T-1$ – one-month lag, $T-2$ – two-month lag variables have been supplemented by one qualitative (binary) variable which refers to the winter or summer half-year, where 1 is the summer and 0 is the winter half-year. After input variable selection, two artificial neural network MLP (Multilayer Perceptron) models and two multilinear regression equations were built. The procedure of creating ANN models have described by Dayhoff and DeLeo in (Dayhoff, DeLeo 2001) whereas MLR by Kutner *et al.* in (Kutner *et al.* 1996). The sets of quantitative variables for the MLP and MLR models are the same, but within both methods, models taking into account the binary variable (Bin+) and those not taking into account the binary variable (Bin–) have been separated. MLR and ANN models were created using Statistica 10 software.

The constructed models were assessed based on four performance measures described below. The most commonly used and easily interpreted is MAPE (mean absolute percentage error) defined as:

$$\mathrm{MAPE} = \left| \frac{WL(t) - \widehat{WL}(t)}{WL(t)} \right| \cdot 100\%, \qquad (1)$$

where $WL(t)$ is water level measured in time t, and $\widehat{WL}(t)$ is the forecasted value of water level. MAPE is an important criterion for the assessment of forecasting models from the perspective of presenting their quality to people familiar with the mean values of the considered time series. This criterion cannot be used when forecasting values should be equal to zero, due to the division by zero. The perfect match exists when MAPE is 0, but no upper value of this criterion exists. The next used performance measure was the so-called mean biased error (MBE), which indicated whether the created model expressed a tendency to over- or underestimate the values of forecasts. It has been calculated as:

$$\mathrm{MBE} = \frac{1}{N}\sum_{t=1}^{N}\left(WL(t) - \widehat{WL}(t)\right), \qquad (2)$$

where N is the sample size. In the case of an ideal match, and when overestimations equal underestimations the value of MBE amounts to 0.

The next criterion was RMSE (root mean square error), which is used to assess the variance of forecast error based on error standard deviation.

$$\mathrm{RMSE} = \sqrt{\frac{1}{N}\sum_{t=1}^{N}\left(WL(t) - \widehat{WL}(t)\right)^2}. \qquad (3)$$

The last performance measure was the coefficient of determination, which ranges from 0 to 1, and for which, values above 0.9 are perceived as very satisfactory and below 0.8 as unsatisfactory (Legates *et al.* 1999). It is expressed as:

$$R^2 = 1 - \frac{\sigma^2\left(\hat{W} - W\right)}{\sigma^2\left(W\right)}, \qquad (4)$$

where σ^2 is set variation.

Due to the particularities of the process of ANN model creation, which requires the division of time series into teaching, validating and testing subsets, MLR and MLP models have been assessed only on the basis of the testing subset.

3. Results and discussion

The obtained results, which are presented in Table 5, are similar for all forecasting models. The most precise forecasts were derived for an ANN model which used the additional binary variable (MLP_B+). It got the most beneficial values of MAPE, RMSE and R2 criteria, while the value of MBE did not diverge far from that of the MLR models. Similar values of individual forecasting criteria have been obtained by other researchers, despite the different characteristics of the lakes investigated (Kisi *et al.* 2012; Noury *et al.* 2014; Young 2015; Yarar *et al.* 2009; Vaziri 1997). This confirms the possibility of an efficient application of ANN to forecasting water level of lakes located in temperate climates (with cold winters) and having a relatively low area (under 100 km²).

Table 5. Values of forecasting performance measures

	MLR_B+	MLR_B–	MLP_B+	MLP_B–
MAPE	2.055%	2.054%	1.838%	2.017%
MBE	0.060	0.061	0.067	–0.245
RMSE	6.654	6.700	5.706	6.374
R2	0.909	0.908	0.934	0.917

The graphical representation of forecasts obtained from the four models shows an approximate congruence to the real values. In Figure 5, the values of the testing set have been sorted in increasing order, in order to depict the model's behaviour in the face of extreme values. This is not related to the trend in water level changes, which over the last 30 years has been in slight decline. Additionally, a horizontal steady line has been marked on the chart and can be used as a benchmark for the values of relative errors. The highest values of deviation from observed water level occurred in the same months. This may suggest that, in forecasting models, not all essential explanatory variables have been considered. Confirmation of this suggestion is provided by the juxtaposition of periods when those deviations were observed. The highest errors occurred during spring months, namely April and May. In this period, the Lake is heavily supplied with water stored in snow and ice sheets. In order to improve forecast quality, knowledge of ice and snow sheet cover thickness in the winter period is essential. The lack of this kind of data for the analyzed area made their application in this study impossible.

It is worth noting that, so far in papers reporting research related to the modeling and forecasting of water level, the issue of snow and ice sheets has gone unnoticed. The vast majority of this research concerned lakes with much greater areas and often located in warm climates. The best example here is Lake Van in Turkey, for which satisfactory forecasting performance of water level was obtained from models based only on water level values from previous periods (Altunkaynak 2007). A set of explanatory variables similar to those presented in this study, with values for: water flow in two supplying rivers (Gilgil and Malewa), evaporation, rainfall and simple time harmonics, was used in case of Lake Naivasha in Africa (Ondimu, Murase 2007).

An investigation into how individual models deal with the forecasting of extreme values appears to be of interest. In those cases, models tend very often to over- or underestimate the real values. From a testing set amounting to 57 samples, five extreme (minimal and maximal) samples were selected. The sum of relative errors among those subsets has been presented in Table 6. In both analyzed methods there is a similar tendency to overestimate

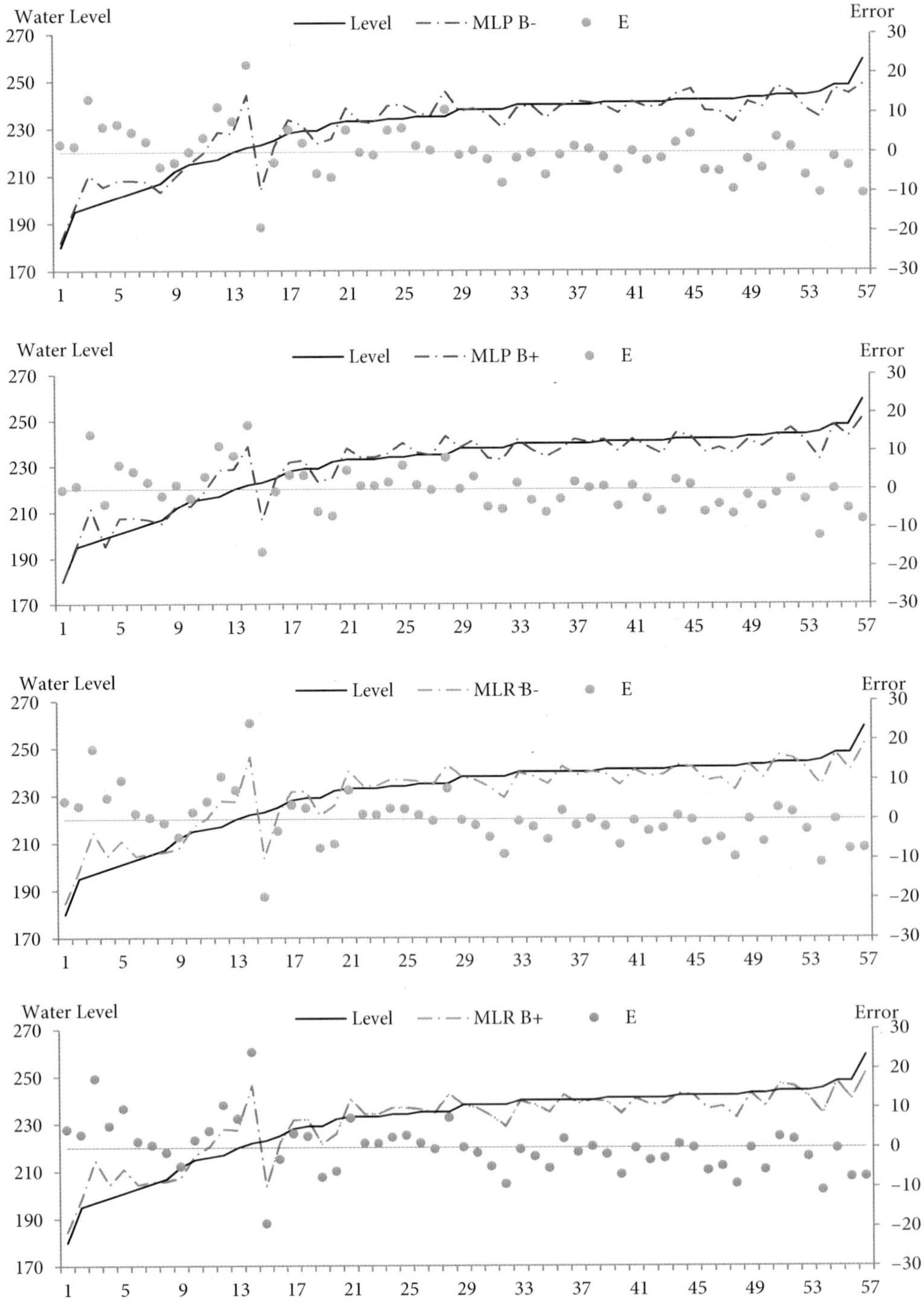

Fig. 5. Observed and forecasted values of water level for various models. Charts have been arranged from the lowest to the highest values of water level along with the forecasted water level and the value of relative errors

the value of forecasts in the case of low water level and vice versa when the water level was high. However, in the case of low water level, the ANN model performed better, as seen in the case of the MLP_B+ model, for which the sum of relative error was over two times smaller than for the MLR models. Also worthy of note is the fact that the MLP_B– model equally under- and overestimated its forecasts. However, in the case of that model, the quality of forecasts was slightly worse than for other models. Based on this comparison, it is possible to state that for a given time series (and testing set) and for the presented input variables, the multiple linear regression models have a tendency to overestimate forecasts for extreme values. In the case of ANN models those values are smaller, and, depending on input set characteristics, they may level out.

Table 6. Sum of relative errors for extreme water level values

Water Level	MLR B+	MLR B–	MLP B+	MLP B–
Low	40.9	40.9	17.8	30.7
High	–28.0	–28.4	–27.2	–31.1

Confirmation of the better forecasting performance of MLP_B+ for extreme values is provided in Figure 6, which presents absolute values of deviation. It is important to note that only in the case of the MLP_B+ model were the obtained values smaller than 18 cm, while in other cases they were over 24 cm.

Conclusions

The attempt to forecast water level in a relatively small lake located in a temperate climate zone by means of MLP and MLR models gave satisfactory results. The values of selected performance measures among analyzed models were approximate and similar to those obtained by other researchers. This is of significance mainly because of the distinct climate conditions and morphological parameters of the lakes considered in all studies previous to this. Small post-glacial lakes located in the Central European Lowlands, to which Lake Serwy belongs, are far more prone to climate factors which determine water level fluctuations. In relation to this, the phenomenon of ice sheet covering lake in winter period and resulting thaw in spring has been brought to attention. Due to the lack of data concerning ice and snow cover in the given area, it was impossible to investigate and consider the additional water supply resulting from the spring thaw. In consequence, the biggest observed forecast error occurred during early spring (April–May).

Taking into consideration the selected performance measures and the absolute deviations from real values, it can be said that the best forecasts were obtained from an ANN model using the additional binary variable and denoted as MLP_B+. The determining factor in the better forecasting performance of this model was the absolute value of deviations from observed values. In the case of MLP_B+ they were much smaller than for the other models.

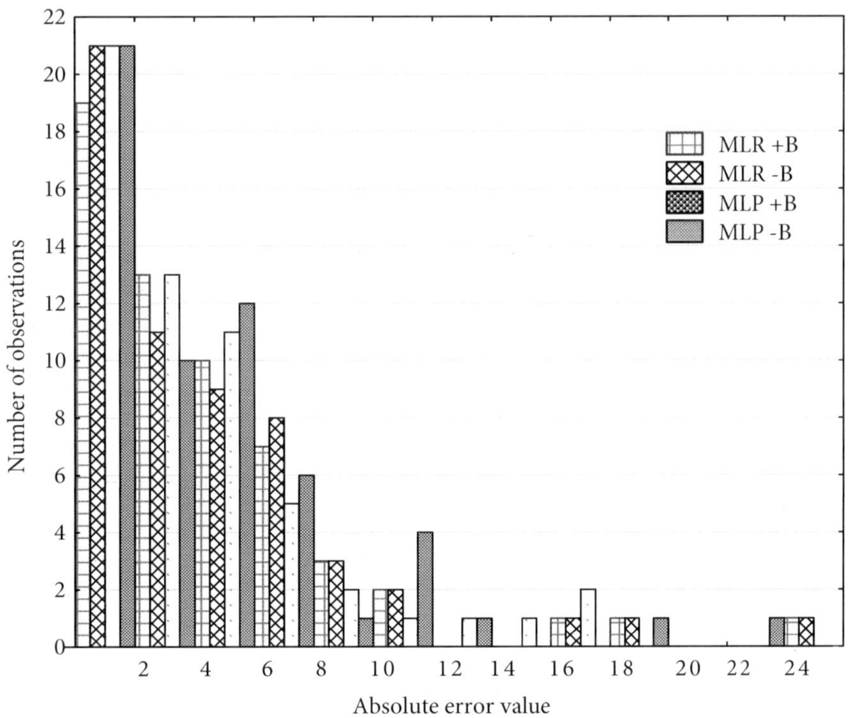

Fig. 6. Histogram of absolute errors for multi-layer perceptron (MLP) and multiple linear regression (MLR) models, with (+B) and without (–B) binary variable

In conclusion, it must be said that ANN models are an efficient tool when it comes to water level fluctuation forecasting, including in the case of relatively small lakes located in temperate climate zones. Their application enables water level forecasting with greater precision and with lower maximal deviations from observed values than the use of multiple linear regression models.

Obtained results indicate several interesting directions for the future research. One should definitely try to compare results of various forecasting models applied to predict water level fluctuation for at least several lakes. Additionally it is also very important to apply other modern forecasting techniques such as those using wavelet transform or fuzzy logic.

Models and results coming from the application of the artificial neural networks in lake water level forecasting may be a crucial parameter in modelling the changes in the natural environment. Such models seems to be of great importance when it comes to investigating various scenarios of water ecosystems operation in the context of observed climate fluctuations. A majority of issues related to the environment protection, especially from the engineering perspective is based upon models and forecasts of changes occurring in the environment which are a consequence of natural and anthropogenic factors. By using artificial neural networks in area of environmental engineering one may get closer to understand the phenomena which drive certain processes like water level fluctuations. Changes in available outflow or water level have huge impact on the operation of hydropower plants in the context of the energy yield as well as their structural performance (Povilaitis, Querner 2008; Šadzevičius *et al.* 2013).

References

Adamowski, J.; Chan, H. F. 2011. A wavelet neural network conjunction model for groundwater level forecasting, *Journal of Hydrology* 407(1): 28–40. https://doi.org/10.1016/j.jhydrol.2011.06.013

Altunkaynak, A. 2007. Forecasting surface water level fluctuations of lake van by artificial neural networks, *Water Resources Management* 21(2): 339–408. https://doi.org/10.1007/s11269-006-9022-6

Angel, J. R.; Kunkel, K. E. 2010.The response of Great Lakes water levels to future climate scenarios with an emphasis on Lake Michigan-Huron, *Journal of Great Lakes Research* 36: 51–58. https://doi.org/10.1016/j.jglr.2009.09.006

Bengtsson, L.; Malm, J. 1997. Using rainfall-runoff modeling to interpret lake level data, *Journal of Paleolimnology* 18(3): 235–248. https://doi.org/10.1023/A:1007982710792

Campisi-Pinto, S.; Adamowski, J.; Oron, G. 2012. Forecasting urban water demand via wavelet-denoising and neural network models. Case study: city of Syracuse, Italy, *Water Resources Management* 26(12): 3539–3558. https://doi.org/10.1007/s11269-012-0089-y

Campolo, M.; Andreussi, P.; Soldati, A. 1999. River flood forecasting with a neural network model, *Water Resources Research* 35(4): 1191–1197. https://doi.org/10.1029/1998WR900086

Chełmicki, W.; Ciszewski, S.; Klimek, M.; Żelazny, M. 2003. Rekonstrukcja wahań zwierciadła wód gruntowych w Puszczy Niepołomickiej za pomocą sieci neuronowych [Reconstruction of groundwater levels in the Niepołomice Forest (S Poland) using neural network], *Przegląd Geograficzny* 75(2): 250–268.

Çimen, M.; Kisi, O. 2009. Comparison of two different data-driven techniques in modeling lake level fluctuations in Turkey, *Journal of Hydrology* 378: 253–262. https://doi.org/10.1016/j.jhydrol.2009.09.029

Clair, T. A.; Ehrman, J. M. 1998. Using neural networks to assess the influence of changing seasonal climates in modifying discharge, dissolved organic carbon, and nitrogen export in eastern Canadian rivers, *Water Resources Research* 34(3): 447–455. https://doi.org/10.1029/97WR03472

Coulibaly, P. 2010. Reservoir Computing approach to Great Lakes water level forecasting, *Journal of Hydrology* 381: 76–88. https://doi.org/10.1016/j.jhydrol.2009.11.027

Daliakopoulos, I. N.; Coulibaly, P.; Tsanis, I. K. 2005. Groundwater level forecasting using artificial neural networks, *Journal of Hydrology* 309(1): 229–240. https://doi.org/10.1016/j.jhydrol.2004.12.001

Dayhoff, J. E.; DeLeo, J. M. 2001. Artificial neural networks, *Cancer* 91(S8): 1615–1635. https://doi.org/10.1002/1097-0142(20010415)91:8+<1615::AID-CNCR1175>3.0.CO;2-L

Hartmann, H. 1990. Climate change impacts on Laurentian Great Lakes levels, *Climate Change* 17: 49–67. https://doi.org/10.1007/BF00149000

Herrera, M.; Torgo, L.; Izquierdo, J.; Pérez-García, R. 2010. Predictive models for forecasting hourly urban water demand, *Journal of Hydrology* 387(1): 141–150. https://doi.org/10.1016/j.jhydrol.2010.04.005

Irvine, K. N.; Eberhardt, A. J. 1992. Multiplicative, seasonal ARIMA models for Lake Erie and Lake Ontario water levels, *Water Resources Bulletin* 28(2): 385–396. https://doi.org/10.1111/j.1752-1688.1992.tb04004.x

Jain, A.; Varshney, A. K.; Joshi, U. C. 2001. Short-term water demand forecast modelling at IIT Kanpur using artificial neural networks, *Water Resources Management* 15(5): 299–321. https://doi.org/10.1023/A:1014415503476

Jones, R. N.; McMahon, T. A.; Bowler, J. M. 2001. Modeling historical lake levels and recent climate change at three closed lakes. Western Victoria, Australia (c. 1840–1990), *Journal of Hydrology* 246(1–4): 159–180. https://doi.org/10.1016/S0022-1694(01)00369-9

Kisi, O.; Shiri, J.; Nikoofar, B. 2012. Forecasting daily lake levels using artificial intelligence approaches, *Computers & Geosciences* 41: 169–180. https://doi.org/10.1016/j.cageo.2011.08.027

Kutner, M. H.; Neter, J.; Wasserman, W.; Nachtsheim, C. 1996. *Applied linear statistical models*. Chicago: Irwin.

Legates, D. R.; Gregory, J.; McCabe, Jr. 1999. Evaluating the use of "goodness-of-fit" measures in hydrologic and hydroclimatic model validation, *Water Resources Research* 35(1): 233–241. https://doi.org/10.1029/1998WR900018

Muleta, M. K.; Nicklow, J. W. 2005. Decision support for watershed management using evolutionary algorithms, *Journal of Water Resources Planning and Management* 131(35): 35–44. https://doi.org/10.1061/(ASCE)0733-9496(2005)131:1(35)

Nayak, P. C.; Rao, Y. S.; Sudheer, K. P. 2006. Groundwater level forecasting in a shallow aquifer using artificial neural network approach, *Water Resources Management* 20(1): 77–90. https://doi.org/10.1007/s11269-006-4007-z

Nicol, J. M.; Ganf, G. G. 2000.Water regimes, seedling recruitment and establishment in three wetland plant species, *Marine and Freshwater Research* 51(4): 305–309. https://doi.org/10.1071/MF99147

Nielsen, D. L.; Chick, A. J. 1997. Flood-mediated changes in aquatic macrophyte community structure, *Marine and Freshwater Research* 48: 153–157. https://doi.org/10.1071/MF96090

Noury, M.; Sedghi, H.; Babazedeh, H.; Fahmi, H. 2014. Urmia lake water level fluctuation hydro informatics modeling using support vector machine and conjunction of wavelet and neural network, *Water Resources* 41(3): 261–269. https://doi.org/10.1134/S0097807814030129

Nowak, M.; Sobota, I. 2015. Artificial neural networks in proglacial discharge simulation: application and efficiency analysis in comparison to the multivariate regression; a case study of Waldemar River (Svalbard), *Geografiska Annaler: Series A, Physical Geography* 97(3): 489–506. https://doi.org/10.1111/geoa.12089

Ondimu, S.; Murase, H. 2007. Reservoir level forecasting using neural networks: Lake Naivasha, *Biosystems Engineering* 96(1): 135–138. https://doi.org/10.1016/j.biosystemseng.2006.09.003

Povilaitis, A.; Querner, E.P. 2008. Possibilities to restore natural water regime in the Žuvintas Lake and surrounding wetlands – modelling analysis approach, *Journal of Environmental Engineering and Landscape Management* 16(3): 105–112. https://doi.org/10.3846/1648-6897.2008.16.105-112

Riis, T.; Hawes, I. 2002. Relationships between water level fluctuations and vegetation diversity in shallow water of New Zealand lakes, *Aquatic Botany* 74: 133–148. https://doi.org/10.1016/S0304-3770(02)00074-8

Šadzevičius, R.; Damulevičius, V.; Skominas, R. 2013. The technical state of earth dams in Lithuania, *Journal of Environmental Engineering and Landscape Management* 21(3): 180–188. https://doi.org/10.3846/16486897.2012.662910

Sajikumar, N.; Thandaveswara, B. S. 1999. A non-linear rainfall–runoff model using an artificial neural network, *Journal of Hydrology* 216(1): 32–55. https://doi.org/10.1016/S0022-1694(98)00273-X

Şen, Z.; Kadioğlu, M.; Batur, E. 2000. Stochastic modeling of the Van Lake monthly level fluctuations in Turkey, *Theoretical and Applied Climatology* 65(1–2): 99–110. https://doi.org/10.1007/s007040050007

Shamseldin, A. Y. 1997. Application of a neural network technique to rainfall-runoff modeling, *Journal of Hydrology* 199(3): 272–294. https://doi.org/10.1016/S0022-1694(96)03330-6

Talebizadeh, M.; Moridnejad, A. 2011. Uncertainty analysis for the forecast of lake level fluctuations using ensembles of ANN and ANFIS models, *Expert Systems with Applications* 38(4): 4126–4135. https://doi.org/10.1016/j.eswa.2010.09.075

Vaziri, M. 1997. Predicting Caspian Sea surface water level by ANN and ARIMA models, *Journal of Waterway, Port, Coastal, and Ocean Engineering* 123(4): 158–162. https://doi.org/10.1061/(ASCE)0733-950X(1997)123:4(158)

Weiher, E.; Keddy, P. A. 1995. The assembly of experimental wetland plant communities, *Oikos* 73: 323–335. https://doi.org/10.2307/3545956

Wiche, G. J.; Vecchia, A. V. 1995. *Lake-level frequency analysis for Devils Lake, North Dakota* [online], [cited 10 January 2016]. Available from Internet: https://pubs.er.usgs.gov/publication/ofr95123

Young, C.-C.; Liu, W.-C.; Hsieh, W.-L. 2015. Predicting the water level fluctuation in an Alpine Lake using physically based, artificial neural network, and time series forecasting models, *Mathematical Problems in Engineering*. Article ID 708204. https://doi.org/10.1155/2015/708204

Yang, Z. P.; Lu, W. X.; Long, Y. Q.; Li, P. 2009. Application and comparison of two prediction models for groundwater levels: a case study in Western Jilin Province, China, *Journal of Arid Environments* 73(4): 487–492. https://doi.org/10.1016/j.jaridenv.2008.11.008

Yarar, A.; Onucyildiz, M.; Copty, N. K. 2009. Modelling level change in lakes using neuro-fuzzy and artificial neural networks, *Journal of Hydrology* 365(3–4): 329–334. https://doi.org/10.1016/j.jhydrol.2008.12.006

Zealand, C. M.; Burn, D. H.; Simonovic, S. P. 1999. Short term streamflow forecasting using artificial neural networks, *Journal of Hydrology* 214(1): 32–48. https://doi.org/10.1016/S0022-1694(98)00242-X

Adam PIASECKI. Mr. Adam Piasecki (MSc) is a PhD student at the Nicolaus Copernicus University in Toruń as well as at the AGH University in Cracow. Currently he is conducting research in area of hydrology and water management. He was a participant of several national and international workshops and conferences as well as author and co-author of many scientific publications.

Jakub JURASZ. Mr. Jakub Jurasz (PhD Eng.) is a research assistant at the AGH University in Cracow. His current research concentrates on the concept of water – energy nexus as well as integration of renewable energy sources to the national power system. Dr Jurasz is an author and co-author of several dozen papers which were published in recognized journals in field of energy.

Rajmund SKOWRON. Mr. Rajmund Skowron (PhD, DSc) is an expert in are of earth sciences. Lifelong research employee of the Nicolaus Copernicus University in Toruń. His research concentrates on limnology as well as the impact of human activities on the natural environment. In his publications he investigates the problem of water temperature as well as glaciation. He was author and co-author of at least 75 scientific contributions from which majority relates physical limnology. Currently he is working on lakes disappearance in Poland as well as long long-term changes in lakes thermal regime.

9

TESTING OF CHLORELLA/SCENEDESMUS MICROALGAE CONSORTIA FOR REMEDIATION OF WASTEWATER, CO_2 MITIGATION AND ALGAE BIOMASS FEASIBILITY FOR LIPID PRODUCTION

Judita Koreivienė[a], Robertas Valčiukas[b], Jūratė Karosienė[a], Pranas Baltrėnas[b]

[a] *Nature Research Centre, Institute of Botany, Žaliųjų Ežerų g. 49, 08604 Vilnius, Lithuania*
[b] *Department of Environmental Protection, Faculty of Environmental Engineering, Vilnius Gediminas Technical University, Saulėtekio al. 11, 10223 Vilnius, Lithuania*

Abstract. Industry, transport and unsustainable agriculture result in the increased quantity of wastewater, release of nutrients and emission of carbon dioxide that promotes eutrophication of water bodies and global climate change. The application of microalgae for phycoremediation, their biomass use for human needs may increase sustainability and have a positive effect on the regional development. The experiments were carried out in order to establish the feasibility of treating the local municipal wastewater with microalgae consortia and their biomass potential for biofuel production. The results revealed that *Chlorella/Scenedesmus* consortium eliminated up to 99.7–99.9% of inorganic phosphorus and up to 88.6–96.4% of inorganic nitrogen from the wastewater within three weeks. The ammonium removal was more efficient than that of nitrate. *Chlorella* algae grew better in diluted, while *Scenedesmus* – in the concentrated wastewater. The consortium treated wastewater more efficiently than a single species. The maximum biomass (3.04 g/L) of algal consortium was estimated in concentrated wastewater. Algae accumulated 0.65–1.37 g of CO_2/L per day in their biomass. Thus, *Chlorella/Scenedesmus* consortium is a promising tool for nutrients elimination from the local wastewater under the climatic conditions specific to Lithuania. However, none of the two species were able to accumulate lipids under the nitrogen starvation conditions.

Keywords: phycoremediation, wastewater, microalgae, biomass, oil accumulation.

Reference to this paper should be made as follows: Koreivienė, J.; Valčiukas, R.; Karosienė, J.; Baltrėnas, P. 2014. Testing of *Chlorella/Scenedesmus* microalgae consortia for remediation of wastewater, CO_2 mitigation and algae biomass feasibility for lipid production, *Journal of Environmental Engineering and Landscape Management* 22(02): 105–114. http://dx.doi.org/10.3846/16486897.2013.911182

Introduction

The demands for energy and freshwater resources are increasing with the growing human population in the world. Industry, transport, unsustainable agriculture increase the quantity of wastewater as well as the release of nutrients and emission of carbon dioxide that promotes eutrophication of the waters and global climate change (Chiu *et al.* 2009; Pittman *et al.* 2011; Olaizola *et al.* 2004; Rawat *et al.* 2011). Photosynthesizing algae are important organisms that can help to control undesirable processes in the ecosystems. Currently, the green technologies are one of the most growing sectors in the world. Algae biomass may be used for alternative application such as biofuel, natural fertilisers production, additives to animal nutrition, etc. (Munoz, Guieysse 2006; Mata *et al.* 2010; Misevičius, Baltrėnas 2011). Applications of algae for human needs are considered to have many benefits such as increasing sustainability, reduction of greenhouse gas emissions, regional development, social structure, agriculture and security of supply (Reijnders 2006).

The wastewater is a suitable resource for microalgae, because it contains large quantities of nitrogen and phosphorus that are the key elements for the algae growth (Li *et al.* 2011; Park *et al.* 2011; Pittman *et al.* 2011). Although large-scale production of algal biofuels using wastewater

Corresponding author: Judita Koreivienė
E-mail: judita.koreiviene@botanika.lt

treatment was first proposed by Oswald and Golueke in 1960 (Park *et al.* 2011 and references therein), the use of microalgae in the wastewater industry is still fairly limited (Pittman *et al.* 2011; Rawat *et al.* 2011). The application of microalgae for phycoremediation and sustainable biofuels production could become economically feasible (Rawat *et al.* 2011). The algal biomass from wastewater treatment systems could be converted to biofuels: by anaerobic digestion to biogas, transesterification of lipids to biodiesel, fermentation of carbohydrate to bioethanol and high temperature conversion to bio-crude oil (Pittman *et al.* 2011; Craggs *et al.* 2011). Under controlled growth conditions, microalgae can accumulate up to 80% of oil in dry cell mass (Pittman *et al.* 2011). Particularly, *Chlorella* species may accumulate 14–63% and *Scenedesmus* 6–55% of oil under various conditions of cultivation (Li *et al.* 2008; Gouveia, Oliveira 2009; Mutanda *et al.* 2011 and references therein). However, lipid accumulation in wastewater grown microalgae is much lower and ranges from low (<10% DW) to moderate (25–30% DW) lipid content (Halim *et al.* 2012).

Various algae species have different ecological requirements for their development. The highly variable composition of wastewater may limit the growth of some algae, whereas the others will flourish (Bhatnagar *et al.* 2011). So, it is essential to select algal species or their consortia capable of growing in particular wastewater under changing climatic conditions that will define the success of the treatment process. Unicellular fast growing green algae are very tolerant to many wastewater conditions and efficient at accumulating nutrients from it (Ruiz-Marin *et al.* 2010). *Chlorella* (*C. ellipsoidea, C. minutissima, C. vulgaris*) and *Scenedesmus* (*S. bijuga, S. quadricauda, S. obliquus*) are the most studied algae for their application in the wastewater treatment (Lau *et al.* 1995; Gouveia, Oliveira 2009; Ruiz-Marin *et al.* 2010; Johnson, Wen 2010; Wang *et al.* 2010; Bhatnagar *et al.* 2011; Praepilas, Kaewkannetra 2011; Yang *et al.* 2011a, b). Pate *et al.* (2011) emphasized that autotrophic microalgae productivity is strongly dependent on environmental conditions most suitable for growth, especially high solar resource and temperatures, which are characteristic of particular geographical region. Thus, the actual values of daily biomass productivity and neutral lipid content that can be achieved in practice will depend on a complex combination of algal strain, cultivation system and local growing conditions.

In Lithuania with the population of about 3 mln., approximately 170 mln. m^3 of domestic-industrial wastewater are produced yearly (http://gamta.lt/cms/index). The release of wastewater around Vilnius city is about one third of the total amount in the country. Study of Gudas, Povilaitis (2013) has revealed that wastewater factor is prominent for water quality in small rivers downstream larger towns. Studies of combined use of microalgae for the treatment and biomass production are very scarce in the country. Makarevičienė *et al.* (2011) tested separately *Chlorella* and *Scenedesmus* species preference to nitrogen resource utilisation and algae biomass as the feedstock for biofuel at temperatures higher than are natural to Lithuania.

The aim of our experiments was to investigate feasibility of *Chlorella/Scenedesmus* microalgae consortia to treat local municipal wastewater and reduce greenhouse gas. We also seeked to evaluate biomass yield of the tested algae and its suitability for biofuel production under the conditions close to climate in Lithuania during summer.

1. Materials and methods

1.1. Algae cultures and growing media for the experiments

The strains of freshwater green algae *Chlorella* sp. and *Scenedesmus* sp. were isolated from Lithuanian lakes. Both cultures were grown in the BG-11 media at 20 °C and 1600 Lux illumination prior to the experiment.

Experimental studies with the mixture of *Chlorella* and *Scenedesmus* cultures were carried out in mechanically cleaned wastewater collected from the water supply and wastewater treatment company "Vilniaus vandenys". Fine soil particles, organic and mineral additives were removed from the solution in laboratory passing the water through the plankton net (pore diameter 20 µm) and additionally filtering through several layers of filter paper. Wastewater was sterilized by autoclave at 120–130 °C 1 atm. pressure for 30 min. for the elimination of bacteria. Bacteria may influence the experiment results of nutrients removal by algae due to decomposition of the organic matter in the solution. Conductivity, pH and chemical analysis of the prepared wastewater media (PO_4, NO_3, NO_2, NH_4 and BOD_5) were performed prior the beginning of the experiment.

The experiment was carried out in 250 ml sterile Erlenmeyer flasks covered with cotton-gauze plugs coated with the plastic. Four types of media with different concentrations of wastewater were used to test the optimal growth conditions for selected green algae species. Wastewater was diluted with distilled water to reach appropriate concentration: M100 – 100% of wastewater; M75 – 75% of wastewater; M50 – 50% of wastewater and M25 – 25% of wastewater in the final solution. BG-11 medium was used as a control. All experiments were carried out in triplicate.

1.2. Experiment design

The generalized experiment scheme is presented in Fig 1. Each variant of 230 ml of prepared growth media was inoculated with 5 ml mixture of *Chlorella/Scenedesmus* species taken from the stock algal cultures. The concentration of algae at the beginning of the experiment was 39×10^3 cells/ml (*Chlorella* sp. 31×10^3 cells/ml and

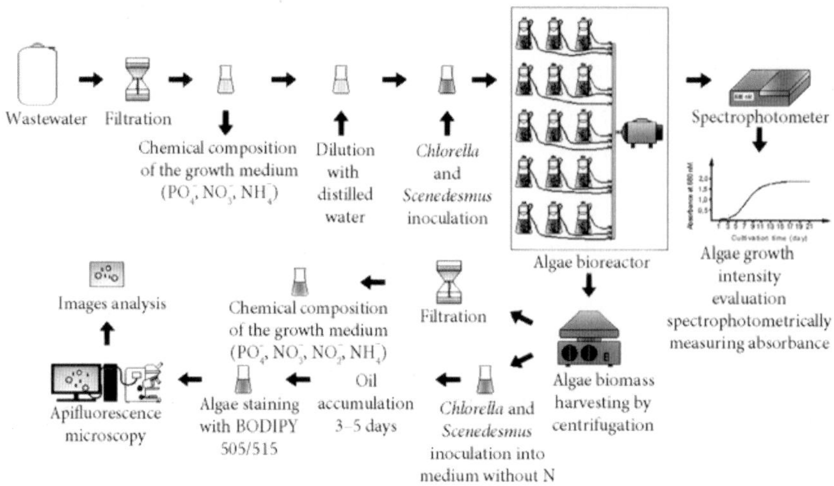

Fig. 1. The general scheme of the experiment

Scenedesmus sp. 8×10^3 cells/ml). Algae were counted in Fuchs-Rosenthal chamber using light microscope.

Climatic conditions close to the natural in Lithuania in June-August were selected to maintain algae in specialized cultivation chamber Percival *Intellus.*

The flasks were incubated at 16:8 day-night cycle with the average 2080 Lux illumination during light period. Day temperature was +20±1 °C and night temperature – +15±1 °C. The mixing of the algae cultures by air pumping (aeration intensity 2 L/min) maintained an equal gas balance and kept algae in the suspension.

Algae growth curves were developed by measuring absorption of samples at 680 nm using spectrophotometer Libra S32PC every second day on the course of 21 days experiment. Algae biomass was evaluated from cell abundance and cell volume calculated based on Olrik *et al.* (1998) recommended formula for ellipsoid with circular cross section algae cells:

$$V = \frac{\pi}{6} \cdot d^2 \cdot l, \quad (1)$$

where: V – the algal cell volume, μm³; d – cell diameter, μm; l – cell length, μm.

Cell volume of 10^6 μm³ was considered as equal to 1 mg of biomass.

1.3. Nutrients and CO_2 removal

For the evaluation of nutrients removal by algae, the chemical analysis of PO_4, NO_3, NO_2, NH_4 concentrations were determined in accordance with standard methods (LST EN ISO 10304; LST EN ISO 14911) in each variant before and after the experiment. After the experiment, the solution was centrifuged 5000 rpm for 1 minute for the separation of algal biomass and media. The remaining water was filtered through 0.9 μm pore diameter filter. Nutrients residues were evaluated. CO_2 consumed and incorporated into algae biomass via photosynthesis was recalculated using formula suggested by Buehner *et al.* (2009):

$$V_{CO_2} = 1.83 \cdot V, \quad (2)$$

where: V_{CO_2} – CO_2 consumed by algae, μm³; V – algae cell volume, μm³.

1.4. Oil accumulation study

Algae start to accumulate oil and other storage resources at critical for their development conditions. The analysis of oil accumulation in the algal cells was performed twice: a) at the end of the experiment when the culture reached stationary phase of growth; b) after growth in nitrogen deprived conditions. In order to create shock conditions and initiation of oil accumulation in the studied green algae cells, 5 ml of algae suspension of each variant was centrifuged (2000 rpm for 5 min.) to separate algae from the growth media. Algae biomass was resuspended into BG-11 medium without nitrogen compounds for three days under the same conditions as during the experiment.

Epifluorescence microscopy was applied to evaluate lipid accumulation in the algae. To stain lipid in the algae cells, Bodipi 505/515 (4,4-difluro-1,3,5,7-tetramethyl-4-bora-3a, 4adiaza-S-indacene; Invitrogen, USA) dyes were added to algae suspension according to Govender *et al.* (2012). Stock solution was prepared by dissolving 1 mg Bodipi 505/515 dyes into 10 ml of DMSO (0.2%). Five μl of prepared dyes was added to 6.25 ml of the sample with *Chlorella*/*Scenedesmus* culture (dye concentration 0.08 μg/ml) and incubated for 5 min. in the dark at 20 °C. Lipids stained with Bodipi 505/515 dyes fluoresce green in algae cells and were investigated with Nikon Eclipse N*i* epifluorescence microscope using a 515 nm emission wavelength filter. Images of algae were taken with Nikon DS-Ri1 camera and analyzed in the NIS-Elements BR software.

1.5. Statistical analysis

All values are expressed as mean ± standard deviation (SD). The data were checked for normality and analyzed by one-way ANOVA followed by Tukey's post-hoc test. The significant differences were considered at $p < 0.05$.

2. Results and discussion

2.1. Characteristics of the wastewater

The physico-chemical parameters of untreated municipal wastewater collected from Vilnius wastewater treatment plant are given in Table 1. The nutrient concentrations of the wastewater fell into the range from medium to strong level according to Rawat *et al.* (2011). Generally, untreated wastewater is characterized by very high concentrations of nutrients and toxic metals (Pittman *et al.* 2011). Total N (TN) and total P (TP) concentrations can be found at values of 10–100 mg/L in municipal wastewater and >1000 mg/L in agricultural effluent (Noue *et al.* 1992). The conductivity values that reach 2.68 mS/cm indicate high amount of various salts. Biochemical Oxygen Demand (BOD_5) shows high amount of easy degradable organic matter in the wastewater. In agreement, light microscopy revealed high abundance of bacteria and scarce number of heterotrophic flagellates, colourless protists.

Table 1. Physico-chemical parameters of municipal wastewater collected from Vilnius wastewater treatment plant

Contaminants	Concentration	Untreated wastewater level*
Total nitrogen, mgN/L	56.5	medium-strong
Total phosphorus, mgP/L	8.3	medium
Inorganic phosphorus, mgP/L	6.1	medium-strong
BOD_5, mgO_2/L	148	week-medium
Oxygen, mg/L	1.41	
Temperature, °C	13.6	
pH	7.74	
Conductivity, mS/cm	2.68	
H_2S, mg/l	0	
Colour	greyish	

* according to Rawat *et al.* (2011).

2.2. Growth rate of microalgae

The mixed culture of *Chlorella* and *Scenedesmus* showed similar pattern of their biomass development under various concentrations of wastewater and considerably differed from the control with BG-11 medium (Fig. 2A). Duration of the lag period was 5 days followed by the exponential growth during the next 8–12 days. Intense photosynthesis processes during the exponential phase also

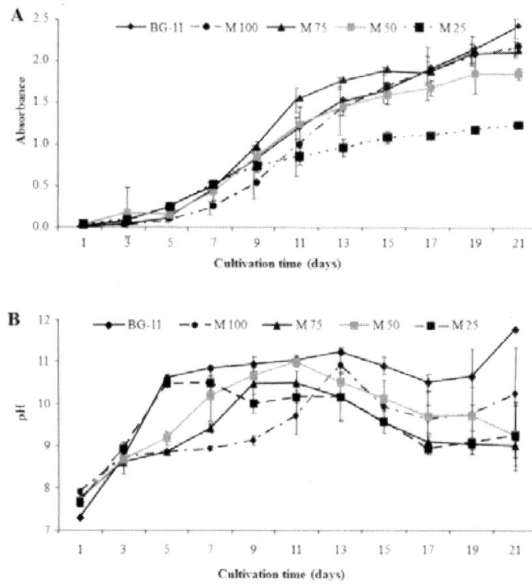

Fig. 2. The growth pattern of *Chlorella/Scenedesmus* algae culture under different wastewater concentrations (A) and the variation of the pH values (B) during the course of experiment

indicated the highest pH values in the 5–13 days (Fig. 2B). Culture slowly entered stationary phase in the 15–19 days of the experiment.

Table 2. Biomass production of *Chlorella* and *Scenedesmus* species based on references

Algae species	Biomass yield, g/L	Reference
Chlorella		
C. pyrenoidosa	2.83 in 2 days	Li *et al.* 2011
C. minutissima	0.073–0.38 in 10 days	Bhatnagar *et al.* 2010
C. vulgaris	0.21	Chinnasamy *et al.* 2009
C. vulgaris	av. 1.5, max 3.0	Gouveia, Oliveira 2009
C. ellipsoidea	0.43	Yang *et al.* 2011a
Chlorella sp.	1.9 in 17 days	Rasoul-Amini *et al.* 2011
Chlorella sp.	1.47–1.71	Wang *et al.* 2010
Scenedesmus		
S. obliquus	1.57±0.67	Abou-Shanab *et al.* 2011
S. obliquus	av. 0.9, max 2.0	Gouveia, Oliveira 2009
S. acutus	0.82	Chaichalerm *et al.* 2012

The biomass yield of mixed algae cultivated in BG-11 medium was the highest 3.7 g/L (Fig. 3). There were little differences of the algae biomass yield between concentrated (mean 2.75–3.04 g/L for M75 and M100) and diluted wastewater (mean 2.1–1.8 g/L for M25 and M50). Moreover, the growth of tested algae in concentrated wastewater was of the similar magnitude as in the chemical media, where the strains were cultured prior the experiment, indicating the wastewater as a perfect source to grow species separately or as their mixture. It seems that our results are similar to biomasses of *Chlorella* and *Scenedesmus* species recorded by the other researchers (Table 2). Nevertheless, it is difficult to compare the results from different studies due to variations of media, culturing time and regime. *Chlorella minutussima* cultured photoheterotrophically built biomass up to 0.38 g/L after 10 days growth (Pittman *et al.* 2011). Strain was defined as a good candidate for high biomass productivity in a municipal wastewater. Biomass yield of our studied *Chlorella/Scenedesmus* culture was 4.5 times higher that recorded by Pittman *et al.* (2011) even in the lowest biomass variant (mean 1.8 g/L) showing good perspectives of the consortia.

2.3. *Chlorella, Scenedesmus* algae density and biomass variations

Control of algae species is still not achieved in wastewater treatment ponds and algal dominance, various species interactions are still poorly understood (Park *et al.* 2011). Our experiment revealed species interactions in the consortia depending on the concentration of nutrients. *Scenedesmus* showed reverse pattern of growth compared to *Chlorella* in the gradient of wastewater concentration ($p < 0.05$) (Fig. 3). Concentrated wastewater (M100) stimulated the growth of *Scenedesmus* algae giving up to 5.5 times higher biomass augmentation compared to the most diluted (M25) variant ($p < 0.05$). *Scenedesmus* cells density increased hundreds to thousands of times by the end of the experiment: up to $4.53 \pm 1.01 \times 10^6$ – $29.23 \pm 3.68 \times 10^6$ cells/mL in M25 and M100 variants, accordingly.

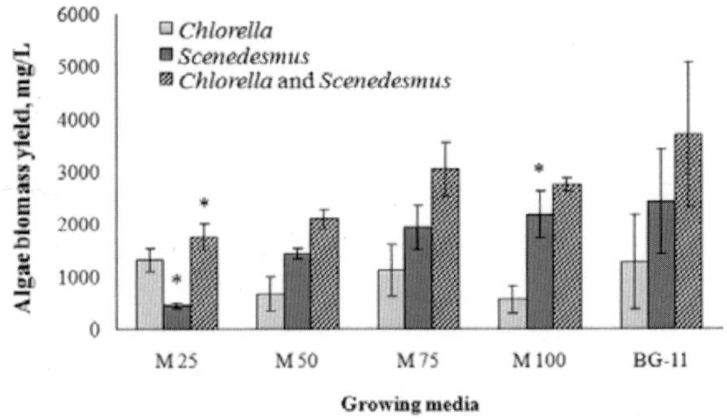

Fig. 3. Biomass yield of *Chlorella* and *Scenedesmus* species in wastewater at different concentrations

* indicate significant differences from the control (BG-11 medium) at $p < 0.05$.

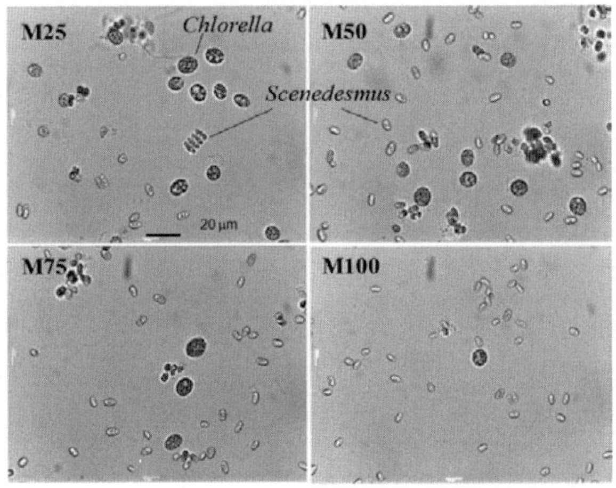

Fig. 4. The mixture of *Chlorella* and *Scenedesmus* species in the stationary growth phase at different concentrations of wastewater

The small algae cells have larger ratio of surface area and volume what enabled a quicker uptake of nutrients (Hein *et al.* 1995). The average volume of *Scenedesmus* cells was nine times lower than that of *Chlorella*. It may determine the faster growth of *Scenedesmus* at higher concentrations of wastewater. Similarly, Ruiz-Marin *et al.* (2010) found that *Scenedesmus obliquus* grew better in municipal wastewater compared to *Chlorella vulgaris*.

Generally, *Chlorella* growth was suppressed by concentrate wastewater. Species growth in most diluted wastewater variants showed the highest values (Figs 3, 4A, B). Species density in M25 and M50 variants reached $2.72 \pm 3.68 \times 10^6$ cells/ml and was three times lower in the most concentrated wastewater ($0.88 \pm 0.28 \times 10^6$ cells/ml). Our results coincide with Ras *et al.* (2011) data for *C. vulgaris* which density varied from 3.0×10^6 to 5.3×10^6 cells/ml. The maximum values were stated as high density cultures.

The increase of *Chlorella* density was lower up to 28–88 times only compared to *Scenedesmus*. Possibly the weakest competition for the resources with *Scenedesmus* in the most diluted variants determined the success of *Chlorella*. Contrary to our results, Johnson, Wen (2010) showed that wastewater was feasible for biomass production of *Chlorella* sp. with pure wastewater producing the highest biomass yield. However, the species of even the same genus have different requirements and growth characteristics (Cha *et al.* 2011).

Algae size is important for the biomass harvesting and desirable products extraction from the cells within the processing of the biomass (Rawat *et al.* 2011). *Chlorella* size varied less if compared to *Scenedesmus* and did not show significant differences between wastewater concentrations as well as before and after nitrogen shock application ($p > 0.05$) (Fig. 5). *Chlorella* mean cell volume increased in all variants from the beginning towards the end of the experiment and at nitrogen deprivation conditions.

Chlorella cells volume was over 30% higher in M50 wastewater variant and coincided with the lowest species culture biomass. It might be because species used a lot of energy to build up large cells slowing down the reproduction processes, what consequently lead to less culture biomass in the variant.

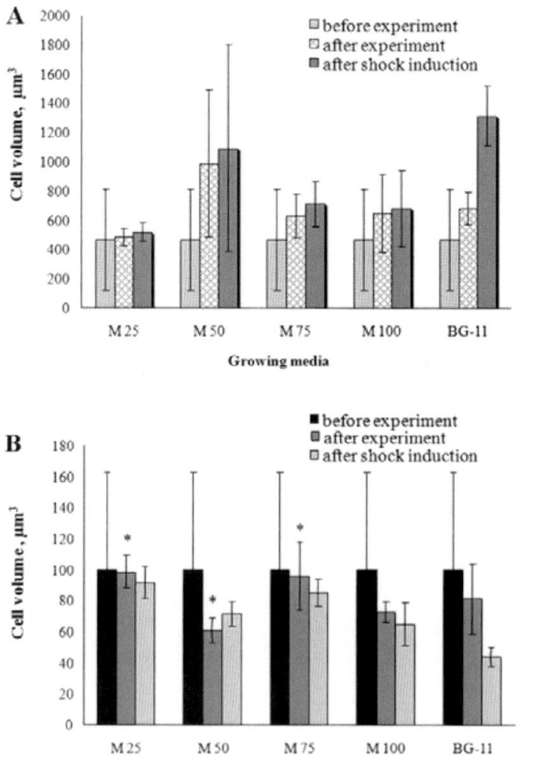

Fig. 5. Variations of *Chlorella* (A) and *Scenedesmus* (B) species cell size in wastewater at different concentrations

* indicate significant differences from the control (BG-11 medium) at $p < 0.05$.

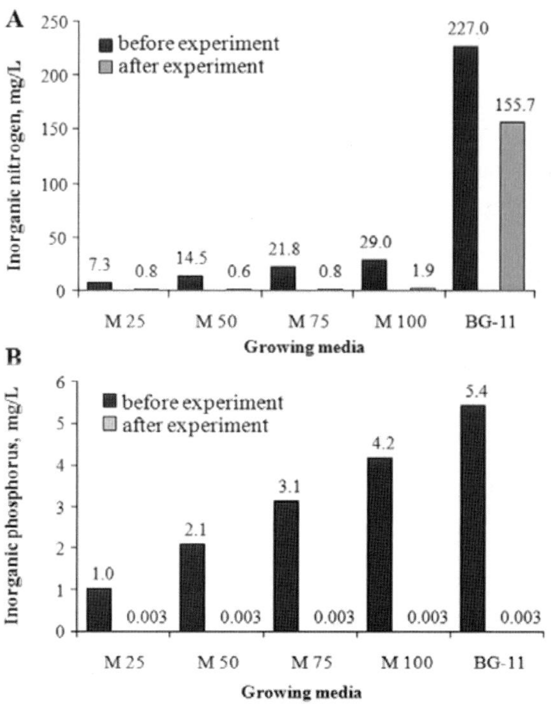

Fig. 6. The inorganic nitrogen (A) and inorganic phosphorus (B) concentrations in the media before and after the experiment

Contrary to *Chlorella*, the mean cell size of *Scenedesmus* diminished during the course of experiment and at nitrogen deprivation conditions without a clear tendency between the variants ($p > 0.05$) (Fig. 5). *Scenedesmus* built large cells and formed typical to species four-celled coenobia in diluted M25 wastewater. It is in agreement with O'Donnell *et al.* (2013) data that increase of nutrients concentration promoted more intense formation of one-celled *Scenedesmus* instead of four-cell coenobia.

2.4. Nutrients and CO_2 removal

Various species of *Chlorella* and *Scenedesmus* can remove from 80% to almost 100% of ammonia, nitrate and phosphate from treated wastewater (Pittman *et al.* 2011), however, the treatment efficiency varies depending on wastewater type and to some extent on climate peculiarities. Phosphorus is particularly difficult to remove from wastewater. Inorganic phosphorus (IP) concentration in the wastewater media variants of the experiment varied from 1.04 to 4.17 mg/L and was slightly higher in the control (BG-11 medium) (Fig. 6).

The experiment revealed that more than 99% of IP was eliminated by growing *Chlorella*/*Scenedesmus* culture within three weeks period (Fig. 7A). The elimination of inorganic nitrogen (IN) compounds from the wastewater was 88.6–96.4% indicating that the main limiting element for algae growth was IP. According to Makarevičienė *et al.* (2011), the removal efficiency of nitrogen and phosphorus of *Chlorella* sp. and *Scenedesmus* sp. reached relatively high values: TN – 91% (for both algae), TP – 94.7% and 95.6%, respectively. Similarly, Woertz *et al.* (2009) laboratory experiments showed nutrient removals of >98% for ammonium and >96% for phosphate with mixed culture microalgae grown on CO_2 supplemented primary wastewater effluent. *Chlorella ellipsoidea* eliminated 99% of nitrogen and 90% of phosphorus from the secondary effluent (Yang *et al.* 2011a). Johnson, Wen (2010) recorded slightly lower values for *Chlorella* sp. removing up to 90% of the TP and 79% of the TN contained within the wastewater.

Usually NH_4-N and PO_4-P predominate as the nutrients in wastewater, thus algae having high productivity and able to utilize these resources are ideal species for the treatment and biomass production (Park *et al.* 2011). IN concentration in the wastewater variants of the experiment media varied from 7.3 to 29.0 mg/L with dominating ammonium (Fig. 6). The concentration of IN in the control media was about ten times higher with domination of nitrate. It should be noted that ammonium from wastewater was eliminated much easier by *Chlorella*/*Scenedesmus* culture compared to nitrates showing that cultures are suitable for application in wastewater treatment. Makarevičienė *et al.* (2011) recorded the best *Chlorella* and *Scenedesmus* biomass productivity using urea as a nitrogen source or modified growing medium BG-11 with decreased concentration of $NaNO_3$. Chaichalerm *et al.* (2012) showed the lowest biomass yield of six chlorophyte species in the BG-11 medium compared to other three tested media (N-8, Kuhl, 3NBBM). They concluded that BG-11 medium has unbalanced concentrations of IN and IP, e.g. nitrogen concentration is high and phosphorous concentration is too low. Conversely, Phukan *et al.* (2011) obtained the highest biomass yield of *Chlorella* sp. (824 mg/L) in BG-11 medium compared to the other tested media (Basal, BBM, Chu-13). Li *et al.* (2008) experimentally showed that nitrate was the most favourable nitrogen source for *Neochloris oleoabundans* among the three tested nitrogen compounds, i.e. sodium nitrate, urea and ammonium bicarbonate. Even the same genus species have different preference for the nutrients. Cha *et al.* (2011) observed different growth pattern of *Chlorella vulgaris* (strain UMT-M1) and *C. sorokiniana* (strain KS-MB2) cultures at various nitrate concentrations.

The results of our experiment showed that the mixture of the culture is suitable for the effective elimination of nutrients from the wastewater. It is in agreement with Bhatnagar *et al.* (2011) that algal consortia grown in wastewater offer high production of renewable biomass for various applications, because in consortia the substantial loss of population of one algae may be compensated by the growth of the other. Chinnasamy *et al.* (2010) demonstrated that a consortium of 15 native algal isolates removed >96% of nutrients in treated wastewater.

Microalgae contain approximately 50% of carbon dry weight, thus about 1.8 kg of CO_2 is required to generate 1 kg of algal biomass (Yang *et al.* 2011b). *Chlorella/Scenedesmus* culture was tested for CO_2 elimination by evaluating biomass increase after three weeks of incubation. The mixture of culture growing on the wastewater incorporated from atmosphere into their biomass on average up to 1.37 g of CO_2/L per day without any significant differences between wastewater concentrations ($p > 0.05$). The values were up to two times lower if compare with the control ($p < 0.05$) (Fig. 7B). Usually room air contains 0.04% of CO_2 (Cheng *et al.* 2006). The experiments of the other researchers indicate that elevated concentrations of CO_2 may enhance the algae biomass yields by several times (Olaizola *et al.* 2004; Ota *et al.* 2009; Yoo *et al.* 2010; Jiang *et al.* 2011). Chinnasamy *et al.* (2009) found that *C. vulgaris* culture fixed 18.3 and 38.4 mg CO_2/L per day at ambient and elevated CO_2 (6%) levels.

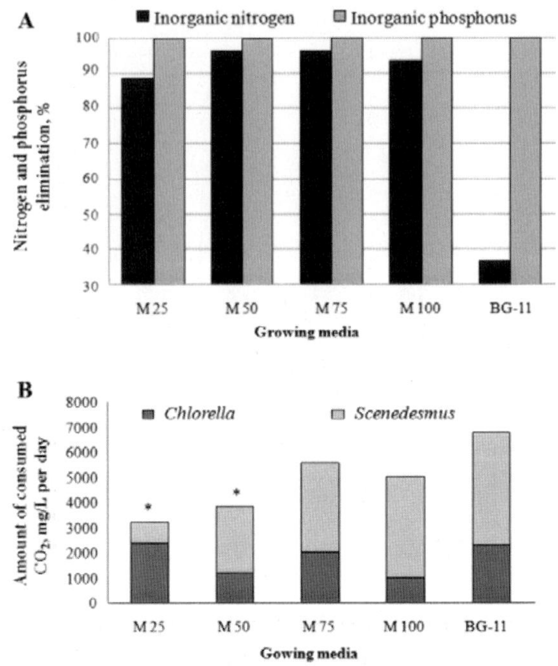

Fig. 7. Elimination of inorganic nitrogen and inorganic phosphorus (A) and the average amount of CO_2 incorporated into algae biomass via photosynthesis (B) at the different concentrations of wastewater

* indicate significant differences from the control (BG-11 medium) at $p < 0.05$.

2.5. Oil accumulation in the algae cells

Stockenreiter *et al.* (2013) found a clear correlation between light use and lipid production in functional diverse algae communities. It is considered as a powerful and cost-effective way to improve biofuel production. Nitrogen starvation was shown as one of the important factors to start oil accumulation by algae (Rodolfi *et al.* 2009; Brennan, Owende 2010; Deng *et al.* 2011; Pruvost *et al.* 2011). However, Stephenson *et al.* (2010) found that the maximal triacylglyceride productivity in *Chlorella vulgaris* was achieved by allowing the cells to deplete the nitrogen naturally instead of transferring cells to a medium without nitrogen. During our experiment period *Chlorella/Scenedesmus* culture reached the stationary phase in all variants of media, however, the nitrogen deprivation conditions were not achieved. The oil accumulation in *Chlorella* and *Scenedesmus* cells was additionally studied after inoculation of the algae biomass into nitrogen starvation conditions.

Under the nitrogen limitation, the growth of algae culture mixture from high concentration wastewater variants was suppressed more compared to diluted variants (Fig. 8). The experiment showed that *Chlorella* species is less susceptible to the N limitation than *Scenedesmus*. In the control variant *Chlorella* biomass increased, whereas *Scenedesmus* diminished. Nitrogen starvation might be not the best method to increase oil accumulation in the cells of particular algae, but application of high light or salt shock may give a positive result. It is also difficult to apply nitrogen starvation conditions at large operational scale to algae biomass grown in wastewater treatment plant.

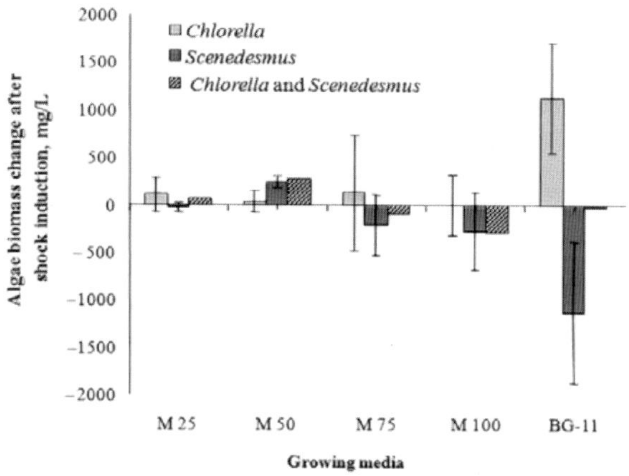

Fig. 8. Biomass yield of *Chlorella* and *Scenedesmus* species in wastewater at different concentrations after three days under nitrogen starvation conditions

Conclusions

1. *Chlorella/Scenedesmus* algae culture effectively eliminated inorganic phosphorus from the wastewater by removing up to 99.7–99.9% and inorganic nitrogen by 88.6 to 96.4% of the initial concentrations after three weeks of growth. Nitrogen effectively was eliminated in the form of ammonium, which predominated in the wastewater. Nitrate elimination was less effective.

2. The maximum biomass of algal culture consortium up to 3.04 g/L was in the concentrated wastewater. Its dilution almost twice reduced algal biomass.

3. The intensity of *Chlorella* algae growth was higher in the more diluted, while *Scenedesmus* algae – in the concentrated wastewater solutions. The consortia of *Chlorella/Scenedesmus* algae treat the wastewater more effectively than the single species, since the chemical composition of wastewater is highly variable over the time.

4. *Chlorella/Scenedesmus* algae culture accumulated on average 0.65–1.37 g of CO_2/L per day in their biomass.

5. The tested *Chlorella* and *Scenedesmus* algae have no ability to accumulate oil in the cells under nitrogen starvation conditions.

6. *Chlorella/Scenedesmus* algae consortium is a promising tool for elimination of nitrogen and phosphorus compounds from local wastewater resources as well as for diminishing CO_2 in the atmosphere under climatic conditions specific to Lithuania.

Acknowledgements

Authors are indebted to V. Ptašekienė for correcting the English. The research was partly supported by the EU (European Regional Development Fund) through the Baltic Sea Region Programme project, Sustainable Uses of Baltic Marine Resources (SUBMARINER No. 055).

References

Abou-Shanab, R. A. I.; Hwang, J.-H.; Cho, Y.; Min, B.; Jeon, B.-H. 2011. Characterization of microalgal species isolated from freshwater bodies as a potential source for biodiesel production, *Applied Energy* 88(10): 3300–3306. http://dx.doi.org/10.1016/j.apenergy.2011.01.060

Bhatnagar, A.; Bhatnagar, M.; Chinnasamy, S.; Das, K. 2010. *Chlorella minutissima* – a promising fuel alga for cultivation in municipal wastewaters, *Applied Biochemstry and Biotechnology* 161(1–8): 523–536. http://dx.doi.org/10.1007/s12010-009-8771-0

Bhatnagar, A.; Chinnasamy, S.; Singh, M.; Das, K. C. 2011. Renewable biomass production by mixotrophic algae in the presence of various carbon sources and wastewaters, *Applied Energy* 88(10): 3425–3431. http://dx.doi.org/10.1016/j.apenergy.2010.12.064

Brennan, L.; Owende, P. 2010. Biofuels from microalgae – a review of technologies for production, processing, and extractions of biofuels and co-products, *Renewable and Sustainable Energy Reviews* 14(2): 557–577. http://dx.doi.org/10.1016/j.rser.2009.10.009

Buehner, M. R.; Young, P. M.; Willson, B.; Rausen, D.; Schoonover, R.; Babbitt, G.; Bunch, S. 2009. Microalgae growth modelling and control for a vertical flat panel photobioreactor, *American Control Conference Proceedings*, 10–12 June, 2009, St. Louis, 2301–2306.

Cha, T. S.; Chen, J. W.; Goh, E. G.; Aziz, A.; Loh, S. H. 2011. Differential regulation of fatty acid biosynthesis in two *Chlorella* species in response to nitrate treatments and the potential of binary blending microalgae oils for biodiesel application, *Bioresource Technology* 102(22): 10633–10640. http://dx.doi.org/10.1016/j.biortech.2011.09.042

Chaichalerm, S.; Pokethitiyook, P.; Yuan, W.; Meetam, M.; Sritong, K.; Pugkaew, W.; Kungvansaichol, K.; Kruatrachue, M.; Damrongphol, P. 2012. Culture of microalgal strains isolated from natural habitats in Thailand in various enriched media, *Applied Energy* 89(1): 296–302. http://dx.doi.org/10.1016/j.apenergy.2011.07.028

Cheng, L.; Zhang, L.; Chen, H.; Gao, C. 2006. Carbon dioxide removal from air by microalgae cultured in a membrane-photobioreactor, *Separation and Purification Technology* 50(3): 324–329. http://dx.doi.org/10.1016/j.seppur.2005.12.006

Chinnasamy, S.; Bhatnagar, A.; Hunt, R. W.; Das, K. C. 2010. Microalgae cultivation in a wastewater dominated by carpet mill effluents for biofuel applications, *Bioresource Technology* 101(9): 3097–3105. http://dx.doi.org/10.1016/j.biortech.2009.12.026

Chinnasamy, S.; Ramakrishnan, B.; Bhatnagar, A.; Das, K. C. 2009. Biomass production potential of a wastewater alga *Chlorella vulgaris* ARC 1 under elevated levels of CO_2 and temperature, *International Journal of Molecular Sciences* 10(2): 518–532. http://dx.doi.org/10.3390/ijms10020518

Chiu, S.-Y; Kao, C.-Y.; Tsai, M.-T.; Ong, S.-C.; Chen, C.-H.; Lin, C.-S. 2009. Lipid accumulation and CO_2 utilization of *Nannochloropsis oculata* in response to CO_2 aeration, *Bioresource Technology* 100(2): 833–838. http://dx.doi.org/10.1016/j.biortech.2008.06.061

Craggs, R. J.; Heubeck, S.; Lundquist, T. J.; Benemann, J. R. 2011. Algae biofuel from wastewater treatment high rate algal ponds, *Water Science and Technology* 63(4): 660–665. http://dx.doi.org/10.2166/wst.2011.100

Deng, X.; Fei, X.; Li, Y. 2011. The effects of nutritional restriction on neutral lipid accumulation in *Chlamydomonas* and *Chlorella*, *African Journal of Microbiology Research* 5(3): 260–270. http://dx.doi.org/10.5897/AJMR10.557

Gouveia, L.; Oliveira, A. C. 2009. Microalgae as a raw material for biofuels production, *Journal of Industrial Microbiology and Biotechnology* 36: 269–274. http://dx.doi.org/10.5897/AJMR10.557

Govender, T.; Ramanna, L.; Rawat, I.; Bux F. 2012. Bodipy staining, an alternative to the Nile Red fluorescence method for the evaluation of intracellular lipids in microalgae, *Bioresource Technology* 114: 507–511. http://dx.doi.org/10.1016/j.biortech.2012.03.024

Gudas, M.; Povilaitis, A. 2013. Factors affecting seasonal and spatial patterns of water quality in Lithuanian rivers, *Journal of Environmental Engineering and Landscape Management* 21(1): 26–35. http://dx.doi.org/10.3846/16486897.2012.696058

Halim, R.; Danquah, M. K.; Webley, P. A. 2012. Extraction of oil from microalgae for biodiesel production: a review, *Biotechnology Advances* 30(3): 709–732. http://dx.doi.org/10.1016/j.biotechadv.2012.01.001

Hein, M.; Pedersen, F. M.; Sand-Jensen, K. 1995. Size-dependent nitrogen uptake in micro- and macroalgae, *Marine Ecology Progress Series* 118: 247–253.

Jiang, L.; Luo, S.; Fan, X.; Yang, Z.; Guo, R. 2011. Biomass and lipid production of marine microalgae using municipal wastewater and high concentration of CO_2, *Applied Energy* 88(10): 3336–3341. http://dx.doi.org/10.1016/j.apenergy.2011.03.043

Johnson, M. B.; Wen, Z. Y. 2010. Development of an attached microalgal growth system for biofuel production, *Applied Microbiology and Biotechnology* 85(3): 525–534. http://dx.doi.org/10.1007/s00253-009-2133-2

Lau, P. S.; Tam, N. F. Y.; Wong, Y. S. 1995. Effect of algal density on nutrient removal from primary settled wastewater, *Environmental Pollution* 89(1): 59–66. http://dx.doi.org/10.1016/0269-7491(94)00044-E

Li, Y.; Horsman, M.; Wang, B.; Wu, N.; Lan, C. Q. 2008. Effects of nitrogen sources on cell growth and lipid accumulation of green alga *Neochloris oleoabundans*, *Applied Microbiology and Biotechnology* 81: 629–636. http://dx.doi.10.1007/s00253-008-1681-1

Li, Y.; Chen, Y.-F.; Chen, P.; Min, M.; Zhou, W.; Martinez, B.; Zhu, J.; Ruan, R. 2011. Characterization of a microalga *Chlorella* sp. well adapted to highly concentrated municipal wastewater for nutrient removal and biodiesel production, *Bioresource Technology* 102(8): 5138–5144. http://dx.doi.org/10.1016/j.biortech.2011.01.091

Makarevičienė, V.; Andrulevičiūtė, V.; Skorupskaitė, V.; Kasperovičienė, J. 2011. Cultivation of microalgae *Chlorella* sp. and *Scenedesmus* sp. as a potentional biofuel feedstock, *Environmental Research, Engineering and Management* 3(57): 21–27.

Mata, T. M.; Martins, A. A.; Caetano, N. S. 2010. Microalgae for biodiesel production and other applications: a review, *Renewable and Sustainable Energy Reviews* 14(1): 217–232. http://dx.doi.org/10.1016/j.rser.2009.07.020

Misevičius, A.; Baltrėnas, P. 2011. Experimental investigation of biogas production using biodegradable municipal waste, *Journal of Environmental Engineering and Landscape Management* 19(2): 167–177. http://dx.doi.org/10.3846/16486897.2011.576456

Munoz, R.; Guieysse, B. 2006. Algal-bacterial processes for the treatment of hazardous contaminants: a review, *Water Resources* 40(15): 2799–2815. http://dx.doi.org/10.1016/j.watres.2006.06.011

Mutanda, T.; Ramesh, D.; Karthikeyan, S.; Kumari, S.; Anandraj, A.; Bux, F. 2011. Bioprospecting for hyper-lipid producing microalgal strains for sustainable biofuel production, *Bioresource Technology* 102(1): 57–70. http://dx.doi.org/10.1016/j.biortech.2010.06.077

Noue de la, J.; Laliberte, G.; Proulx, D. 1992. Algae and waste water, *Journal of Applied Phycology* 4: 247–254. http://dx.doi.org/10.2166/wst.2011.100

O'Donnell, D. R.; Fey, S. B.; Cottingham, K. L. 2013. Nutrient availability influences kairomone-induced defenses in *Scenedesmus acutus* (Chlorophyceae), *Journal of Plankton Research* 35(1): 191–200. http://dx.doi.org/10.1093/plankt/fbs083

Olaizola, M.; Bridges, T.; Flores, S.; Griswold, L.; Morency, J.; Nakamura, T. 2004. Microalga removal of CO2 from flue gases: CO_2 capture from a coal combustor, *Biotechnology and Bioprocess Engineering* 8: 360–367.

Olrik, K.; Blomquist, P.; Brettum, P.; Cronberg, G.; Eloranta, P. 1998. *Methods for quantitative assessment of phytoplankton in freshwaters*, Part 1, Naturvardsverket Forlag, Stockholm.

Ota, M.; Kato, Y.; Watanabe, H.; Watanabe, M.; Sato, Y.; Smith, R. L.; Inomata, H. 2009. Fatty acid production from a highly CO_2 tolerant alga, *Chlorocuccum littorale*, in the presence of inorganic carbon and nitrate, *Bioresource Technology* 100(21): 5237–5242. http://dx.doi.org/10.1016/j.biortech.2009.05.048

Park, J. B. K.; Craggs, R. J.; Shilton, A. N. 2011. Wastewater treatment high rate algal ponds for biofuel production, *Bioresource Technology* 102(1): 35–42. http://dx.doi.org/10.1016/j.biortech.2010.06.158

Pate, R.; Klise, G.; Wu, B. 2011. Resource demand implications for US algae biofuels production scale-up, *Applied Energy* 88(10): 3377–3388. http://dx.doi.org/10.1016/j.apenergy.2011.04.023

Phukan, M. M.; Chutia, R. S.; Konwar, B. K.; Kataki, R. 2011. Microalgae *Chlorella* as a potential bio-energy feedstock, *Applied Energy* 88(10): 3307–3312. http://dx.doi.org/10.1016/j.apenergy.2010.11.026

Pittman, J. K.; Dean, A. P.; Osundeko, O. 2011. The potential of sustainable algal biofuel production using wastewater resources, *Bioresource Technology* 102(1): 17–25. http://dx.doi.org/10.1016/j.biortech.2010.06.035

Praepilas, D.; Kaewkannetra, P. 2011. Effects of wastewater strength and salt stress on microalgal biomass production and lipid accumulation, World Academy of Science, *Engineering and Technology* 60: 1163–1168.

Pruvost, J.; Van Vooren, G.; Le Gouic, B.; Couzinet-Mossion, A.; Legrand, J. 2011. Systematic investigation of biomass and lipid productivity by microalgae in photobioreactors for biodiesel application, *Bioresource Technology* 102(1): 150–158. http://dx.doi.org/10.1016/j.biortech.2010.06.153

Ras, M.; Lardon, L.; Bruno, S.; Bernet, N.; Steyer, J.-P. 2011. Experimental study on a coupled process of production and anaerobic digestion of *Chlorella vulgaris*, *Bioresource Technology* 102(1): 200–206. http://dx.doi.org/10.1016/j.biortech.2010.06.146

Rasoul-Amini, S.; Montazeri-Najafabady, N.; Ali Mobasher, M.; Hoseini-Alhashemi, S.; Ghasemi, Y. 2011. *Chlorella* sp.: a new strain with highly saturated fatty acids for biodiesel production in bubble-column photobioreactor, *Applied Energy* 88(10): 3354–3356. http://dx.doi.org/10.1016/j.apenergy.2010.12.040

Rawat, I.; Ranjith Kumar, R.; Mutanda, T.; Bux, F. 2011. Dual role of microalgae: phycoremediation of domestic wastewater and biomass production for sustainable biofuels production, *Applied Energy* 88(10): 3411–3424.

Reijnders, L. 2006. Conditions for the sustainability of biomass based fuel use, *Energy Policy* 34(7): 863–876. http://dx.doi.org/10.1016/j.enpol.2004.09.001

Rodolfi, L.; Zittelli, G. C.; Bassi, N.; Padovani, G.; Biondi, N.; Bonini, G.; Tredici, M. R. 2009. Microalgae for oil: strain selection, induction of lipid synthesis and outdoor mass cultivation in a low-cost photobioreactor, *Biotechnology and Bioengineering* 102(1): 100–112. http://dx.doi.org/10.1002/bit.22033

Ruiz-Marin, A.; Mendoza-Espinosa, L. G.; Stephenson, T. 2010. Growth and nutrient removal in free and immobilized green algae in batch and semi-continuous cultures treating real wastewater, *Bioresource Technology* 101(1): 58–64. http://dx.doi.org/10.1016/j.biortech.2009.02.076

Stephenson, A. L.; Dennis, J. S.; Howe, C. J.; Scott S. A.; Smith A. G. 2010. Influence of nitrogen-limitation regime on the production by *Chlorella vulgaris* of lipids for biodiesel feedstocks, *Biofuels* 1(1): 47–58. http://dx.doi.org/10.4155/bfs.09.1

Stockenreiter, M.; Haupt, F.; Graber, A.-K.; Seppälä, J.; Spilling, K.; Tamminen, T.; Stibor, H. 2013. Functional group richness: implications of biodiversity on light and lipid yield in microalgae, *Journal of Phycology* 49(5): 838–847. http://dx.doi.org/10.1111/jpy.12092

Wang, L.; Min, M.; Li, Y.; Chen, P.; Chen, Y.; Liu, Y.; Wang, Y.; Ruan, R. 2010. Cultivation of green algae *Chlorella* sp. in different wastewaters from municipal wastewater treatment plant, *Applied Biochemistry and Biotechnology* 162(4): 1174–1186. http://dx.doi.org/10.1007/s12010-009-8866-7

Woertz, I. C.; Fulton, L.; Lundquist, T. J. 2009. Nutrient removal & greenhouse gas abatement with CO_2-supplemented algal high rate ponds, in *WEFTEC Annual Conference Proceedings*, 12–14 October, 2009, Orlando Florida, 7924–7936.

Yang, J.; Li, X.; Hu, H.; Zhang, X.; Yu, Y.; Chen, Y. 2011a. Growth and lipid accumulation properties of a freshwater microalga, *Chlorella ellipsoidea* YJ1, in domestic secondary effluents, *Applied Energy* 88(10): 3295–3299. http://dx.doi.org/10.1016/j.apenergy.2010.11.029

Yang, J.; Xu, M.; Zhang, X. Z.; Hu, O.; Sommerfield, M.; Chen, Y. 2011b. Life-cycle analysis on biodiesel production from microalgae: water footprint and nutrients balance, *Bioresource Technology* 102(1): 159–165. http://dx.doi.org/10.1016/j.biortech.2010.07.017

Yoo, C.; Jun, S.-Y.; Lee, J.-Y.; Ahn, C.-Y.; Oh, H.-M. 2010. Selection of microalgae for lipid production under high levels carbon dioxide, *Bioresource Technology* 101(1): 71–74. http://dx.doi.org/10.1016/j.biortech.2009.03.030

Judita KOREIVIENĖ. Researcher at the Institute of Botany of the Nature Research Centre. Doctor of Biomedicine (Botany) sciences of Vilnius University and Institute of Botany, 2005. Author of 15 scientific papers. Scientific interests: cyanobacteria and algae flora, morphology, ecology and distribution; trophic interactions in hydroecosystems, harmful algae blooms and cyanotoxins; invasive species; cyanobacteria and algae isolation; algae application in biotechnology.

Robertas VALČIUKAS. Master of Environment Engineering, Vilnius Gediminas Technical University. Scientific interests: environmental protection, application of algae in the environment engineering.

Jūratė KAROSIENĖ. Researcher at the Institute of Botany of the Nature Research Centre. Doctor of Biomedicine (Botany) sciences of Vilnius University and Institute of Botany, 2008. Author of 13 scientific papers. Scientific interests: algae and cyanobacteria diversity, distribution, ecology, species morphology, invasive species, harmful algae blooms, phytobentos functional activity, statistical analysis.

Pranas BALTRĖNAS. Dr Habil. Prof. and Head of the Department of Environmental Protection, Vilnius Gediminas Technical University (VGTU). Doctor Habil. of Science (air pollution), Leningrad Civil Engineering Institute (Russia), 1989. Doctor of Science (Air Pollution), Ivanov Textile Institute (Russia), 1975. Employment: Professor (1990), Associate Professor (1985), Senior Lecturer (1975), Vilnius Civil Engineering Institute (VISI, now VGTU). Publications: author of 13 monographs, 24 study guides, over 320 research papers and 67 inventions. Honorary awards and membership: prize-winner of the Republic of Lithuania (1994), a corresponding Member of the Ukrainian Academy of Technological Cybernetics, a full Member of International Academy of Ecology and Life Protection. Probation in Germany and Finland. Research interests: air pollution, pollutant properties, pollution control equipment and methods.

10

EVALUATION WITH STREAM CHARACTERISTICS OF DOWNSTREAM FLOOD PROBLEMS AFTER DAM CONSTRUCTION

Tamer Bagatur, Nizamettin Hamidi

Civil Engineering Department, Engineering Faculty, Dicle University, 21280 Diyarbakir, Turkey

Abstract. The flood may cause serious loss of life and property. The evaluation of flood events is necessary for the planning and design of engineering projects. In a dam, flood problems should be considered for both upstream and downstream conditions. The characteristics of flood problems can be derived from observation of dam sites. This paper aims to evaluate the impact of flood hazards from spillway based on downstream conditions after dam construction. Also, the paper discusses downstream flood problems depending on stream characteristics in an example dam site. In results, three reasons are determined for downstream floods problems: control problems of spillway gates, the naturally developing vegetation through channel flow line and incompetence of stream cross-section for carrying of flood flows. Therefore, downstream conditions should be investigated before dam construction. This investigation can be considered in the planning phase. In addition, the cross-section and longitudinal profile of stream should be controlled via maintenance of stream after dam construction. In stream cross-section problems (such as vegetation, incompetence and irregularities), some modifications (e.g. straighten, deepen, and widen) can be applied with engineering works.

Keywords: environmental impact assessment, flood damages, dam downstream condition, stream cross-section, environmental sustainability, environmental politics.

Reference to this paper should be made as follows: Bagatur, T.; Hamidi, N. 2014. Evaluation with stream characteristics of downstream flood problems after dam construction, *Journal of Environmental Engineering and Landscape Management* 22(02): 96–104. http://dx.doi.org/10.3846/16486897.2013.852555

Introduction

Flooding is the consequence of the lack of stream capacity to carry huge volumes of runoff induced by heavy precipitation or snow melt. Worldwide, floods are responsible for the fatalities and the economic losses. Therefore, people have fought to decrease the negative impacts of floods (Julien 2002). For example, Guzzetti *et al.* (2005) evaluated flood and landslide risk in Italy.

Water resources projects are very large national projects requiring huge dimensions. Hence, the planning and design of such systems must be performed with great care. The management and control of water resources requires the conception, planning, and execution of designs to use of the water or avoid damages from too much water. More alternatives must be considered in the planning phase.

There are around 40 000 large dams (>15 m high) worldwide (Jackson *et al.* 2001). A dam is the most important structure in water resources projects. The dams are probably built for the purposes of irrigation, energy, flood control, and water supply. In almost all dams, provision must be made for the safe discharge of excess water by the construction of a spillway. A spillway is that component of a dam which serves to evacuate the flood wave from the reservoir without damaging the dam structure (Chow 1959). It may be controlled or uncontrolled (or gated) type. If flow cross-section of river and stream is insufficient, cross-section will not carry of dam discharges. Thus, downstream floods can be observed. These floods negatively impact environmental and agricultural purposes. All water resources projects (dam, irrigation channel, hydroelectric power plant, etc.) are planned for the future. But, downstream effects of dam projects are rarely considered in project appraisal. Therefore, the dam projects must be evaluated for both upstream and downstream flood conditions. Fitzhugh *et al.* (2011) investigated

Corresponding author: Tamer Bagatur
E-mail: tbagatur@dicle.edu.tr

the impact of dams on flood flows in the United States. Downstream conditions must be investigated before dam construction for downstream flood risk. Also, the cross-section and longitudinal profile of stream must be controlled via maintenance of stream after dam construction. Because, maintenance and operation are important processes after the dam construction works. Non-uniform cross-sectional profiles, meanders, riffles and pools, and vegetation increase the heterogeneity of flow depths and velocities. The water depths and velocities are significant on stream channel hydraulics (Yen 2002).

The objective of this study is to evaluate causes and effects of downstream flood problems after dam construction. In addition, this paper recommends some technical approaches for protecting from downstream flood impacts. These approaches include stream channel geometry and roughness characteristics after dam construction. As an example dam site, this paper also describes some technical solutions for downstream flood problems on agriculture areas of the Goksu dam in Southeast of Turkey. The Goksu flood event is different from that presented in textbooks.

1. Background

1.1. Dam planning

Planning is an important step in the development of the water resources project. Project evaluation is developed based on economic, financial, environmental, and social considerations. Therefore, the planning and design of water resources projects must be performed with great care. The poor forecasts or estimates may lead to poor planning. Thus, the consequences of poor planning can be arrived large problems. In dam planning and design, both upstream and downstream conditions must be considered and investigated for negative impacts on flood-threatened lands with respect to environmental and agricultural purposes. These investigations should be aimed at predicting the structural problems.

1.2. GIS technique for water resources

In nowadays, computer simulation, analysis and its different applications (for example, GIS) are a popular way for evaluating of water resources projects. Geographic Information System (GIS) technology has been used to develop solutions for water resource engineering problems. The subjects of GIS can be summarized such as assessing water quality, determining water availability, preventing flooding, understanding the natural environment, managing water resources on a local or regional scale and mapping of river channel plan and bank erosion probability. Overton (2005) used GIS in modeling floodplain inundation on a regulated river. Shahapure *et al.* (2010) studied on flood simulation using GIS. In recent years, GIS has become an excellent tool for spatial date storage, visualization, and analysis. Thus, the downstream conditions of dam projects can be evaluated with GIS technique before and after dam construction.

1.3. The human role in changing stream channels

The extrinsic and intrinsic human roles affecting channels have been thought of as role in changing river channels. Direct consequences of human activities affect river channels through engineering works including canalization, dam construction, diversion and culvert. Gregory (2006) evaluated the human role in changing river channels in detail. There are many negative side effects of the dam construction on the downstream river's regime. Particularly, changing of river cross-section is most evident changes of size, shape and composition. Moreover, the overall channel plan form, the network and the ecology can be changed.

1.4. Flood problem

Various types of open channel problem such as flood occur in engineering practice. A flood is the result of runoff from rainfall and melting snow in quantities too great to be confined in the low-water channels of streams. Precipitation and stream flow vary widely from year to year. Flood problems involve a unique set of physical conditions and constraints, which can be resolved by the careful coordination the various disciplines. The physical facts of most flooding situations are against the planner who attempts to provide absolute flood control. Therefore, dam projects should be evaluated for floods depending on downstream conditions. The downstream conditions include checking of cross-section and longitudinal profile of stream. Therefore, the cross-section and longitudinal profile have a significant effect on downstream floods.

1.5. Spillway and radial gates

Operation of a storage reservoir is governed by the inflow. During normal periods of river flow, the reservoir will be maintained at the normal pool level for the operation of water supply purposes. If extremely high flows are can be drawn to such an elevation that the minimum expected flood flow will be sufficient to restore the active storage to its maximum level. A spillway is that component of a dam which serves to evacuate the flood wave from the reservoir without damaging the dam structure (Chow 1959; Chanson 2004). Shortly, a spillway is the safety valve for a dam. It may be controlled or uncontrolled. A controlled spillway is provided with crest gates. Gates permit the operator to control the release of water downstream for various purposes. The relation between storage and outflow for a reservoir with spillway is dependent on the number of gate. Storage in reservoir may modify the shape of a flood wave more markedly than an equivalent length of natural channel. The construction of a dam usually leads to control variations in reservoir water level depending on climatic conditions. Thus, downstream flood control may be made with reservoir storage and gradually-varied spillway releases.

1.6. Stream cross-section and flow line

Producing a cross-section of a river channel is fundamental to all river studies. A cross-section refers to the section of a channel taken transverse to the direction of the flow. Cross-section and longitudinal profile analysis to determine the best of several alternatives is required in planning most projects. The channel may consist of a main channel carrying normal discharges. The floodplain carries flow in excess of the channel capacity.

1.7. Roughness of stream channel and vegetation

In natural rivers and streams channel topography, bank vegetation, and in-stream woody debris may have a great influence on hydraulics. Vegetation is an important roughness element in channels. Reduced flow velocity around vegetation or woody debris leads to deposition of the fine sediment that is essential for bank cohesion. The accumulation of debris and excavation materials in the stream beds decreases the water carriage capacity of the bed and causes the floods. Hydraulically, debris act as large roughness elements that provide a varied flow environment, reduce average velocity, and locally elevate the water-surface profile. This can significantly increase flood travel time. Azami *et al.* (2004) examined the effects of a dam reservoir flood control on riparian plant communities. Nikora *et al.* (2008) examined vegetation effects on hydraulic resistance in small streams. Also, they suggested simple quantitative relationships for predicting and quantifying these effects using the plant characteristics.

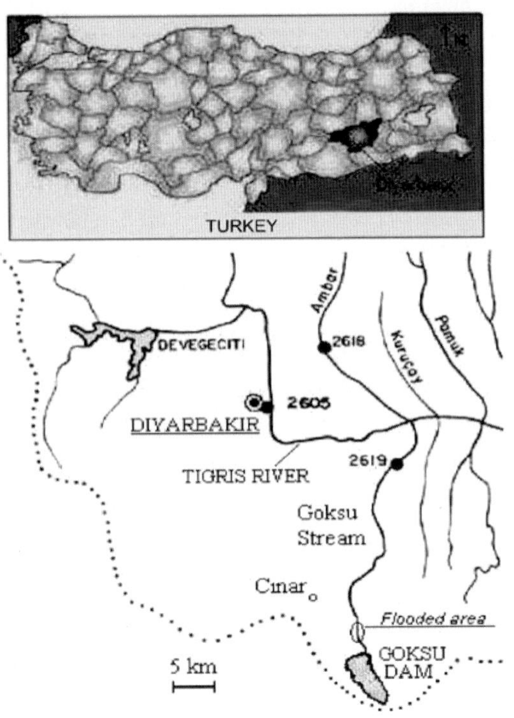

Fig. 1. Location map of study area

1.8. Hydraulic analysis: water-surface profile and standard step method

The hydraulic analysis of natural channels is similar to that of engineered channels. Hydraulic analysis of the flow in streams and rivers is performed using computer programs employing the varied flow equations. Computer programs, such as HEC-RAS (developed by the U. S. Army Crops of Engineers Hydrologic Engineering Center) and MIKE 11 (developed by DHI Water and Environment in Denmark), based on the preceding concepts are available for the computation of water-surface profiles. Water surface profiles can be examined standard step method. This method is easily adapted to prismatic channels with high precision. A reasonably accurate determination of the water-surface profile is possible by this method if the increments of depth are small. Billa *et al.* (2011) used as input into a MIKE 11 GIS and hydrological modeling process for flood inundation mapping of the Langat River Basin in Malaysia. Uniform flow rarely occurs in natural streams because of changes in depth, width, and slope along the channel. Therefore, the engineer is often concerned with non-uniform (varied) flow in open channels. One method of solving problems of non-uniform flow utilizes a step-by-step procedure.

1.9. Environmental, economic and social impacts

Analysis of the environmental and social impacts of the various project alternatives is important studies in dam construction. Large dams are often criticized because of their negative environmental and social impacts: changes in water and food security, increases in communicable diseases, and the social disruption caused by construction and involuntary resettlement. The impact of hydraulic engineering works and human activities on hydrology of stream basin has become a hot issue along with the economy and social development.

2. Materials and methods

2.1. Study area and dam site description

This study was conducted in the Goksu Stream Basin, which is located in the province of Diyarbakir (37° 41′ N, 40° 26′ E, in the southeastern Anatolia region of Turkey). This basin has a drainage area of 734 km^2. Goksu stream is 33 km long, flows directly into the Tigris River. A diagram of the study field is shown in Figure 1. Also, Table 1 shows the main characteristics of study area and dam site.

Table 1. The main characteristics of study area

Basin name:	Dicle Basin
Stream name:	Goksu Stream
Location:	Diyarbakir/Turkey
Drainage area	734 km^2
Annual average rainfall	487.6 mm
Dam name:	Goksu Dam
Construction year:	1987–1991
Dam volume:	1.632 million m^3
Dam height	46 m
Dam crest elevation	702 m
Irrigation Area:	35.82 km^2
Reservoir volume at normal water surface elevation:	62 million m^3
Reservoir area at normal water surface elevation:	4 km^2

2.2. Goksu dam and spillway structure

Figure 2 illustrates cross-section of Goksu dam and spillway structure. The dam was built for irrigation purposes as earth-fill type. The length of dam crest and crest width is 673.50 m and 10.00 m, respectively.

The spillway structure of dam is controlled with radial gates. The spillway has four radial gates (Fig. 3). The dimensions of radial gates are 7.50 m (horizontal) and 11.00 m (vertical). The length of spillway crest is 31.80 m. Spillway design discharge (Q_{10000}) is 2215 m^{3}s^{-1}.

Radial gate flow is a type of variable-area orifice flow, which may be either free or submerged. The general equation for spillway flow through an undershot gate can be derived from the Bernoulli equation and expressed as:

$$Q_{gate} = C_d \cdot w \cdot B \cdot \sqrt{2gH}, \qquad (1)$$

where: Q_{gate} = discharge (spillway gate flow); C_d = coefficient of discharge, typically 0.5–0.7; w = vertical opening of gate; B = gate width; g = gravitational constant; and H = a head term.

The relation between storage and outflow for a reservoir with spillway gates is dependent on the number of gate. For four radial gates of Goksu dam, spillway total discharge (Q_{sp}) can be calculated as:

$$Q_{sp} = Q_{gate1} + Q_{gate2} + Q_{gate3} + Q_{gate4}. \qquad (2)$$

2.3. Characteristics of Goksu streamflow records

Values of discharge are necessary for determination of streamflow characteristics. Also, they are part of the information routinely collected at streamflow gaging stations. Discharge data may be either instantaneous values or may be mean values. These valuables can be obtained by averaging flows through specified increments of time such as daily, monthly, annually, or for all or part of the period of record. Observation period of stream flow characteristics is from 1960 to 1987. In these years, Figure 4 shows variation of maximum discharges for 28 water years at the downstream gage. Maximum discharges are observed from 10 to 60 m^{3}s^{-1}. Mean annual maximum discharge is estimated 19 m^{3}s^{-1}.

Figure 5 illustrates the long-term mean monthly discharges for 20 years. The long-term mean annual discharge of stream was 2.86 m^{3}s^{-1} before dam construction.

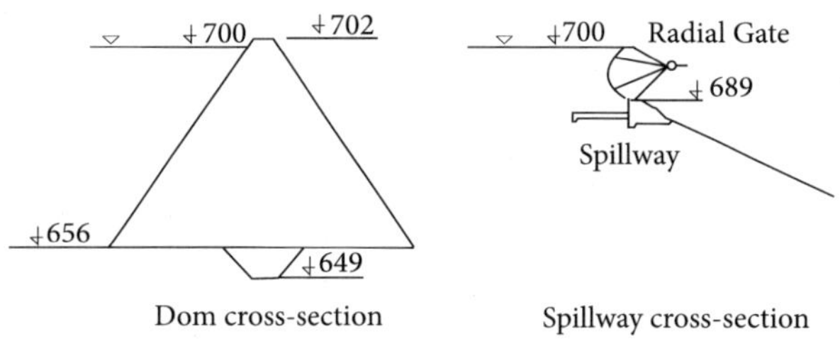

Fig. 2. Goksu dam and gated spillway structure

Fig. 3. Gated spillway structure of Goksu dam

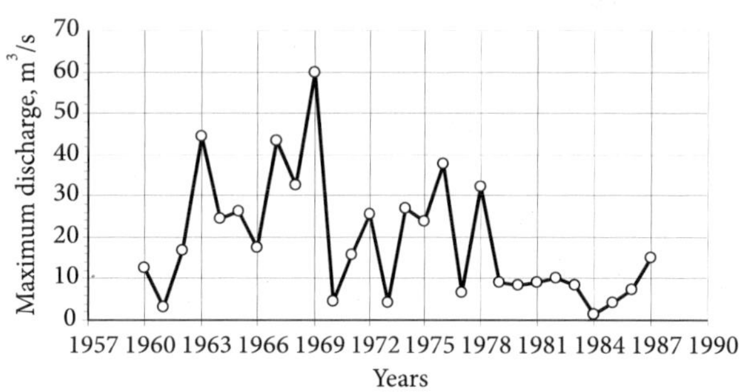

Fig. 4. Variation of maximum flow values of Goksu stream in different years (dam construction was started in 1987 and completed in 1995)

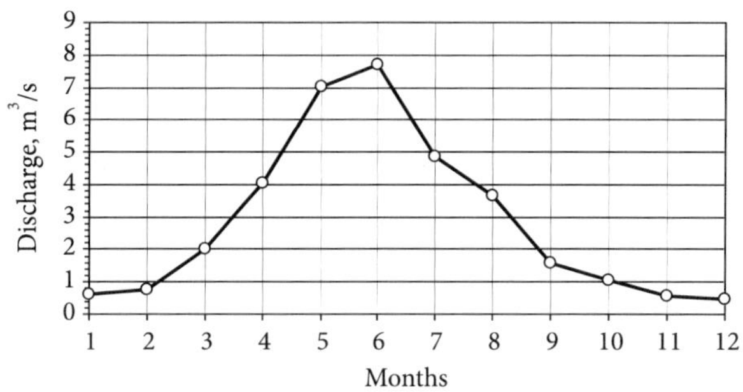

Fig. 5. Long-term mean monthly discharge of Goksu stream (observation period is from 1968 to1987)

2.4. "Zero" cross-section and changes in stream channel line

Rivers and stream carry material in the form of solid particles. These particles may alternatively be deposited on the river bed (sedimentation) and entrained into the moving water (erosion). Although design engineers are aware of the extreme hydrology of the catchments, it appears that they ignore the catchments erosion and sediment-load motion taking place during extreme floods. Thus, cross-section and flow line of the stream change with sedimentation and erosion in time. Moreover, channel cross-sections can not carry flood capacity of the stream in flood time. For downstream flood management, the cross-section and longitudinal profile of stream are the most important scientific knowledge. In Goksu basin, the stream cross-sections and flow line were surveyed after 2004 flood. Some cross-sections of stream are filled with bed material. The cross-section of the stream is nearly flat in flooded areas. Thus, "zero" cross-section is observed. The channel of Goksu stream has undergone significant hydrologic and morphologic changes in the relatively short period. For evaluating the effects of vegetation, riverbank erosion and stability on flood, the bed and bank material is important for the resulting depth and width. The channel of Goksu stream is entirely alluvial in composition with banks of sand and clay and a bed of fine to medium gravel with dense vegetation. The predominant geomorphic process is over bank flooding and deposition.

2.5. Roughness of stream channel

In natural rivers and streams, channel topography, bank vegetation, and in-stream woody debris may have a great influence on hydraulics and physical habitat. Particularly, the naturally developing vegetation has a higher hydraulic roughness. Therefore, depth-discharge relationships should be established for the varying roughness conditions.

The Manning formula is used to estimate flow in open channel situations (Chow 1959). The Manning formula states:

$$V = \frac{1}{n} \cdot R^{2/3} \cdot S^{1/2}, \tag{3}$$

where: V is the cross-sectional average velocity (m/s); n is the Manning coefficient; R is the hydraulic radius (m); S is the slope of the water surface or the linear hydraulic head loss.

The influence of roughness has been cataloged in the form of a roughness coefficient based on data obtained from field and laboratory observations. Typical values of the roughness coefficient, commonly called Manning coefficient (n), are presented in Table 2 (Chow 1959).

The values of Manning coefficient (n) include the effect of all hydraulic factors. For example, they include channel blockage by excessive aquatic vegetation.

In Goksu stream, the most striking visual change was from a sparse growth of streamside grasses to a dense growth of riparian woody vegetation. In such areas covered by dense vegetation, the vegetation characteristics play an important role on the magnitude of flow parameters such as velocities and depths.

Table 2. Variation with bed profile, vegetation and material type of Manning roughness coefficient in natural river channels

Bed profile	Vegetation	Manning coefficient (n)	
		Sand and gravel	Coarse gravel
Uniform	None	0.020	0.030
Uniform	Some	0.040	0.050
Undulating	None	0.030	0.040
Undulating	Some	0.050	0.060
Highly irregular	None	0.055	0.070
Highly irregular	Extensive	0.080	0.090

The behavior of grass in an open channel lining is complicated by the fact that grass stems bend as flow depth and shear stress increase. Vegetation will grow among the stones, after sediment has had an opportunity to be deposited between the stones. The aquatic vegetation growth resulted in a large increase in the roughness coefficient. According to Table 2, Manning coefficient (n) can be determined as 0.060–0.080 for Goksu stream due to dense vegetation and irregular bed profile.

3. Results and discussion

3.1. Downstream floods

In Goksu dam, the flood events have not observed from 1987 to 2003. In the time of flood of 2004, reservoir water was released from spillway gates due to enhancing of water elevation in reservoir. Thus, agricultural areas are flooded with released water. However, floods in some parts of the basin have lead to investigate of downstream conditions such as erosion, sedimentation, the stream line and cross-sections. The cross-sections were investigated the probability of the flood storage capacity between 0–33 km. As a result, from the spillway toe to 33th km of Goksu stream, it has been found to be more sensitive to a probable inundation. Inundation of agriculture areas supports the findings obtained by the study. Figure 6 shows stream channel banks covered by dense vegetation in flooded areas. The water depth increases due to stream channel banks covered by dense vegetation. Figure 7 shows stream channel configuration after dam construction and flood areas. Observations from agricultural areas close to the stream embankments mentioned flood depths of 1 to 2 m over land. Agriculture areas (62 hectares) suffered badly from flood conditions. Approximately 95% of the Goksu catchment is agricultural land, 65 % of which is cultivated for arable crops.

Fig. 6. Goksu channel banks covered by dense vegetation

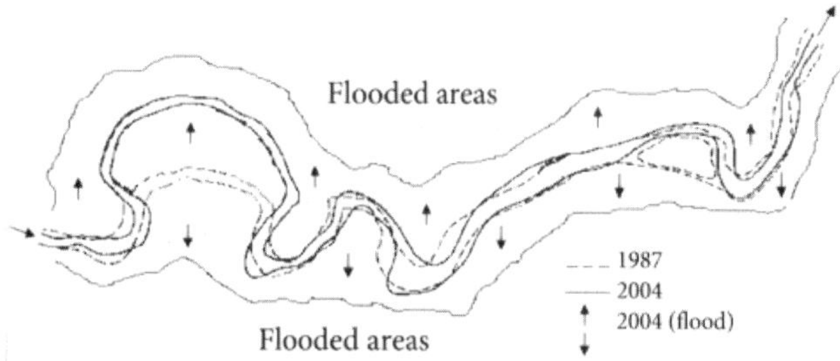

Fig. 7. Stream channel configuration after dam construction and flood areas (canal configuration in 1987 (dashed line) and 2004 (solid line) of Goksu stream)

3.2. Spillway releases and reservoir water level

In dam structures, the flood water is discharged through a radial gated spillway. Radial gates allow controlled discharges from reservoir. In this study, the field data were collected in February-March 2004. In these months, radial gates of Goksu dam were opened due to flood water. Figure 8 shows variation of total discharge (Q_{sp}) versus time of released water from spillway gates in February 2004 flood. The spillway released flood water from 5 to 30 m^3s^{-1} within 25 days in February 2004 flood.

The 2004 flood event extended from February 5 to March 23, a period of 48 days. Figure 9 shows release discharges from spillway gates in March 2004 flood. The spillway gates were opened for release water flood from 22 to 8 m^3s^{-1} discharge within 23 days in March 2004 flood. Water elevation in reservoir is shown depending on days in Figure 10. On 5 February 2004, reservoir water level increased with flood time. Also, reservoir water level increased with flood time on 24 February 2004. Figure 11 shows Variation between reservoir water elevation and days in March 2004 flood. The reservoir water elevation was arranged for stable conditions with controlled spillway releases.

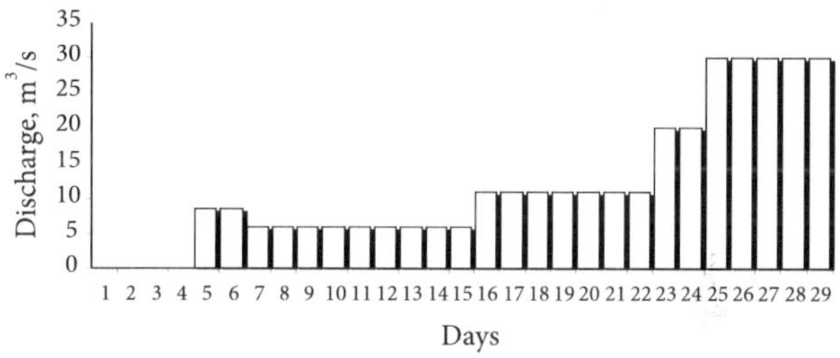

Fig. 8. Release discharges from gates of Goksu dam in the time of flood (February 2004)

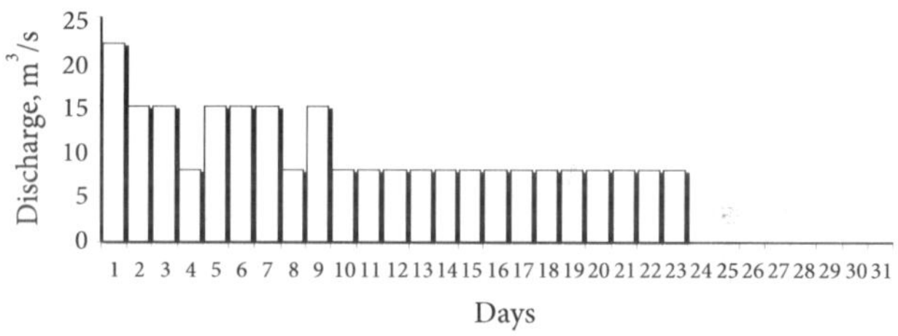

Fig. 9. Release discharges from gates of Goksu dam in the time of flood (March 2004)

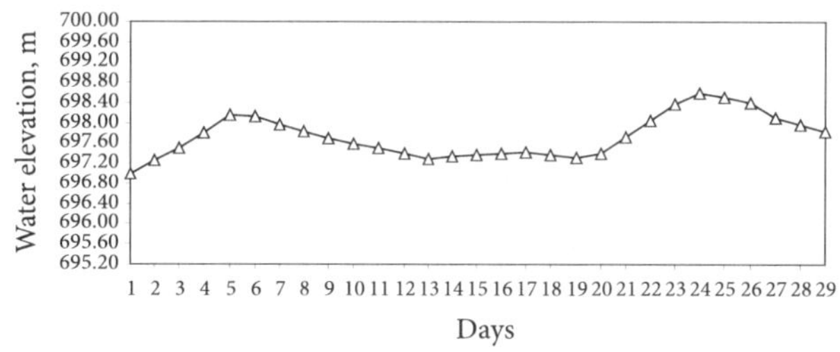

Fig. 10. Variation between reservoir water elevation and days in Goksu dam (February 2004)

3.3. Some considerations for safety of downstream floods

Flood analysis studies are the multidisciplinary research field. The flood risk maps should be produced. Flood risk analysis and management can significantly contribute to safety and life living in flood areas at along streams and rivers.

The commonly accepted measures for reducing flood damage are as follows: a) Reduction of peak flow by reservoirs; b) Confinement of the flow within a predetermined channel by levees, flood walls, or a closed conduit; c) Reduction of peak stage by increased velocities resulting from channel improvement; d) Diversion of flood waters through a flood bypass, which may return the water to the same channel at a point downstream or deliver it to another channel or different watershed; e) Reduction of flood runoff by land management; f) Temporary evacuation of

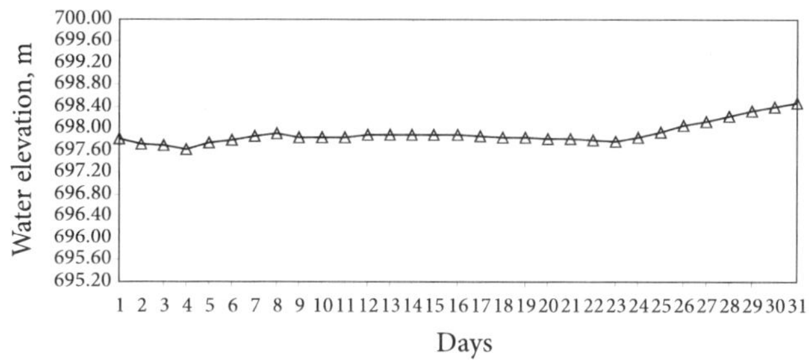

Fig. 11. Variation between reservoir water elevation and days in Goksu dam (March 2004)

flood threatened areas on the basis of flood warnings; g) Flood plain management.

Some technical considerations can be suggested for safety of downstream floods as:

- Use of intensive structural technologies such as reshaping of the land surface.
- Deepening and widening the original cross-section is highly effective in increasing the flood carrying capacity of rivers.
- Water surface profile should be calculated before dam construction. For this, standard step method can be used.

These technical considerations present special requirements for evaluating flood damage reduction.

3.4. Engineering works for solution of downstream flood problem

Learning from practical cases is an essential factor in engineering design. Engineering practice is often a combination of theory and practical constraints. Downstream flood problems can be solved by the careful coordination based on past experience and analysis. In flood management, both structural measures and more economic nonstructural safety measures should be used. Flood-mitigation works often utilize a combination of measures such as the cross-section and longitudinal profile of stream. Natural channels may be improved for flood by regulating the flow in storage reservoirs, dredging, contraction works, bank stabilization, straightening, and snag removal. Also, the stream is straightened and deepened for flood capacity.

During the regulation process of the channels of rivers and streams, their natural bends were eliminated. Thus, the cross-sectional profile was formed artificially. Rivers regulated in such a way became comparatively straight channels arranged in the valley of a former meandering stream (Baublys 2006).

The local hydraulic conditions are determined by flow resistance and geometry of the channel. Engineering works aim to reconstruct longitudinal changes in hydraulic geometry and flow processes during the passage of flood.

Flood capacity can be increased by enlarging the cross section and straightening the alignment, smoothing the banks to reduce the value of Manning's coefficient (n), and through the use of levees. Engineering works on streams and rivers serve a number of purposes: increasing the flow capacity, stabilizing the alignment, deepening the channel for navigation, and preventing bank erosion and sloughing.

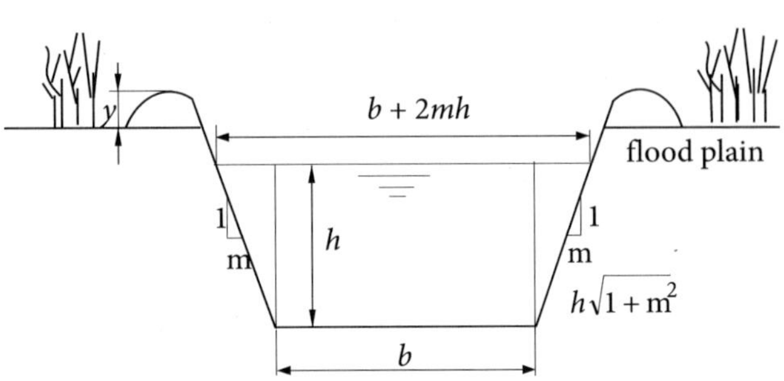

Fig. 12. New cross-section in Goksu dowstream

Deepening and widening of the original cross-section is highly effective in increasing of the flood carrying capacity. In Goksu stream, flood capacity was increased by enlarging of the cross section and straightening of flow direction. Thus, the configuration of streams and rivers can be modified by dredging. These works are mainly to assure the safety of persons who are downstream of the dam.

3.5. Modification and new channel design: trapezoidal channel cross-section

Floods in some parts of the basin have lead to investigate and revise the cross-sections. Cross-section analysis is used to determine the conveyance and storage of a stream channel and over bank areas. In results of spillway releases, cross-section investigation and observation of flooded areas, it was be noticed that the original cross-section of stream had the conveyance problem due to dense vegetation, sedimentation and erosion. Therefore, new cross-section is considered deepening and widening of the original cross-section as stream modification. The main purpose of channel rehabilitation in the Goksu stream is for flood control. Thus, the new cross-section will serve as the main corridor for the flood flow.

The stream channel is 33 km long and has a surface width (top width) range of one to five meters. The gradient through the study reach is 0.0014. Required new cross-sections are typically designed as trapezoidal section (Fig. 12). Stream section requirements are defined by the hydraulic calculation.

Wet area of trapezoidal cross-section (A):

$$A = (bh + mh^2), \qquad (4)$$

where: A is cross-sectional area (m^2); b is channel width; h is flow depth; m is bank slope.

Wet perimeter of trapezoidal cross-section (P):

$$P = (b + 2h\sqrt{1+m^2}). \qquad (5)$$

Hydraulic radius (R):

$$R = \frac{(bh + mh^2)}{(b + 2h\sqrt{1+m^2})}. \qquad (6)$$

The new channel cross-section may be able to minimize damage to crops and property within the flood plain of the stream. Cross-section parameters must be obtained at sufficient intervals to define the flow carrying capacity of the stream. The value of river bed width (b) is calculated according to the analysis and routing of flood flows.

The design channel width (b) refers to channel width in the straight reach between two bends corresponding to the design discharge (Q). The design channel width (b) may be determined by calculation from equations. From continuity and Manning equation:

$$Q = A \cdot V = A \cdot \left(\frac{1}{n} R^{2/3} S^{1/2} \right), \quad (7)$$

where: Q is discharge of channel (m^3s^{-1}); A is cross-sectional area (m^2); V is the cross-sectional average velocity (ms^{-1}); n is the Manning coefficient; R is the hydraulic radius (m); S is the slope of the water surface or the linear hydraulic head loss.

Table 3 shows typical values of Manning roughness coefficient for man-made channel. The Manning coefficient (n) of new channel cross-section was selected as 0.030 according to Table 3.

Table 3. Typical values of Manning roughness coefficient in forman-made channel (Chow 1959)

Number	Channel material	n
1	Smooth earth	0.018
2	Firm gravel	0.023
3	Natural channels in good condition	0.025
4	Natural channels with stones and weeds	0.035
5	Very poor natural channels	0.060
6	Gravel beds, straight	0.025
7	Gravel beds plus large boulders	0.040
8	Earth, straight, with some grass	0.026
9*	Earth, winding, no vegetation	0.030
10	Earth , winding with vegetation	0.050

* selected value in present study.

After construction of Goksu dam, the average channel capacity has reduced due to stream ecology, habitat and the physical characteristics. The bed alternates in elevation through a series of pools and riffles. Also, the river bed generally composed of sand having a median size of 2 mm. The discharge capacity has been enhanced by a new channel design. The design discharge of the trapezoidal cross-section is selected according to mean annual maximum discharge of the Goksu stream (estimated 19 m^3s^{-1}).

The hydraulic characteristics of trapezoidal the cross-section are calculated with Equations (3), (4), (5), (6) and (7). Table 4 summarizes the hydraulic characteristics of the trapezoidal cross-section.

The new channel is about 10 m wide at the top, and this is the width of the flow at bankful stage. The bottom width of a trapezoidal channel having side slopes of 1 vertical on 1.5 horizontal is 4 m. The depth of water is designed about 2 m.

As a part of the modification project (study) reach (29 km) was excavated to replace the straightened channel in 2005 by DSI (Local State Hydraulic Work of Turkey). Figure 13 and Figure 14 show the new cross-section and flow line modification after engineering works.

Table 5 shows the characteristics of engineering works as a part of the modification project.

Table 4. The hydraulic characteristics of the new trapezoidal cross-section in the Goksu downstream

Parameter	Value
Discharge, Q (m^3s^{-1})	19
Manning coefficient, n	0.030
Bank slope, m	1.5
Stream slope, S	0.0014
Flow depth, h (m)	2
Width, b (m)	4
Area, A (m^2)	14
Perimeter, P (m)	12.94
Hydraulic Radius, R (m)	1.08
Velocity, V (ms^{-1})	1.31
Floodplain height, y (m)	1.5

Table 5. Characteristics of engineering works

Description	Value
Stream channel length	33 km
Modified channel length	29 km
Estimated work distance in day	250 m/day^{-1}
Finishing time of engineering works	$\frac{29000\ m}{250\ m/day} = 116\ day$
Excavated volume in the stream line	300 000 m^3

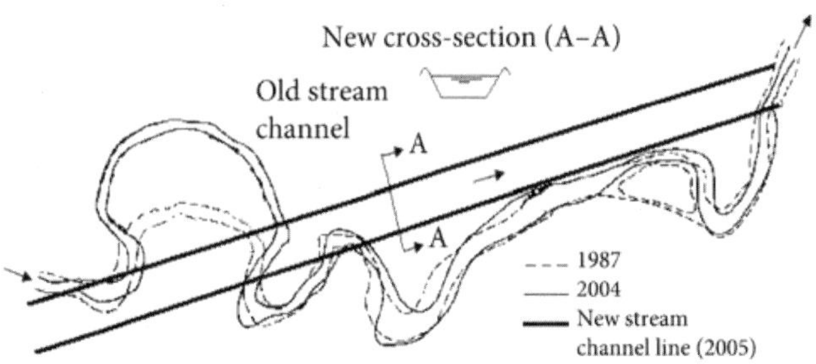

Fig. 13. Cross-section and flow line modification after flood event in Goksu dowstream

Fig. 14. View of new trapezoidal cross-section in Goksu downstream (November 2011)

Conclusions

The evaluation of flood events is necessary of the planning and design of dam projects. The flood problems should be considered for both upstream and downstream conditions in dam. The characteristics of these problems can be derived from observation and investigation of dam sites.

In results, three reasons are determined for downstream floods problems in dam: control problems of spillway gates, the naturally developing vegetation through channel flow line and incompetence of stream cross-section for carrying of flood flows.

For downstream flood management, the cross-section and longitudinal profile of stream are the most important scientific knowledge. Therefore, downstream conditions should be investigated before dam construction. In addition, the cross-section and longitudinal profile of stream must be controlled via maintenance of stream after dam construction. In incompetence of cross-section geometry and longitudinal profile, some modifications (e.g. straighten, deepen and widen) can be applied. Thus, the stream is straightened and deepened for flood capacity.

Also, different safety margins may be derived considering the local conditions. Hence, environmental management is enhanced with defending of environmental and agricultural fields from spillway floods.

The case study showed that both flow resistance and cross-section geometry are vital factors to determine local hydraulics. The present investigation may assist engineering works associated with the planning, project, operation and maintenance of the dam site.

Acknowledgment

The authors would like to thank the DSI (Local State Hydraulic Works) for their assistance in providing data for this study.

References

Azami, K.; Suzuki, H.; Toki, S. 2004. Changes in riparian vegetation communities below a large dam in a monsoonal region: Fatuse Dam, Japan, *River Research Applications* 20(5): 549–563. http://dx.doi.org/10.1002/rra.763

Baublys, R. 2006. Investigation of deformation processes in channels of regulated streams, *Journal of Environmental Engineering and Landscape Managemen* 14(3): 141–148.

Billa, L.; Mansor, S.; Mahmud, A. R. 2011. Pre-flood inundation mapping for flood early warning, *Journal of Flood Risk Management* 4(4): 318–327. http://dx.doi.org/10.1111/j.1753-318X.2011.01115.x

Chanson, H. 2004. *The hydraulics of open channel flow*. Oxford, UK: Butterworth-Heinemann. 650 p.

Chow, V. T. 1959. *Open channel hydraulics*. New York: McGraw-Hill. 680 p.

Fitzhugh, T. W.; Vogel, R. M. 2011. The impact of dams on flood flows in the United States, *River Research. Applications* 27(10): 1192–1215. http://dx.doi.org/10.1002/rra.1417

Gregory, K. J. 2006. The human role in changing river channels, *Geomorphology* 79(3–4): 172–191. http://dx.doi.org/10.1016/j.geomorph.2006.06.018

Guzzetti, F.; Stark, C. P.; Salvati, P. 2005. Evaluation of flood and landslide risk to the population of Italy, *Environmental Management* 36(1): 15–36. http://dx.doi.org/10.1007/s00267-003-0257-1

Julien, P. Y. 2002. *River mechanics*. Cambridge, UK: Cambridge University Press. 434 p. http://dx.doi.org/10.1017/CBO9781139164016

Jackson, R. B.; Carpenter, S. R.; Dahm, C. N.; McKnight, D. M.; Naiman, R. J.; Postel, S. L.; Running, S. W. 2001. Water in a changing world, *Ecological Applications* 11: 1027–1045. http://dx.doi.org/10.1890/1051-0761(2001)011[1027:WIACW]2.0.CO;2

Nikora, V.; Larned, S.; Nikora, N.; Debnath, K.; Cooper, G.; Reid, M. 2008. Hydraulic resistance due to aquatic vegetation in small streams: field study, *Journal of Hydraulic Engineering* 134(9): 1326–1332. http://dx.doi.org/10.1061/(ASCE)0733-9429(2008)134:9(1326)

Overton, I. C. 2005. Modelling floodplain inundation on a regulated river: integrating GIS, remote sensing and hydrological models, *River Research Applications* 21(9): 991–1001. http://dx.doi.org/10.1002/rra.867

Shahapure, S. S.; Eldho, T. I.; Rao, E. P. 2010. Coastal urban flood simulation using FEM, GIS and remote sensing, *Water Resorces Management* 24(13): 3615–3640. http://dx.doi.org/10.1007/s11269-010-9623-y

Yen, B. C. 2002. Open channel flow resistance, *Journal of Hydraulic Engineering* 128(1): 20–39. http://dx.doi.org/10.1061/(ASCE)0733-9429(2002)128:1(20)

Tamer BAGATUR. Dr Habil., Associate Prof., Department of Civil Engineering, Dicle University. Doctor Habil. of Science (Civil Engineering), Dicle University, 2009; Doctor of Science (Civil Engineering), Firat University, 2000. Employment: Associate Professor (2009), Senior Research Worker (2000), Junior Research Worker (1993), Engineer (1990). Publications: over 36 research paper. Research interests: fluid mechanics, river flows and process, interaction between hydraulic structures and the environment.

Nizamettin HAMIDI. Dr Habil., Assistant Prof., Department of Civil Engineering, Dicle University. Doctor Habil. of Science (Civil Engineering), Dicle University, 1995; Doctor of Science (Civil Engineering), Istanbul Technical University, 1994. Employment: Assistant Professor (1995), Senior Research Worker (1994), Junior Research Worker (1984), Engineer (1982). Publications: over 30 research paper. Research interests: hydraulic, hydrology, fluid mechanics, river flows and process, interaction between hydraulic structures and the environment.

11

FLOOD PATTERN CHANGES IN THE RIVERS OF THE BALTIC COUNTRIES

Diana SARAUSKIENE[a], Jurate KRIAUCIUNIENE[a], Alvina REIHAN[b], Maris KLAVINS[c]

[a]*Laboratory of Hydrology, Lithuanian Energy Institute, Breslaujos g. 3, 44403 Kaunas, Lithuania*
[b]*Institute of Environmental Engineering, Tallinn University of Technology, Ehitajate tee 5, 19086 Tallinn, Estonia*
[c]*Faculty of Geography and Earth Sciences, University of Latvia, Raina Blvd. 19, 1586, Riga, Latvia*

Abstract. Estimation of both the frequency and variation of spring floods is a key issue for the assessment and management of flood risks. Changes in river floods in Estonia, Latvia and Lithuania have been investigated in few national studies. However, there are no studies of the changes of flood patterns by using a common methodology for the rivers of this region. In this study flood pattern changes in the rivers of the Baltic countries were estimated applying trend and frequency analysis for the periods of 1922–2010, 1922–1960, 1961–2010 and 1991–2010, i.e. for the whole spring flood data sets, periods before and after 1960 (this year was considered as the beginning of the remarkable climate change), as well as for the two past decades. A comparative study of five probability distributions was performed in order to estimate which distribution at best represents statistical characteristics of the flood data. The results showed that maximum discharges of spring floods decreased over the whole studied period. Only some insignificant positive trends of maximum discharges were found in the last time period in continental and transitional rivers. Generalized extreme value distribution provided the best approximation to the maximum discharge data series of the rivers of Baltic countries for the whole observation period.

Keywords: Baltic countries, rivers, environmental monitoring, maximum discharge, spring floods, trends, probability distributions.

Introduction

Flooding is the most pervasive natural hazard. According to EU Directive 2007/60/EC on the assessment and management of flood risks, floods have the potential to cause fatalities, displacement of people and damage to the environment, to severely compromise economic development and to undermine the economic activities of the Community. It is feasible and desirable to reduce the risk of adverse consequences, especially for human health and life, the environment, cultural heritage, economic activity and infrastructure associated with floods.

Recently, significant attention is paid to analysis of this very special hydrological regime phase. Since the 1990s extensive and repeated floods across Europe have been bringing a lot of damage and loss (Arduino *et al.* 2005; Barredo 2009; De Wrachien *et al.* 2011). That shows failure of contemporary flood prevention approaches to control natural processes, despite the significant funding that is spent for flood prevention measures.

For evaluation of the risk of flooding it is important to predict the time of its occurrence and magnitude of floods heights and also it is necessary to ensure the reliability of various hydro technical structures. This analysis should neither underestimate nor overestimate magnitude of such an event. At the same time climate change is expected to increase flood frequency and its magnitude through increased precipitation. Several studies showed that seasonality of changes varies with location: increases are strongest in the warm season in the USA; while in Europe changes are most notable in the cool season (Groisman *et al.* 2004; Haylock, Goodess 2004; Villarini *et al.* 2011). Research in the Baltic countries showed that maximum discharge of spring floods tends to decrease in the most regions (Apsite *et al.* 2011; Kriauciuniene *et al.* 2012; Latkovska *et al.* 2012; Reihan *et al.* 2012), however, there some questions may arise in the context of climate change: if trends observed in flood series continue in the future or if the historical data are quite good to represent the future flood events?

Corresponding author: Diana Sarauskiene
E-mail: eko@mail.lei.lt

Many studies showed not clear increase of magnitude of flood events but what is promised that flood events can be more frequent. In the Northern Europe, Dankers and Feyen (2008) reported a considerable reduction of 10–40% in 100-year discharges in Finland, much decreases in northern Sweden and north-western Russia by the end of the century due to decrease in snow accumulation; however Lehner *et al.* (2006) evaluated that the 100-year floods in the same areas will occur more frequently by the 2070s.

To protect inundated areas and provide floods risk assessment the flood frequency analysis (FFA) is used as the most well-known procedure to ensure the reliable estimation of expected floods and their frequency. Main spring flood parameters, such as maximum discharge, flow duration, are usually used in FFA analysis. These parameters have already been investigated in the Baltic States (Gailiusis *et al.* 2001; Dumbrauskas *et al.* 2008; Klavins *et al.* 2009, etc.). The annual peak series with more than 50 observations for 40 streams in Estonia, Latvia and Lithuania were studied previously, however, there is no common procedure for the FFA in all these three countries and therefore the results of the same analysis can differ from one another. In Lithuania and in Latvia usually normal and Gumbel distribution laws were used to calculate the parameters, but Estonian hydrologists preferred log-Pearson type III distribution. Therefore, choice of the distribution most suitable to the recorded sample series is important from these aspects.

Gailiusis *et al.* (2001) calculated parameters of probability curve for the most of Lithuanian rivers. Recently, Sarauskiene and Kriauciuniene (2011) examined different probability distributions for spring flood frequency analysis in Lithuania for different periods with a focus on probability distribution fitting to the actual data in spring flood time. Their studies showed that the log-Pearson type III and generalized extreme value distributions are the most suitable for the Lithuanian rivers. However, such analysis is not performed yet in Estonia and Latvia.

The aim of this study is to analyse and to compare the patterns of spring flood data in the selected rivers from different hydrological regions in the Baltic countries applying trend analysis, calculating anomalies and estimating the best fit of probability distribution for the selected periods.

1. Data

The Baltic countries are north-eastern region of Europe containing the countries of Estonia, Latvia, and Lithuania, on the eastern shores of the Baltic Sea. These countries cover relatively small area, although hydro-meteorological differences across the States are significant. The territory of the Baltic States is divided into nine hydrological regions (3 regions in each country) (Fig. 1a) (according to Reihan *et al.* 2012). The regions of western Lithuania (W-LT), Latvia (W-LV) and Estonia (W-ES), the territory which is close to the Baltic Sea, belong to the marine climate zone and the main source of river feeding is precipitation. South-eastern Latvia (SE-LV) and Lithuania (SE-LT) together with eastern Estonia (E-ES) are the continental part of the Baltic States. The rivers of this

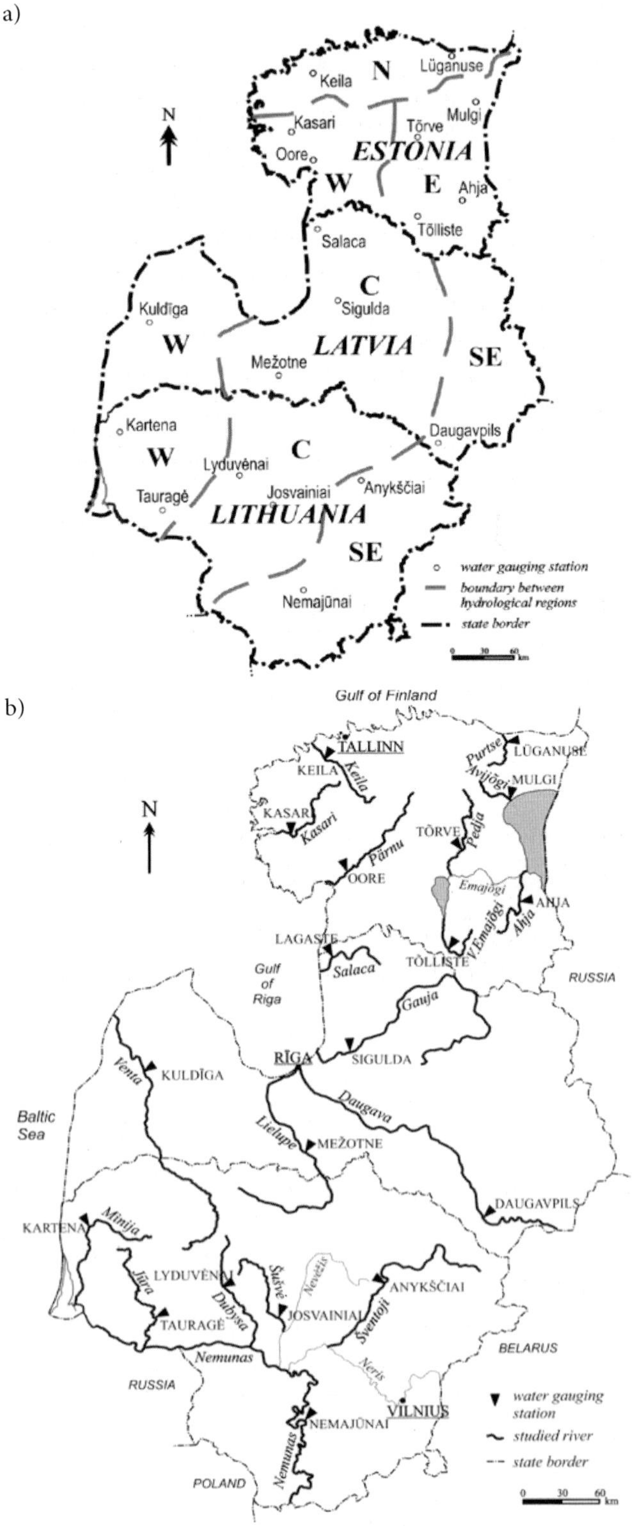

Fig. 1. Hydrological regions, 19 water gauging stations (a) and studied rivers across the Baltic States (b)

territory have prevailing snowmelt and subsurface feeding and the annual discharge of these rivers is distributed rather equally. The patterns of the other hydrological regions (N-ES, C-LV and C-LT) of the Baltic States are of a more individual character (Fig. 1a), the authors called these rivers transitional.

Annual maximum discharge data for the present study were obtained from Estonian Meteorological and Hydrological Institute, Latvian Environmental, Geology and Hydrometeorological Centre and Lithuanian Hydrometeorological Service. Annual maximum discharge time series at 19 water gauging stations across the Baltic States were used (Table 1, Fig. 1b). The rivers were selected according to availability and quality of data, i.e. there were used long data series and the rivers that are unaffected by significant upstream regulation. Average duration of observations was 83 years. The catchments of selected rivers at water gauging stations (WGS) have a different size (Table 2). The rivers were chosen so as to characterize the regime of the hydrological regions to which they belong.

Table 1. Summary of the used data

Country	Number of WGSs	Station-years of data
Estonia	8	632
Latvia	5	478
Lithuania	6	470
The Baltic States	19	1580

In order to evaluate flood frequency changes over time and to investigate climate change impact on spring flood events the analysis was made for different periods: 1922–2010, 1922–1960, 1961–2010 and 1991–2010.

2. Methods

The calculation of trend statistics of spring flood data series was based on the nonparametric Mann-Kendall test (Gilbert 1987). This procedure (Mann-Kendall test) is used since missing values are allowed and the data do not need to conform any particular distribution. In this study the significance of trend was tested by using three different α levels of significance: $\alpha = 0.05$, $\alpha = 0.01$ and $\alpha = 0.001$.

There are many probability distributions that are selected for flood projections in different countries all over the world. However, there is no one specific universal distribution for modelling flood data. In addition, some later conclusions from the various analyses of outputs for both real and modelled data showed that by employing several alternative distributions to the time series for a catchment one can hardly get the same distribution for samples for another duration (Mitosek *et al.* 2006; Kidson, Richards 2005).

In this study five widely used probability distributions were suggested and compared. Gumbel (EV1), generalized extreme value (GEV), log-Pearson type III (LP3), three-parameter lognormal (LN3) and generalized logistic (GLO) (Table 3) were selected for the analysis of statistical characteristics of observed flood data of the rivers in the Baltic States.

Table 2. Description of the river data

River	WGS	Hydrological region	Catchment area, km^2	Record period (years)	Mean discharge (m^3/s)	Annual maximum discharge (m^3/s)
		Continental type rivers				
Ahja	Ahja	E-ES	909	1960–2010	29.49	111
Avijogi	Mulgi	E-ES	366	1955–2010	25.41	55.7
V. Emajõgi	Tõlliste	E-ES	1050	1922–2010	63.62	198
Pedja	Tõrve	E-ES	776	1925–2010	52.04	200
Daugava	Daugavpils	SE-LV	64500	1922–2010	2552	6930
Šventoji	Anykščiai	SE-LT	3600	1928–2010	143.42	423
Nemunas	Nemajūnai	SE-LT	42800	1922–2010	919.15	3460
		Marine type rivers				
Venta	Kuldīga	W-LV	8320	1922–2010	589.44	1300
Minija	Kartena	W-LT	1230	1925–2010	118.73	287
Jūra	Tauragė	W-LT	1690	1926–2010	206.78	510
Kasari	Kasari	W-ES	2640	1925–2010	206.06	703
Pärnu	Oore	W-ES	5150	1922–2010	319.74	810
		Transitional type rivers				
Keila	Keila	N-ES	635	1923–2010	39.18	144
Purtse	Lüganuse	N-ES	784	1923–2010	53.77	156
Lielupe	Mežotne	C-LV	9390	1922–2010	647.23	2430
Gauja	Sigulda	C-LV	8510	1940–2010	335.30	870
Salaca	Lagaste	C-LV	3220	1926–2010	173.26	457
Dubysa	Lyduvėnai	C-LT	1070	1933–2010	66.35	184
Šušvė	Josvainiai	C-LT	1100	1940–2010	80.86	312

Generalized extreme value and Gumbel distributions are extreme value distributions widely used for modelling of extreme or rare events such as extreme floods and snowfalls, high wind speeds, extreme temperatures. The Gumbel (EV1) distribution, also known as the extreme value type I distribution, is unbounded (defined on the entire real axis). The generalized extreme value (GEV) distribution is a flexible three-parameter model that combines the Gumbel, Fréchet and Weibull maximum extreme value distributions.

The log-Pearson type III (LP3) is mostly used for annual maximum floods. In the USA, it is a standard distribution for flood frequency analysis.

The generalized logistic (GLO) distribution is also found to perform pretty well for modelling floods in some countries.

The three-parameter lognormal (LN3) distribution is popular in the studies of the frequency analysis of floods as well.

Parameter estimation for the different probability distributions was performed using methods of L-moments (for EV1, GEV, GLO), moments (for LP3) and maximum likelihood (for LN3). Selection of the corresponding distribution was achieved through goodness-of-fit comparison based on the Anderson-Darling test. Product of MathWave Technologies (EasyFit 5.4 with the built-in distribution viewer StatAssist) was used for probabilistic modelling of flood data of the selected rivers.

3. Results

3.1. Spring flood discharge anomalies

The main sources of spring floods in the Baltic rivers include accumulation of snow during wintertime, hard winters (indicated by sum of negative temperatures during winter), and intensive rainfalls that usually associated with increase of air temperature during the spring season. As far as the Baltic countries are situated in lowlands with extensive river network, the floods are one of the major sources of natural hazards. The flood impacts are aggravated due to location of major cities of the region either in the lowest reaches of biggest rivers or at rivers. Historically several major floods with high damages caused were recorded. Flood threat since 18th century initiated building of hydrotechnical structures to reduce flood risks and, especially, in the last century number of events of major floods has reduced (Table 4); however as factors affecting flood recurrence also climate change and increase of urban areas can be mentioned. Previous studies (Jaagus 2006; Kriauciuniene *et al.* 2012) indicate positive air temperature anomalies throughout a year and higher precipitation amounts in the winter season in the recent

Table 3. Probability density functions of the used distributions

Distribution type	Probability density function	Parameters
EV1	$f(x)=\frac{1}{\sigma}\exp(-z-exp(-z))$	σ – continuous scale parameter (σ>0); μ – continuous location parameter; $z\equiv\frac{x-\mu}{\sigma}$
GEV	$f(x)=\begin{cases}\frac{1}{\sigma}\exp(-(1+kz)^{-\frac{1}{k}}(1+kz)^{-1-\frac{1}{k}} & k\neq 0\\ \frac{1}{\sigma}\exp(-z-\exp(z)) & k=0\end{cases}$	k – continuous shape parameter; σ – continuous scale parameter (σ>0); μ – continuous location parameter; $z\equiv\frac{x-\mu}{\sigma}$
LP3	$f(x)=\frac{1}{x\lvert\beta\rvert\Gamma(\alpha)}\left(\frac{\ln(x)-y}{\beta}\right)^{\alpha-1}exp\left(-\frac{\ln(x)-y}{\beta}\right)$	α – continuous parameter (α >0); β – continuous parameter (β≠0); γ – continuous parameter
LN3	$f(x)=\frac{exp\left(-\frac{1}{2}\left(\frac{\ln(x-y)-\mu}{\sigma}\right)^{2}\right)}{(x-\gamma)\sigma\sqrt{2\pi}}$	σ – continuous parameter (σ>0); μ – continuous parameter; γ – continuous location parameter
GLO	$f(x)=\begin{cases}\frac{(1+kz)^{-1-1/k}}{\sigma(1+(1+kz)^{-1/k})^2} & k\neq 0\\ \frac{\exp(-z)}{\sigma(1+\exp(-z))^2} & k=0\end{cases}$	k – continuous shape parameter; σ – continuous scale parameter (σ>0); μ – continuous location parameter

decades in the territory of the Baltic countries. Significant changes in land use are also happening: urban areas are growing up; movement of people from rural to urban areas is obvious.

Table 4 shows the numbers of spring floods with probability equal or less than 10% in the studied rivers. The number of spring floods of the selected probability in the whole studied period comprised from 5 to 9. The comparison of two periods of the same length, 1922–1960 and 1961–1999, showed significant decrease of the spring floods (of the selected probability); in 7 rivers no cases of such floods were recorded. The exception is two Lithuanian rivers of the marine type, the Minija and the Jura, where the number of the spring floods increased (to 5 cases in 1961–1999). In the last two decades a single spring flood or no floods with ≤10% probability occurred. The reason of flood decreasing could be warmer winters, smaller thickness and duration of snow cover in the last year period.

Decrease of spring flood discharges can be observed using other methods of analysis, like estimation of anomalies. Anomalies of annual maximum discharge data were calculated as a ratio, expressed by %, between the flood discharges of a particular year and the mean maximum spring discharge of the long-term period (Figs 2–4).

Flood discharge anomalies in the continental rivers (Fig. 2) showed that annual maximum discharges have decreased from 1961. In the period of 1922–1960 there were many high positive anomalies (the greatest flood discharges in the studied period: in 1926 – 284% in the Pedja, in 1931 – 211% in the V.Emajogi, 172% in the Daugava and 195% in the Sventoji, in 1958 – 276% in the Nemunas and in 1960 – 276% in the Ahja). These great anomalies were caused by exceptionally big floods in the period. From 1961 the amount of positive and negative anomalies was almost the same. In the last two decades the most spring flood maximum discharges were less than the mean value of the long term period.

Flood discharge anomalies in the marine rivers (Fig. 3) indicate more even distribution of maximum flood discharges than in continental rivers. A significant decrease of spring maximum discharges can be noticed from 1971 to 1975 that could be caused by the decreased precipitation amounts from the seventh decade (Reihan *et al.* 2012). Only in the Minija the biggest flood was observed in 1994 (anomaly of 142%), whereas in other rivers the greatest positive anomalies calculated till 1970: in 1931 – 241% in the Kasari and 153% in the Parnu, in 1951 – 121% in the Venta, in 1968 – 147% in the Jura.

In transitional rivers (Fig. 4), like in marine rivers, much bigger than the average spring flood discharges were observed till 1970; the greatest floods also occurred in this period, whereas very low values of spring floods were recorded in 1971–1977 (possibly also due to the decreased precipitation amounts) and since then negative flood anomalies dominated.

Table 4. Number of big spring floods (probability of ≤10%) in the rivers of continental (a), marine (b) and transitional type (c)

a)

Periods	Ahja	Avijogi	V. Emajogi	Pedja	Daugava	Sventoji	Nemunas
1922–2010			10	9	8	8	11
1922–1960			10	9	7	8	9
1961–1999	3	2	0	0	1	0	2
1991–2010	1	1	0	0	0	0	0

b)

Periods	Kasari	Parnu	Venta	Minija	Jura
1922–2010	8	8	9	9	8
1922–1960	7	8	8	4	3
1961–1999	1	0	1	5	5
1991–2010	0	0	0	1	1

c)

Periods	Keila	Purtse	Lielupe	Gauja	Salaca	Dubysa	Susve
1922–2010	8	9	9	5	9	9	6
1922–1960	8	8	9	4	7	5	4
1961–1999	0	1	0	0	1	4	2
1991–2010	0	0	0	1	1	0	0

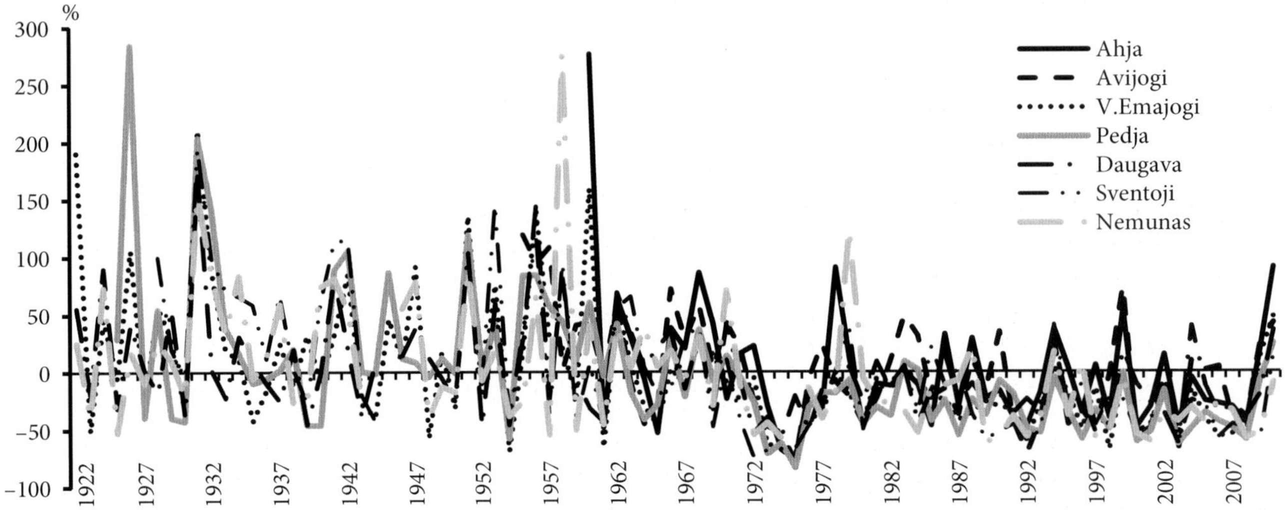

Fig. 2. Annual maximum spring flood anomalies (%) from the long-term average in the continental rivers

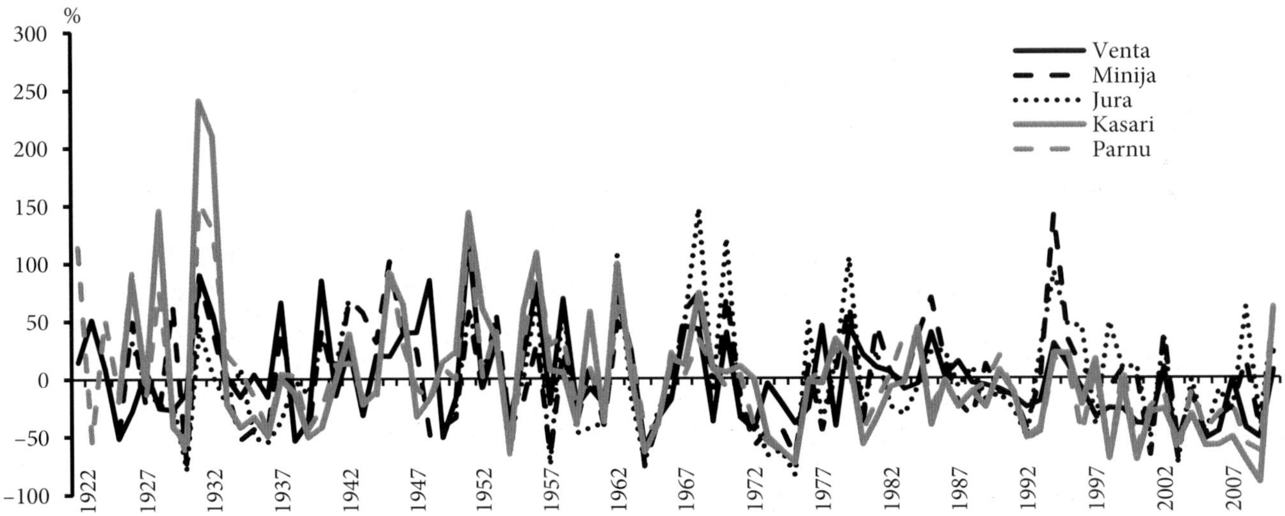

Fig. 3. Annual maximum spring flood anomalies (%) from the long-term average in the marine rivers

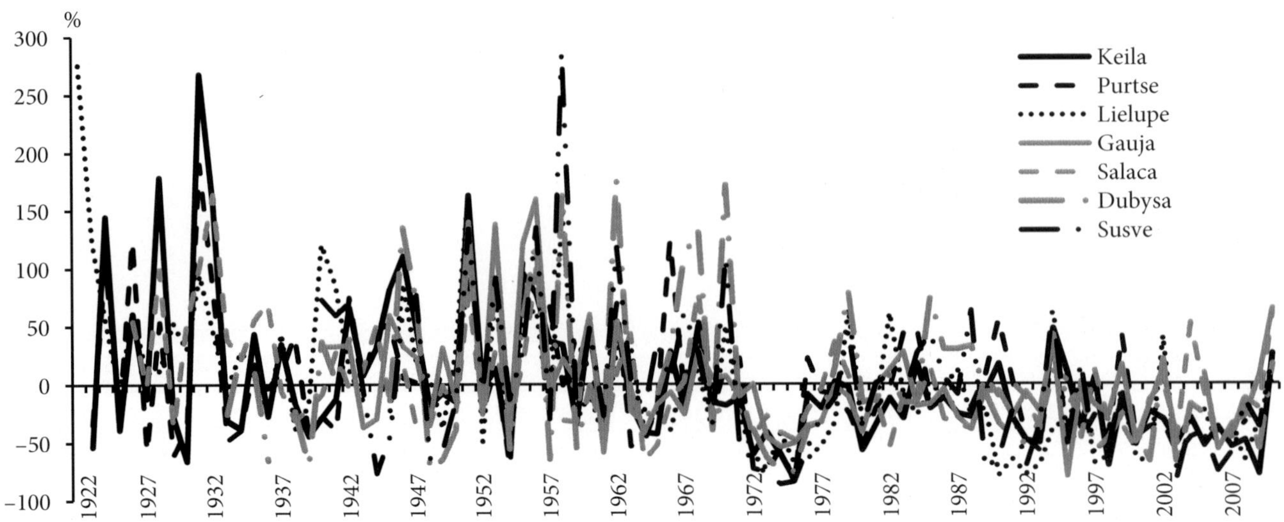

Fig. 4. Annual maximum spring flood anomalies (%) from the long-term average in the transitional rivers

3.2. Trend analysis

Trend analysis was performed for the whole spring flood data sets (1922–2010), periods before and after 1960, i.e. the beginning of the remarkable climate change (BACC Author Team 2008): 1922–1960 and 1961–2010, as well as for the two past decades (1991–2010). The trend was considered as statistically significant at the 5% level if the test statistics was greater than 1.65 or less than −1.65.

In the longest analysed period (1922–2010) the studied flood discharge time series had negative trends, except in the Jura and the Minija, both marine type rivers and both with the smallest catchment area in this group. Their trends can differ from other results since in small watersheds influence of local factors on the value of spring flood is more significant than other factors such as climate change for example. The Dubysa (transitional type river), where only poor insignificant trend was detected, was also the exception from the general decreasing tendencies (Fig. 5). Four (out of five rivers that have measurements for this period) continental rivers showed trend at $\alpha = 0.001$ level, i.e. the detected trends were very significant. Trends of the same significance were found in two transitional rivers: the Keila and the Lielupe as well. The reason of such character of floods may be the fact that in all continental rivers the highest (or the second highest) floods occurred in 1931, i.e. in the beginning of the studied period, and later in most cases maximum flood discharges were getting less and less. In spring of 1931 Estonia suffered damaging floods, 2% of its land was under water (RPV USSR 1972). Table 4 (b) and Figure 3 can explain why the data of two marine rivers in Lithuania, the Minija and the Jura, do not have clear trend: they both had high spring floods (of probability ≤10%) before and after 1961, moreover, the most spring flood discharge peaks were measured in the last four decades of 20th century.

The analysed above long-term period was split into two separate time slices: 1922–1960 and 1961–2010. Analysis showed (Fig. 5) no significant trend (neither negative nor positive) in maximum flood discharges in 1922–1960, except for two transitional rivers: the Lielupe and the Salaca in Latvia, where Mann-Kendall test displayed significant negative trends at $\alpha = 0.05$ level. And since there is no trend in other rivers it can be concluded that in the studied river basins flood regime (as well as natural and anthropogenic factors that influence it) was relatively stable in this time period.

Starting from 1961 (Fig. 5) the decreasing tendencies of flood discharges in the most of rivers were detected, however only in the continental type rivers: the Pedja, the Daugava and the Nemunas, and marine rivers: the Kasari, the Parnu and the Venta, changes were more pronounced and can be defined as significant.

In the last 20-year period only two marine type rivers: the Kasari and the Venta had significant negative trends at $\alpha = 0.05$ level in the flood data sets (Fig. 5). Also it should be mentioned that some opposite (positive) insignificant trends were observed in the selected rivers of continental type, whereas for the studied group of transitional rivers no tendencies could be noticed. The recent study of temperature and precipitation changes in the Baltic States (Kriauciuniene *et al.* 2012) showed that in the last two decades in the eastern regions, i.e. continental part, amount of precipitation in winter increased the most. This precipitation increase together with higher temperatures in the whole region in this period could give such consequences – higher floods in the rivers of continental type, since more rainfall or snowmelt water reached the rivers.

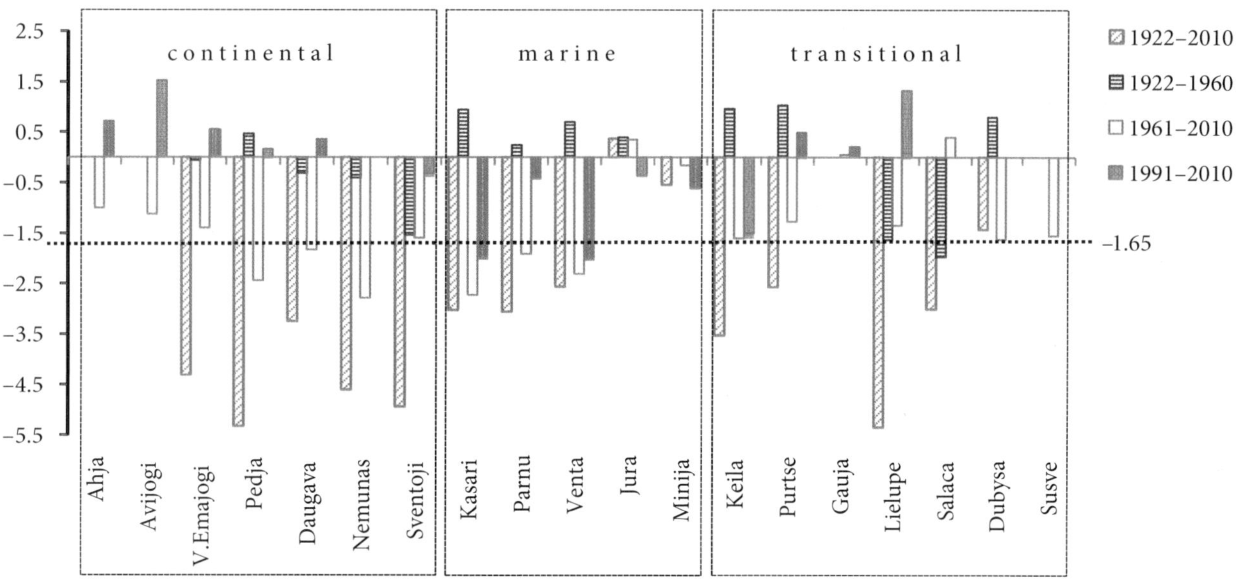

Fig. 5. Trends (according to trend statistics) of annual maximum discharges of spring floods in rivers of different type (continental, marine and transitional) in 1922–2010, 1922–1960, 1961–2010 and 1991–2010

3.3. Flood frequency analysis

In order to assess the spring flood risk, the probability of this event has to be identified.

Annual maximum discharge (spring flood) data of the selected rivers have been fitted to 5 probability distribution (PD) models. The aim of the goodness of fit test is to measure the "distance" between the data and the tested distribution, and compare that distance to some threshold value. If the distance (the test statistic) is less than the threshold value (the critical value), the fit is considered good. The critical values depend on the sample size and the significance level chosen ($\alpha = 0.05$). The distribution with the lowest statistic value is considered as the best fitting model.

The PDs were ranked according to the results of Anderson-Darling (AD) test in order to find out the dominant PD for all studied rivers. The best fitting PD (i.e. the distribution with the lowest statistic value) received rank 1, whereas the worst got rank 6.

The PD fitting was accomplished for the same observation periods as the trend analysis (1922–2010, 1922–2060, 1961–2010 and 1991–2010).

The fitting of annual maxima series of different hydrological regions showed that flood distribution approximations differed slightly (Table 5).

In the longest studied period (1922–2010) GEV distribution provided the best approximation to spring flood data in marine rivers, whereas in other studied rivers maximum discharges were mostly GEV or LP3 distributed. In Figure 6 visualisation of the probability distribution fitting

Table 5. Summed fitting ranks according to different distribution functions

River type (number of used data series)	Probability distribution				
	GLO	GEV	LP3	LN3	EV1
			1922–2010		
Continental (7)	22	14	15	19	35
Marine (5)	17	9	14	14	21
Transitional (7)	25	15	14	19	32
			1922–1960		
Continental (4)	15	8	7	13	17
Marine (5)	19	10	10	13	23
Transitional (6)	26	11	8	21	24
			1961–2010		
Continental (7)	29	10	17	18	31
Marine (5)	21	8	9	13	24
Transitional (7)	22	11	22	21	29
			1991–2010		
Continental (7)	26	13	17	16	33
Marine (5)	19	6	13	17	20
Transitional (7)	21	14	18	20	32

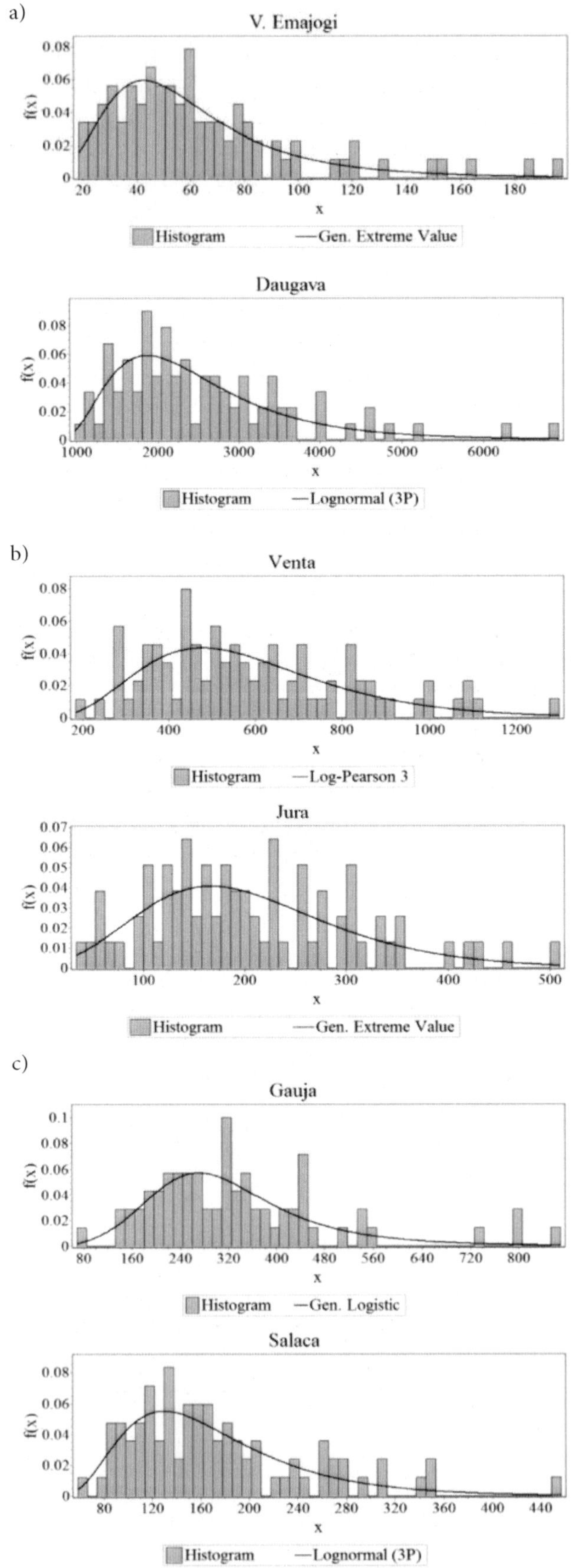

Fig. 6. Results of theoretical probability density function (PDF; displayed as a continuous curve) fitting to the empirical PDF (displayed as a histogram) for: a) continental rivers; b) marine rivers; and c) transitional rivers (in x axis – annual maximum discharge, m³/s, in y axis – probability density function)

to the actual spring flood data of the whole observation period is presented; it shows the probability density functions of the distributions which best fit to the spring flood data (the cases, where AD test statistics value was the lowest, i.e. less than 0.2).

In the period of 1922–1960 GEV and LP3 models showed the best (almost the same) correspondence to the flood data time series. The data sets after 1960 seemed to be the most similar to GEV distribution patterns; the used goodness of fit test gave very high ranks for this PD when fitting continental and transitional rivers data.

Maximum discharges were definitely GEV distributed in marine, continental and transitional zone of the studied territory in the last two decades as well.

Performed analysis showed that GEV distribution provides the best approximation to spring flood data in all the Baltic countries for all observation periods; LP3 and LN3 were the next best fitted PD; other models, GLO and EV1, were shown to perform poorly (Table 6). Therefore GEV distribution can be employed to estimate the occurrence probability of a given flood event in the studied rivers. This may help a lot in accurate and safe design of bridges, embankments, dams and other hydro technical structures.

Table 6. Final results of goodness of fit test (ranking of probability distributions)

	GEV	LP3	LN3	GLO	EV1
Overall summing	129	165	204	262	321
Final rank	I	II	III	IV	V

4. Discussion

For the analysis of flood pattern changes a group of rivers was selected from each the Baltic State. Each river group consisted of rivers representing three hydrological regions: marine, continental and transitional. The aim was to find out if there were any significant differences concerning spring flood patterns among the rivers of the separate groups in different time periods.

During the whole studied time period in the most of investigated rivers decrease of annual maximum spring flood discharges was observed. The greatest spring discharges were more frequent in the period of 1922–1960. The comparison of the periods of 1922–1960 and 1961–1999 showed significant decrease of spring floods; only in two Lithuanian rivers of the marine type, the Minija and the Jura, number of the spring floods increased. In the last two decades a single spring flood or no floods of ≤10% probability occurred.

Calculated flood discharge anomalies revealed a clear decrease of floods exceeding mean values since 1961 as well. Beginning from 1961 the amount of positive and negative anomalies was almost the same. The period of 1963–1977 was the driest period of the rivers runoff for all regions of the Baltic countries; that might be the reason why a significant decrease of spring maximum discharges can be noticed from 1971 to 1975–1977, since then negative flood anomalies dominated in the most of the studied rivers. The analysis of trend magnitude performed by Reihan *et al.* (2007) showed that the greatest amount of change of meteorological parameters occurred during the period of 1961–2003. During this period the average winter temperature raised by 3 °C and precipitation increase was 43 mm. These climate changes influenced a river discharge for the winter season. In such a way redistribution of spring runoff occurred, i.e. spring floods tended to start earlier because of the warmer winters and earlier snowmelt, and they became smaller.

In the last two decades the most spring flood maximum discharges were less than the mean value of the long term period.

Detected negative trends in spring flood data for the period of 1961–2010 confirmed the described tendencies. Decrease of the maximal discharge level and reduction of extreme yearly discharges recently in the Baltic rivers was estimated by other studies (e.g. Reihan *et al.* 2012) as well.

Already mentioned marine rivers the Minija and the Jura deviated from the other studied rivers the most. Big floods in these rivers were recorded in both periods before and after 1961; that may explain the absence of trends in the studied periods as well as in the whole period of 1922–2010. Catchments of these rivers are situated in Western Lithuania, this region is characterised by the highest amount of annual precipitation (735–810 mm) and exactly precipitation comprises the greatest part of a river feeding (53%) there. Moreover the catchments of the Minija and the Jura are small: the smaller a catchment, the shorter its response to changes of meteorological conditions is (Bagdziunaite-Litvinaite *et al.* 2011).

Long spring flood data series enable to use it for flood frequency analysis and make projection of floods of a certain probability. The analysis demonstrated that floods in the selected rivers may be most often represented by either GEV or LP3 distributions. These probability distribution models may be used for estimation of the design floods in the rivers of the studied area, and these models are among the most commonly applied in many countries (Abida, Ellouze 2008; Gubareva 2011) as well.

The observed variability of spring flood discharges in time is definitely climate driven, although the changing scale of urbanization plays significant role as well. Growing urban areas change infiltration capacities of water into the soil, consequently changing the runoff patterns within river catchments. It is difficult to get accurate and reliable data on land use in the studied territory, but other studies (like Theobald *et al.* 2009) show a significant impact of the land use changes on a maximum discharge formation. Other potential drivers of flood changes such as increased water use and deforestation (Preti *et al.* 2011) should not be forgotten as well. Combination of all these factors together makes estimation and prediction of flood events very complicated.

Conclusions

1. The performed analysis of big spring floods (probability of ≤10%), calculated anomalies and trend analysis of maximum annual discharges showed no significant differences in spatial patterns among the different groups of rivers, but indicated obvious changes in their temporal behaviour, i.e. the decrease of spring floods in the studied rivers in the period of 1922–2010.

2. The performed probabilistic modelling of the flood discharge data and probability distribution fitting revealed that generalized extreme value and log-Pearson type III distributions represent the most valid model for the studied river data.

3. The detected changes of spring flood maximum discharges over time are definitely mainly climate driven.

Acknowledgements

A part of research described in this paper was supported by COST action ES0901 European procedures for flood frequency estimation and partly by Environmental protection and technology programme No 3.2.0802.11-0043 of European Regional Fund within the project EstKliima.

References

Abida, H.; Ellouze, M. 2008. Probability distribution of flood flows in Tunisia, *Hydrology and Earth System Science* 12: 703–714. http://dx.doi.org/10.5194/hess-12-703-2008

Apsite, E.; Bakute, A.; Elferts, D.; Kurpniece, L.; Pallo, I. 2011. Climate change impacts on river runoff in Latvia, *Climate Research* 48: 57–71. http://dx.doi.org/10.3354/cr01004

Arduino, G.; Reggiani, P.; Todini, E. 2005. Recent advances in flood forecasting and flood risk assessment, *Hydrology and Earth System Sciences* 9: 280–284. http://dx.doi.org/10.5194/hess-9-280-2005

BACC Author Team 2008. *Assessment of climate change for the Baltic Sea basin.* Berlin Heidelberg: Springer. http://dx.doi.org/10.1007/978-3-540-72786-6

Bagdziunaite-Litvinaitiene, L.; Litvinaitis, A.; Saulys, V. 2011. Patterns of river runoff change considering the size of the basin, *Journal of Environmental Engineering and Landscape Management* 19(14): 326–334. http://dx.doi.org/10.3846/16486897.2011.634057

Barredo, J. I. 2009. Normalized flood losses in Europe: 1970–2006, *Natural Hazards and Earth System Sciences* 9: 97–104. http://dx.doi.org/10.5194/nhess-9-97-2009

Dankers, R.; Feyen, L. 2008. Climate change impact on flood hazard in Europe: an assessment based on high-resolution climate simulations, *Journal of Geophysical Research* 113, D19105. http://dx.doi.org/10.1029/2007JD009719

Dumbrauskas, A.; Bagdziunaite-Litvinaitiene, L.; Vyciene, G. 2008. Trend detection in hydrological series of main Lithuanian rivers, in *The 7th International conference "Environmental Engineering"*: Selected Papers, 22–23 May, 2008, Vilnius, Lithuania, 2: 508–514.

De Wrachien, D.; Mambretti, S.; Schultz, B. 2011. Flood management and risk assessment in flood-prone areas: measures and solutions, *Irrigation and Drainage* 60: 229–240. http://dx.doi.org/10.1002/ird.557

Gailiusis, B.; Jablonskis, J.; Kovalenkoviene, M. 2001. *Lietuvos upės. Hidrografija ir nuotėkis* [The Lithuanian rivers. Hydrography and runoff]. Kaunas: LEI. 792 p.

Gilbert, R. O. 1987. *Statistical methods for environmental pollution monitoring.* New York: Van Nostrand Reinhold. 336 p.

Groisman, P. Y.; Knight, R. W.; Karl, T. R.; Easterling, D. R.; Sun, B.; Lawrimore, J. H. 2004. Contemporary changes of the hydrological cycle over the contiguous United States: trends derived from in situ observations, *Journal of Hydrometeorology* 5: 64–85. http://dx.doi.org/10.1175/1525-7541(2004)005<0064:CCOTHC>2.0.CO;2

Gubareva, T. S. 2011. Types of probability distributions in the evaluation of extreme floods, *Water Resources* 38(7): 962–971. http://dx.doi.org/10.1134/S0097807811070074

Haylock, M. R.; Goodess, C. M. 2004. Interannual variability of extreme European winter rainfall and links with mean large-scale circulation, *International Journal of Climatology* 24: 759–776. http://dx.doi.org/10.1002/joc.1033

Jaagus, J. 2006. Climate changes in Estonia during the second half of the 20th century in relation with changes in large-scale atmospheric circulation, *Theoretical and Applied Climatology* 83(1–4): 77–88. http://dx.doi.org/10.1007/s00704-005-0161-0

Kidson, R.; Richards, K. S. 2005. Flood frequency analysis: assumptions and alternatives, *Progress in Physical Geography* 29(3): 392–410. http://dx.doi.org/10.1191/0309133305pp454ra

Klavins, M.; Briede, A.; Rodinov, V. 2009. Long term changes in ice and discharge regime of rivers in the Baltic region in relation to climatic variability, *Climatic Change* 95: 485–498. http://dx.doi.org/10.1007/s10584-009-9567-5

Kriauciuniene, J.; Meilutyte-Barauskiene, D.; Reihan, A.; Koltsova, T.; Lizuma, L.; Sarauskiene, D. 2012. Variability in temperature, precipitation and river discharge in Baltic States, *Boreal Environment Research* 17: 150–162.

Latkovska, I.; Apsīte, E.; Elferts, D.; Kurpniece, L. 2012. Forecasted changes in the climate and the river runoff regime in Latvian river basins, *Baltica* 25(2): 143–152. http://dx.doi.org/10.5200/baltica.2012.25.14

Lehner, B.; Döll, P.; Alcamo, J.; Henrichs, T.; Kaspar, F. 2006. Estimating the impact of global change on flood and drought risk in Europe: a continental, integrated analysis, *Climatic Change* 75: 273–299. http://dx.doi.org/10.1007/s10584-006-6338-4

Mitosek, H. T.; Strupczewski, W. G.; Singh, V. P. 2006. Three procedures for selection of annual flood peak distribution, *Journal of Hydrology* 323(1–4): 57–73. http://dx.doi.org/10.1016/j.jhydrol.2005.08.016

Preti, F.; Forzieri, G.; Chirico, G. B. 2011. Forest cover influence on regional flood frequency assessment in Mediterranean catchments, *Hydrology and Earth System Sciences* 15: 3077–3090. http://dx.doi.org/10.5194/hess-15-3077-2011

Reihan, A.; Koltsova, T.; Kriauciuniene, J.; Lizuma, L.; Meilutyte-Barauskiene, D. 2007. Changes in water discharges of the Baltic States rivers in the 20th century and its relation to climate change, *Nordic Hydrology* 38(4–5): 401–412. http://dx.doi.org/10.2166/nh.2007.020

Reihan, A.; Kriauciuniene, J.; Meilutyte-Barauskiene, D.; Kolcova, T. 2012. Temporal variation of spring flood in rivers of the Baltic States, *Hydrology Research* 43(4): 301–314. http://dx.doi.org/10.2166/nh.2012.141

RPV SSSR (Surface water resources of USSR). 1972. Protasjeva, M. S.; Eipre, T. F. (Eds.). Leningrad: Gidrometeoizdat, 106–119 (in Russian).

Sarauskiene, D.; Kriauciuniene, J. 2011. Flood frequency analysis of Lithuanian rivers, in *Environmental Engineering: 8th International Conference*, 19–20 May 2011, Vilnius, Lithuania, 2: 666–671.

Theobald, D. M.; Goetz, S. J.; Norman, J. B.; Jantz, P. 2009. Watersheds at risk to increased impervious surface cover in the conterminous United States, *Journal of Hydrologic Engineering* 14(4): 362–368. http://dx.doi.org/10.1061/(ASCE)1084-0699(2009)14:4(362)

Villarini, G.; Smith, J. A.; Serinaldi, F.; Ntelekos, A. A. 2011. Analyses of seasonal and annual maximum daily discharge records for central Europe, *Journal of Hydrology* 399(3–4): 299–312. http://dx.doi.org/10.1016/j.jhydrol.2011.01.007

Diana SARAUSKIENE. Dr, Senior Research Associate, Laboratory of Hydrology, Lithuanian Energy Institute, Lithuania. Bachelor of Science (biology), Vytautas Magnus University, 1994; Master of Science (environmental management), Vytautas Magnus University, 1996; Doctor of Physical Sciences (geography), Lithuanian Energy Institute, 2002. Publications: author of 13 scientific publications. Research interests: climate impact on river runoff, floods and thermal regime, research of thermal regime of water bodies, ice formation.

Jurate KRIAUCIUNIENE. Dr, Head of Laboratory of Hydrology, Lithuanian Energy Institute. Doctor of Natural sciences (geography) of Lithuanian Energy Institute, 1998. Publications: over 60 scientific publications. Research interests: modelling of hydrodynamics, waves, sediment and advection-dispersion processes; assessment of anthropogenic and natural impact on the hydrological regime of water bodies; climate change and water recourses.

Alvina REIHAN. Dr, Hydrologist, Assoc. Prof. of Department of Environmental Engineering, Tallinn University of Technology. Doctoral degree in Civil Engineering. Publications: over 20 scientific publications. Research interests: climate change impact on surface water resources and water quality, river runoff modelling, water balance, water resources estimation and management and scenarios for water management.

Maris KLAVINS. Dr hab. chem., Professor at the University of Latvia, Faculty of Geography and Earth Sciences, Department of Environmental Sciences. He obtained his scientific degree in chemistry of biologically active compounds at the Moscow State University in 1986 but a habilitation degree at the University of Latvia in 1994. He is a member of the Academy of Sciences of Latvia. Research interests are related to studies of natural organic matter, wetlands and bogs, surface water hydrology and environmental pollution problems.

12

GEP PREDICTION OF SCOUR AROUND A SIDE WEIR IN CURVED CHANNEL

Fevzi Onen

Civil Engineering Department, Engineering Faculty, Dicle University, 21280 Diyarbakir, Turkey

Abstract. Side-weirs have been widely used in hydraulic and environmental engineering applications. Side-weir is known as a lateral intake structure, which are significant parts of the distribution channel in irrigation, land drainage, and urban sewerage system, by flow diversion device. Local scour involves the removal of material around piers, abutments, side-weir, spurs, and embankments. Clearwater scour depth based on five dimensional parameters: approach flow velocity (V_1/V_c), water head ratio ($(h_1-p)/h_1$), side-weir length (L/r), side-weir crest height (b/p) and angle of bend θ. The aim of this study is to develop a new formulation for prediction of clear-water scour of side-weir intersection along curved channel using Gene Expression Programming (GEP) which is an algorithm based on genetic algorithms (GA) and genetic programming (GP). In addition, the explicit formulations of the developed GEP models are presented. Also equations are obtained using multiple linear regressions (MLR) and multiple nonlinear regressions (MNRL). The performance of GEP is found more influential than multiple linear regression equation for predicting the clear-water scour depth at side-weir intersection along curved channel. Multiple nonlinear regression equation was quite close to GEP, which serve much simpler model with explicit formulation.

Keywords: side-weir, clear-water scour, gene expression programming, environmental processes modelling.

Introduction

A side-weir is an overflow weir set into the side of a channel. Side-weirs are used for water level control, diverting excess water into relief channel during floods, as storm overflows from urban sewage systems and as head regulators of distributaries. The hydraulic behavior and the discharge coefficient of side-weirs for the different type of weirs, main channel, flow and bed conditions have been studied by many researchers such as: Muslu (2002); Cosar, Agacıoglu (2004); Emiroglu *et al.* (2010); Agaccioglu *et al.* (2012). Only a few investigators have been considered for side-weirs flow on an alluvial bed such as: Fares (2000); Agacıoglu, Onen (2005); Onen, Agacıoglu (2007a, b).

As can be seen from Figure 1, at sub-critical flow conditions the water surface profile along the weir plane increases from the upstream end towards the downstream end of the side-weir. The flow characteristics are very complicated due to the helicoidally flow pattern when the side-weir is located in a curved channel. In a curved channel, the inertia of the water keeps the water flowing in a straight path. The water near the surface, which has a velocity somewhat greater than the average velocity, posseses

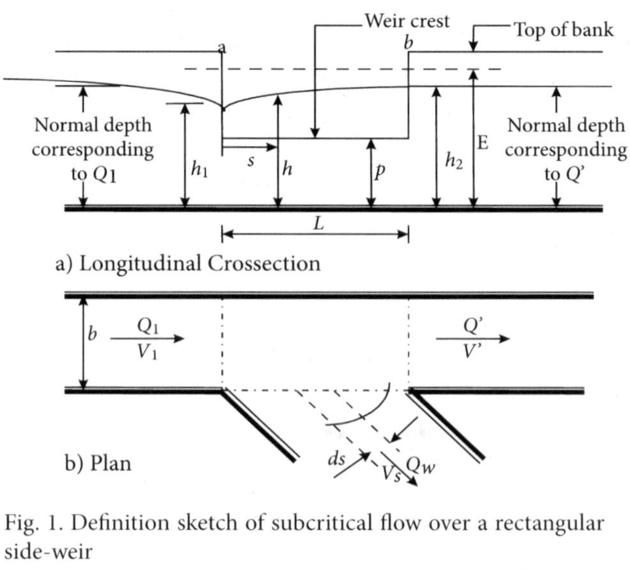

Fig. 1. Definition sketch of subcritical flow over a rectangular side-weir

Corresponding author: Fevzi Onen
E-mail: fonen@dicle.edu.tr

greater inertia than water near the bottom, which has a velocity less than the average. This greater inertia directs the higher-velocity water toward the outer channel bank, while the slower water moves inward because of the lateral pressure gradient set up by the outward-moving water at the surface. This causes a spiral flow that may persist for some distance downstream. The path of the thread of maximum velocity in an 180° curved channel also deviates from its normal course towards the inner bank at an upstream section of the bend. The path gradually shifts outward from 30° to 60°, and then follows the other bank to about a 150°. After that section, it again returns toward its normal course (Agaccioglu *et al.* 2007).

Scours around the structures such as the bridge piers, the obstacles and lateral intakes are a natural phenomenon caused by the flow of water in rivers and streams. Around these hydraulic structures, scouring introduces the possibilities of a reduction in the support given to foundations of these structures. Scour depths can be seen to depend on the properties of the bed material in the stream (grading, layering, particle shape and size, alluvial or cohesive), the foundation geometries and the properties of flow (Agacıoglu, Onen 2005). According to many investigators, the depth of scour around the structure is related to flow velocity or intensity. Melville and Chiew (1999) mentioned that clear-water scour occurs for mean flow velocities, V up to the threshold velocity for bed sediment entrainment, V_c, i.e. $V \leq V_c$, while live bed scour occurs for $V > V_c$. The maximum depth of scour occurs at $V = V_c$.

Under clear water conditions, the local depth of scour in uniform sediment increases almost linearly with flow velocity to a maximum value at the threshold velocity ($V = V_c$). For mean velocity larger than this value (V_c), the depth of scour fluctuates periodically with time and about an equilibrium value due to the effect of bed form migration. In addition, Chiew (1984) gave H_{dmax}/D versus V/V_c with d_{50} in Figure 2 for bridge piers.

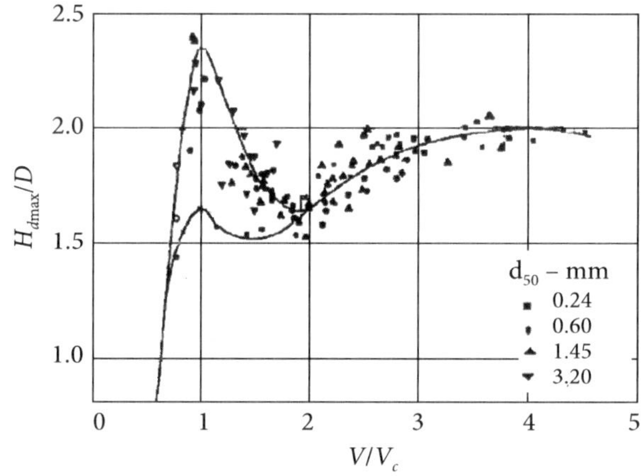

Fig. 2. H_{dmax}/D versus V/V_c with d_{50} (Chiew 1984)

The dimensionless parameters of the equilibrium depth of scour at a cylindrical pier of diameter D in uniform sediment are given by Melville and Chiew (1999) the following:

$$\frac{d_{se}}{D} = f\left(\frac{V}{V_c}, \frac{y}{D}, \frac{d_{50}}{D}, \frac{t}{t_e}\right), \tag{1}$$

where: d = local scour depth at time t; d_{se} = equilibrium depth of scour; t_e = time to develop the equilibrium scour depth; V/V_c = flow intensity; y/D = flow shallowness and d_{50}/D = sediment coarseness.

Genetic programming (GP) is a new technique used in the field of water resources engineering. Babovic and Keijzer (2002) applied GP to rainfall-runoff modelling. Harris *et al.* (2003), studied velocity predictions in compound channels with vegetated floodplains using genetic programming. Dorado *et al.* (2003) studied prediction and modelling of the rainfall-runoff transformation of a typical urban basin using artificial neural network (ANN) and GP. Giustolisi (2004) determined Chezy resistance coefficient in corrugated channels by using GP. Khorchani and Blanpain (2005) investigated development of a discharge equation for side-weirs using artificial neural network. Aytek and Kişi (2008) studied a genetic programming approach to suspended sediment modelling. Guven and Gunal (2008) studied a genetic programming approach for prediction of local scour downstream hydraulic structures. Guven and Kisi (2010) investigated linear genetic programming for time series modelling of daily flow rate. Azamathulla *et al.* (2008) studied genetic programming to predict ski-jump bucket spill-way scour. Guven and Aytek (2009) studied new approach for stage-discharge relationship: Gene-expression programming. Eldrandaly (2009) studied integrating gene expression programming and geographic information systems for solving a multi-site land use allocation problem. Azamathulla *et al.* (2010) applied GP to predict bridge pier scour. Bilhan *et al.* (2011) applied neural Networks for estimation of discharge capacity of triangular labyrinth side-weir located on a straight channel. Kisi *et al.* (2012) investigated prediction of lateral out flow over triangular labyrinth side weirs under subcritical conditions using soft computing approaches. Onen (2013) predicted penetration depth in a plunging water jet using soft computing approaches. Bagatur and Onen (2013) studied a predictive model on air entrainment by plunging water jets using GEP and ANN.

The objective of this current study is to develop a new formulation technique for prediction of Clear-water scour of side-weir intersection along curved channel using GEP. For this aim, experimental data are used. The performance of the GEP models was compared with multiple linear regressions and multiple nonlinear regressions.

2. Experimental set-up

Experiments to study the development of scour hole around side-weir, a model located on a rectangular flume were conducted in a Plexiglas flume at the Hydraulic Laboratory of Yildiz Technical University, Istanbul, Turkey. The main and discharge flume located between two straight channel are central angle of 180° with 2.95 m and 3.40 m radius to centreline, respectively. The bend is connected with an upstream straight reach of 14 m and a downstream straight reach of 3 m. The main flume is 0.40 m wide and a 0.55 m deep with a well-finished aluminium bottom. The discharge collection flume is 0.55 m deep, 0.50 m wide and situated parallel to the main flume. Details of the flume are shown in Figure 3.

Uniform bed materials were placed on as a 20 cm thick layer in the flume with a bed slope of 0.001. The bed of flume was firstly mixed to prevent sediment grading and layering, and then flattened before each run. During the experiments, the upstream valve was slowly adjusted without causing any disturbance to bed material until the desired flow condition was obtained in the flume. No sediment inflow was allowed from upstream due to clear-water conditions. Three different lengths (25, 40 and 50 cm) and three different weir crest height (7, 12 and 17 cm) of the side weir were located at the 30°, 60°, 90°, 120° and 150° section of bend. The dimensions of side-weirs and the ranges of the parameters tested are given in Table 1. Side-weirs were sharp edged and fully aerated on the overflowed side. The water depth is controlled by a sluice gate located at the end of the main flume to produce the uniform flow. Water is supplied from an overhead tank with a constant head and measured by a calibrated 90° V notched weir (Q_1) at the beginning of the main channel. $Q_1 = \left(\frac{h}{5.11}\right)^{2.447}$ was used for measuring the flow with 90° V notched weir (h, cm). Another calibrated 90° V notched weir measured the side-weirs flow rate (Q_w), which was located at the downstream end of the collection channel. $Q_w = 0.0134 \times h^{2,5896}$ was used for side-weirs flow rate. Then, the downstream discharge (Q') in the main channel according to the side-weir location was determined as follows:

$$Q' = Q_1 - Q_w . \tag{2}$$

The measurements of water surfaces and bed levels were determined using point gauges capable of reading to the nearest 0.1 mm. The mean particle size of the bed material (d_{50}) is 1.15 mm. The critical velocity of initiation of motion, (V_c) was determined using Shields diagram with a series of test run and given the following:

$$V_c = 0.1556 \log(4809 h_1) , \tag{3}$$

where: h_1 = flow depth at the upstream end of side-weir at the flume centre as in m and V_c is in m/s. Clear water scour experiments were done in the duration between 300–720 minutes (Agaccioglu, Onen 2005).

The changes in the main channel bottom topography around side-weirs are briefly mentioned according to flow visualization and the obtained bottom topography at 30° section of the bend as given in Figure 4, when the side-weir has L = 40 cm and p = 7 cm. The reduction in velocities and shear stresses causes to the development of stagnation along the inner bank and separation at the upstream corner of the side overflow at the downstream end of the side-weir in the main channel. They cause scour at the centre of reverse flow area. The accumulation in front of the upstream of the side-weir and the sand bar behind of the scour hole is also observed. The location of maximum scour hole moves toward the downstream with an increasing of the dimensionless parameters of approach flow intensity V_1/V_c as well as the water head ratio $(h_1-p)/h_1$ the weir crest height b/p and the weir length L/r. As mentioned below when the approach flow velocity or remaining momentum towards the downstream direction in

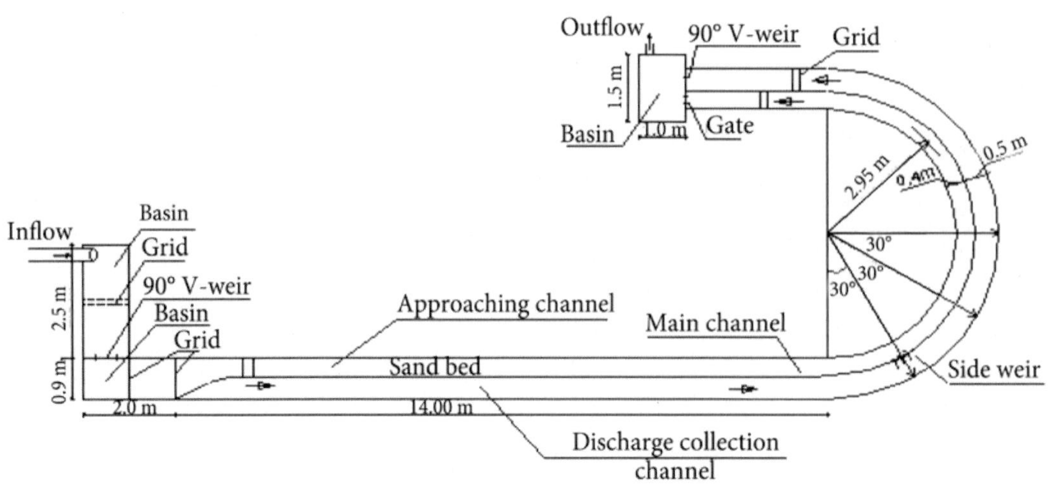

Fig. 3. Plan of test flume

the main channel increases then reverse flow area and the location of scour move toward the downstream end of the side-weir (Agaccıoglu *et al.* 2007).

The proposed GEP model formulations are trained with experimental data taken from experimental study (Table 1). The statistical parameters such as; the arithmetic mean (x_{mean}), standard deviation (S_x) and uncertainty (u) reflects the distribution properties of any data set. The uncertainty in measurements (x_{mean} = 0.382, S_x = 0.31, $u = S_x/n^{0.5}$, % $u = ((u/x_{mean})\times100)$ is about ±0.0231 (6.05%) for Clear-water scour of side-weir intersection along curved channel.

3. Dimensionless parameters

The variables that affect the phenomenon of scour depth, H_d, produced by a side-weir located on a channel having an alluvial bed are the mean upstream velocity of flow, V_1;

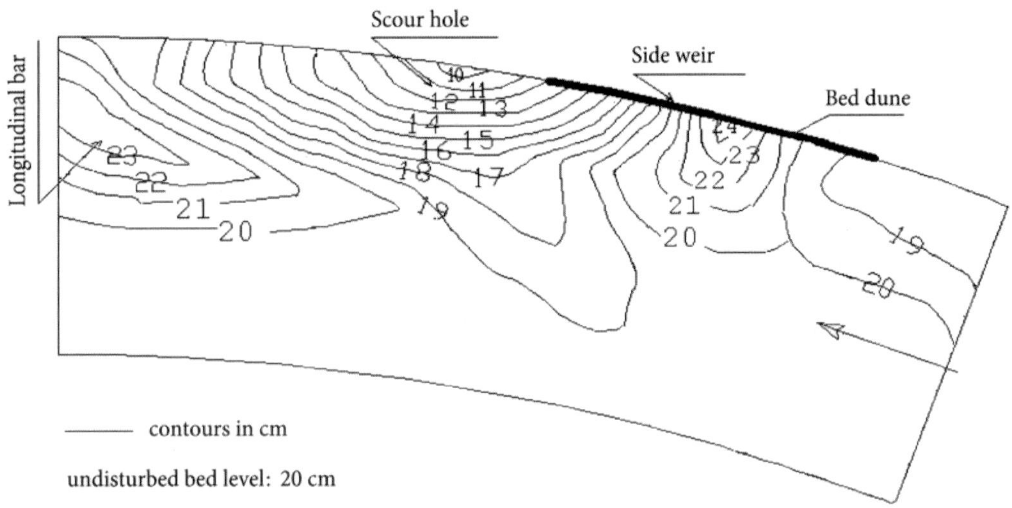

Fig. 4. Bed topography at a side-weir intersection when $V_1/V_c = 0.98$

Table 1. The dimensions of side-weirs and the ranges of parameters tested for curved channel

θ	p (cm)	L (cm)	h_1 (cm)	$(h_1-p)/h_1$	b/p	L/r	V_1 (m/s)	V_c (m/s)	V_1/V_c
(1)	(2)	(3)	(4)	(5)	(6)	(8)	(9)	(10)	(11)
30°	7	25	15–21	0.50–0.70	5.71	0.085	0.25–0.47	0.44–0.47	0.57–1.0
	12	25	21–27	0.45–0.60	3.33	0.085	0.31–0.49	0.47–0.49	0.66–1.0
	17	25	27–30	0.35–0.45	2.35	0.085	0.34–0.49	0.47–0-49	0.69–1.0
	7	40	15–21	0.50–0.70	5.71	0.136	0.25–0.49	0.44–0.47	0.57–1.0
	12	40	21–27	0.45–0.60	3.33	0.136	0.31–0.49	0.47–0.49	0.58–1.0
	17	40	27–30	0.35–0.45	2.35	0.136	0.34–0.49	0.47–0.49	0.63–1.0
	7	50	14–16	0.50–0.60	5.71	0.170	0.25–0.45	0.44–0.46	0.60–0.98
	12	50	20–25	0.40–0.55	3.33	0.170	0.31–0.48	0.45–0.48	0.66–1.0
	17	50	27–30	0.35–0.45	2.35	0.170	0.34–0.49	0.47–0.49	0.69–1.0
60°	7	40	15–20	0.50–0.65	5.71	0.136	0.25–0.47	0.44–0.47	0.58–1.0
	12	40	21–27	0.45–0.60	3.33	0.136	0.28–0.48	0.46–0.48	0.60–1.0
90°	7	40	15–20	0.50–0.65	5.71	0.136	0.25–0.47	0.44–0.47	0.58–1.0
	12	40	21–27	0.45–0.60	3.33	0.136	0.31–0.48	0.47–0.49	0.66–1.0
120°	7	25	15–21	0.50–0.70	5.71	0.085	0.25–0.47	0.44–0.47	0.57–1.0
	12	25	21–27	0.45–0.60	3.33	0.085	0.31–0.48	0.47–0.49	0.66–1.0
	17	25	27–30	0.40–0.45	2.35	0.085	0.34–0.48	0.48–0-49	0.69–1.0
	7	40	15–21	0.50–0.65	5.71	0.136	0.25–0.47	0.44–0.47	0.57–1.0
	12	40	21–27	0.45–0.55	3.33	0.136	0.27–0.48	0.47–0.49	0.58–1.0
	17	40	27–30	0.37–0.45	2.35	0.136	0.34–0.48	0.48–0.49	0.63–1.0
	7	50	14–16	0.50–0.62	5.71	0.170	0.27–0.45	0.44–0.45	0.60–1.0
	12	50	20–25	0.40–0.55	3.33	0.170	0.31–0.48	0.46–0.48	0.66–1.0
	17	50	27–30	0.35–0.45	2.35	0.170	0.34–0.49	0.47–0.49	0.69–1.0
150°	7	40	15–20	0.50–0.65	5.71	0.136	0.27–0.47	0.44–0.47	0.57–1.0
	12	40	21–27	0.45–0.60	3.33	0.136	0.31–0.48	0.47–0.49	0.66–1.0

the critical mean flow velocity for entrainment of bed sediment, V_c; the upstream depth of flow, h_1; the width of side-weir, L; the crest height of side-weir, p; the width of channel, b; the radius of curved channel, r; the angle of the bend, θ. Dimensionless analysis defines the equilibrium depth of scour at the side-weir intersection on an alluvial channel through the following function:

$$\frac{H_d}{p} = f(V_1/V_c, (h_1 - p)/h_1, b/p, L/r, \theta), \qquad (4)$$

where: dimensionless parameters on the right-hand side V_1/V_c, $(h_1-p)/h_1$, b/p, L/r, and θ represent, respectively, approach flow velocity, water head ratio, side-weir crest height, side-weir length, and the angle of bend (Agaccioglu *et al.* 2007).

In this study, the dimensionless parameters in equation were used as input and output parameters in GEP modelling. Table 1 reports the ranges of input and output parameters.

4. Multiple linear regressions

Multiple linear regressions (MLR) is a method used to model the linear relationship between a dependent variable and one or more independent variables. The dependent variable is sometimes also called the predict, and the independent variables the predictors. MLR is based on least squares: the model is fit such that the sum-of-squares of differences of observed and predicted values is minimized. The general Multiple linear regression (MLR) models with response y and terms $x_1,\ldots, x_i$ have the form:

$$y = b_0 + b_1 x_1 + ... + b_i x_i, \qquad (5)$$

where: b_i = model coefficients; y = predicted variable; x_i = independent parameters. The model coefficients (b_i) were calculated by least square regression. Values of root mean square error (RMSE) were used to find the fitting degrees of linear models with experimental data. The best model selection process stopped when the lowest RMSE value was reached. Multiple linear regression analysis yielded the following equations:

$$H_d/p = 2.2 + 1.67V_1/V_c + 0.000056L/r + 1.83(h_1-p)/h_1 + 0.085b/p; \quad R^2 = 0.84; \qquad (6)$$

$$H_d/p = -1.68 + 1.86V_1/V_c - 0.0058L/r - 0.164(h_1- p)/h_1 + 0.125b/p; \quad R^2 = 0.81; \qquad (7)$$

$$H_d/p = -1{,}59 + 1.85V_1/V_c - 0.0054L/r - 0.139 (h_1- p)/h_1 + 0.127b/p - 0.118\sin\theta; \quad R^2 = 0.82; \qquad (8)$$

Eqn (6) contains four independent variables and used data of 30°. Eqn (7) contains four independent variables and used data from 30° to 150°. Eqn (8) contains five independent variables and used data from 30° to 150°.

5. Multiple non-linear regression

Multiple non-linear functions of interest are as follows:

$$Y = a\, X_1^{\,b}\, X_2^{\,c}\, X_3^{\,d}. \qquad (9)$$

Nonlinear relationships between the dependent variables and the independent variables based on multivariate power function were considered. The following equations were derived:

$$H_d/p = 0.87(V_1/V_c')^{2.7}(L/r)^{0.356}(h_1- p)/h_1^{\,3.07}(b/p)^{0.58}; \quad R^2 = 0.94; \qquad (10)$$

$$H_d/p = 0.18(V_1/V_c)^{3.94}(L/r)^{-0.12*}(h_1-p)/h_1^{\,0.16}(b/p)^{1.27}; \quad R^2 = 0.87; \qquad (11)$$

$$H_d/p = 0.158(V_1/V_c)^{3.9}(L/r)^{-0.107*}(h_1-p)/h_1^{\,0.154} \times (b/p)^{1.28}\sin\theta^{-0.263}; \quad R^2 = 0.88. \qquad (12)$$

Eqn (10) obtained for 30°, Eqn (11) and Eqn (12) obtained from 30° to 150°.

6. Gene expression programming (GEP)

Ferreira proposed a new algorithm based on genetic algorithms (GA) and genetic programming and incorporates both the simple, linear chromosomes of fixed length similar to the ones used in genetic algorithms and the ramified structures of different sizes and shapes similar to the parse trees of genetic programming. So, the phenotype of GEP is made of the same kind of ramified structures used in genetic programming, but the ramified structures created by GEP are the expression of a totally autonomous genome. The main aim of GEP find a mathematical function principal using a set of data presented. For the mathematical equation the GEP process performed the symbolic regression by means of the most of the genetic operators of GA. The brief flowchart of GEP is given in Figure 5. Initially, chromosomes of each individual of the population are randomly generated. Then chromosomes are mean and each individual is appreciated based on fitness function and selected to reproduce with modification. The individuals of new generation are subjected to some developmental processes such as expression of the genomes, confrontation of the selection environment and reproduction with modification. The process is repeated for a certain number of generations or until a good solution is found (Ferreira 2002).

The two main players of GEP are the chromosomes and expression trees (ETs). The expression of the genetic information's encoded in the chromosome. As in nature, the process of information decoding is named translation and this translation implies a code and a set of rules. The genetic code of GEP is very simple; a one-to-one relation between the symbols of the chromosome and the nodes they represent in the trees. The rules determine the spatially organization of nodes in the expression trees and the type of interaction between sub-ETs. Thus, there are two languages in GEP; the language of the genes and the language of expression trees and, thanks to ETs structures which are determined from simple rules and their interactions. It is possible to at once infer the expression tree given the sequence of a gene and vice versa. This is termed Karva language (Ferreira 2002). The GEP genes have two main parts called the head and tail. The head includes some mathematical operators, variables and constants (+, –, *, /, sin, cos, 1, a, b, c) and they are used to encode a mathematical expression. The tail includes variables and constants (1, a, b, c) named as terminal symbols. In GEP, some operators such as selection, transposition, mutation and cross-over (recombination) are used to modify chromosomes for the next generation.

7. GEP model development

To generate the mathematical function for the prediction of the clear-water scour of a side-weir intersection along curved channel was the main aim of development of GEP models. For that reason, a development of GEP model was realized. The GEP model has five input parameters (approach flow velocity, water head ratio, side-weir crest height, side-weir length and angle of bend). The parameters of GEP models were presented in Table 2.

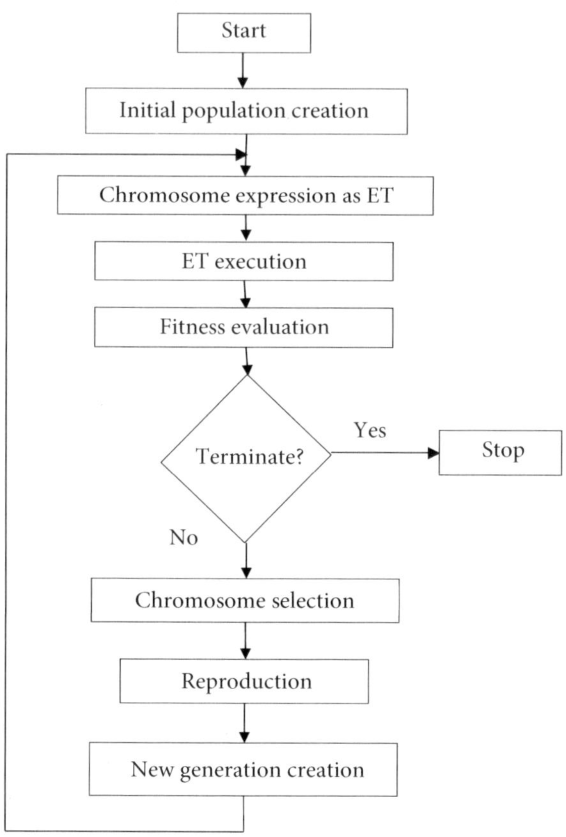

Fig. 5. GEP algorithm

Table 2. Parameters of the GEP model

Chromosomes	30
Fitness function error type	R^2
Number of the genes	3
Head size	8
Linking function	*
Function set	+, -, *, /, √, x^2
Mutation rate	0.044
One-point recombination rate	0.3
Two-point recombination rate	0.3
Inversion rate	0.1
Transposition rate	0.1

There are five major preliminary steps to solve a problem using GEP. These are the determination of: (1) the set of terminals; (2) the set of functions; (3) the fitness measure; (4) the values of numerical parameters and qualitative for controlling the run; and (5) the criterion for designating a results and terminating a run (Ferreira 2006).

The first major step in preparing to employ the GEP paradigm is to identify the set of terminals to be used in the individual computer programs. The major steps of terminal sets contain the independent variables of the problem, the state variables of the system and the functions with no arguments. The terminal set used in present study is $\left\{V_1/V_c,\ L/r,\ (h_1-p)/h_1,\ {}^{b}\!/_{p},\ \theta\right\}$.

The second major step is to define the set of functions and arithmetic operations. We used the four basic arithmetic operators (+, -, *, /), and some basic mathematical functions (√, x^2). The third major step is fitness measure, which specifies the way of evaluating how good a given problem solves a particular problem. In the present study, root mean square error (RMSE) of the training set is taken as fitness function. The terminals and the functions are the components of the programs that form the junctions in the tree. The four major steps is the selection of certain parameters to control the runs. The control parameters contain the size of the population, the rate of crossover, etc. the last step is the determination of the criteria to terminate the run. If the RMSE between the model results got reasonably close to zero, then the model was considered acceptable (Kisi *et al.* 2012).

We used a length of the head, h = 8 and three genes per chromosome. In this case, we linked the sub-ETs by multiplication.

Figure 6 shows the ET of the formulation which actually is:

$$\text{pow}((d[0]\times G1C0),2)\times(((d[4]-d[2])/(d[0]+G2C0))\times d[0])\times((((d[4]-G3C1)-(d[3]-d[2]))/((d[2]\times d[2])\times d[4]))+d[0]), \quad (13)$$

where: the constants in the formulation are G1C0 = −0.989; G2C0 = −5.78 and G3C1 = 2.86 and the actual parameters are d[0] = V_1/V_c, d[1] = L/r, d[2] = b/p, d[3] = $(h_1 - p)/h_1$ and d[4] = sin. After putting the corresponding values, the final equation becomes for 30°–150° of bend:

$$H_d/p = \left[-0.989\frac{V_1}{V_c}\right]^2 \left[\frac{sin\theta-(b/p)}{(V_1/V_c)-5.78}\left(\frac{V_1}{V_c}\right)\right]\times \left[\frac{sin\theta-(h_1-p)/h_1+(b/p)-2.86}{(b/p)^2\times sin\theta}+\frac{V_1}{V_c}\right]. \qquad (14)$$

It should be noted that the proposed GEP formulation is valid for the ranges of the training set given in Table 2. The prediction of the proposed GEP formulation versus actual experimental values for training and testing sets are given in Figures 7 and 8. Eqn (14) contains five independent variables and used data from 30° to 150°. Figure 7 shows how good GEP learned the nonlinear relation between parameters and Figure 8 proves the high generalization capacity of the proposed GEP formulation. Eqn (15) contains four independent variables and used data from 30° to 150°.

$$H_d/p = \left[\frac{V_1}{V_c}-\frac{(V_1/V_c)+2.3+(b/p)^{0.5}}{3.3(b/p)}\right]\times \left[\frac{-2.09}{6.44+(V_1/V_c)+(b/p)-(h_1-p)/h_1}+\frac{V_1}{V_c}\right]^2\times \left[b/p\right]^{0.5}; \qquad (15)$$

$$H_d/p = \left[\frac{V_1}{V_c}-\left(\left(V_1/V_c+7.75-7.75/L/r\times(V_1/V_c)(h_1-p)/h_1\right)\right)^2\right]; \left[(L/r-V_1/V_c)\frac{b}{p}+(L/r)^2-(V_1/V_c\times\frac{b}{p}+74.13)\right]\times \left[\frac{(h_1-p)/h_1}{L/r}\times\frac{(h_1-p)/h_1}{64-6.44\times b/p}\right]. \qquad (16)$$

Eqn (15) contains four independent variables and used data from 30° to 150°. Eqn (16) contains four independent variables and used data for 30° of bend.

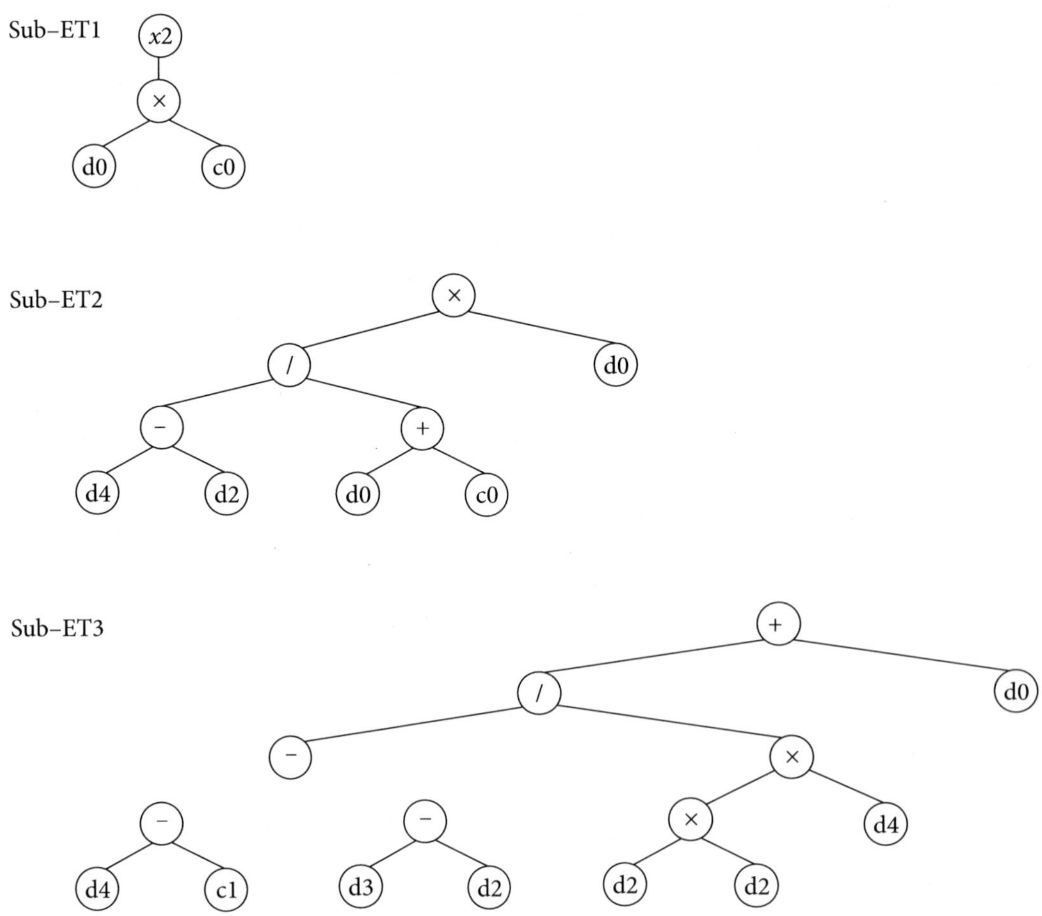

Fig. 6. Expression tree (ET) for the proposed GEP formulation

8. Comparison of GEP with regression model

The statistical performances of both GEP and regression models are summarized in Table 3. As far as Table 3 is concerned, satisfactory agreement between the model predictions and experimental data is observed for most of models. In statistics, the overall error performances of the relationship between two groups can be interpreted from coefficient of correlation (R) values. If a proposed model gives $R > 0.8$, there is strong correlation between measured and prediction values for overall the data available in the database. In addition to this, the statistical performance of any model is evaluated in terms of some error criteria such as root-mean-square (RMSE), which is significant criteria as well as R value, since sometimes a model with high R^2 value may exhibit high RMSE (Taskiran 2010).

Both multiple linear regressions and multiple nonlinear regressions are applied between H_d/p with V_1/V_c, L/r, $(h_1-p)/h_1$, b/p for 30°. It is applied between H_d/p with V_1/V_c, L/r, $(h_1-p)/h_1$ and b/p for 30°–150°. Then it is applied between H_d/p with V_1/V_c, L/r, $(h_1-p)/h_1$, b/p and θ for 30°–150°. We obtained six regression formulations (Eqns (6), (7), (8), (10), (11) and (12)). In the GEP model, first five independent parameters (V_1/V_c, L/r, $(h_1-p)/h_1$, b/p, θ) used for 30°–150° of bend, four independent parameters (V_1/V_c, L/r, $(h_1-p)/h_1$, b/p) used for 30°–150° of bend and then four independent parameters (V_1/V_c, L/r, $(h_1-p)/h_1$,

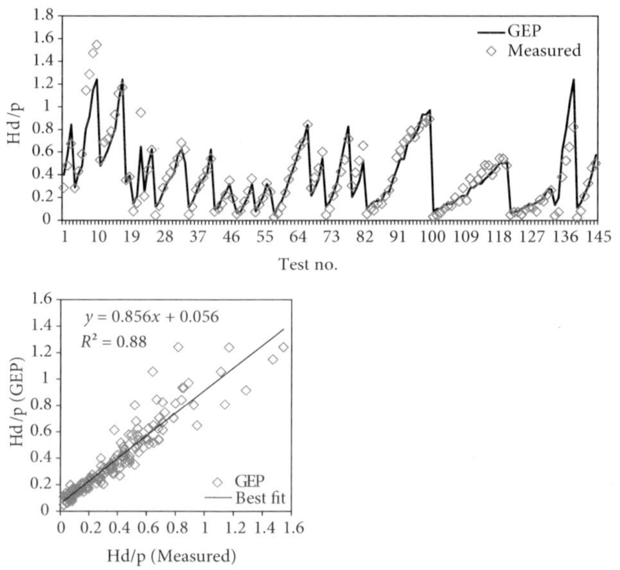

Fig. 7. GEP estimations of H_d/p versus measured ones for train set (model 1)

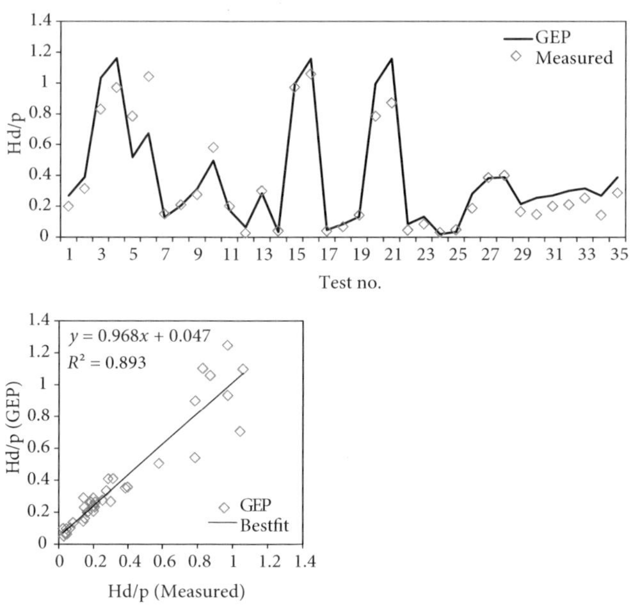

Fig. 8. GEP estimations of H_d/P versus measured ones for test set (model 1)

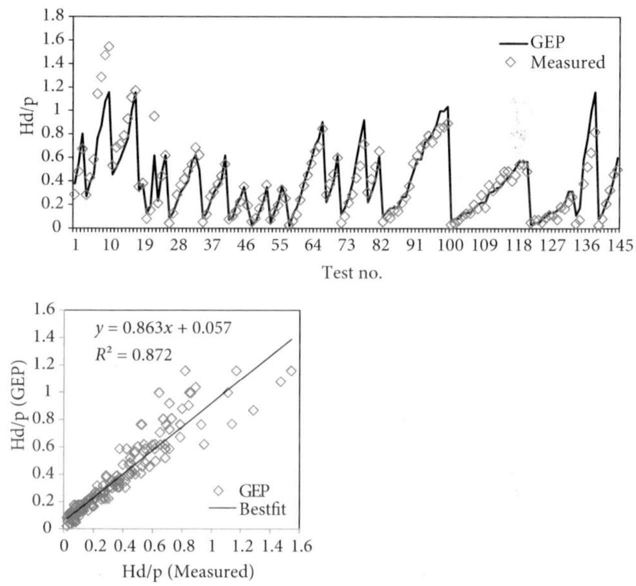

Fig. 9. GEP estimations of H_d/P versus measured ones for train set (model 2)

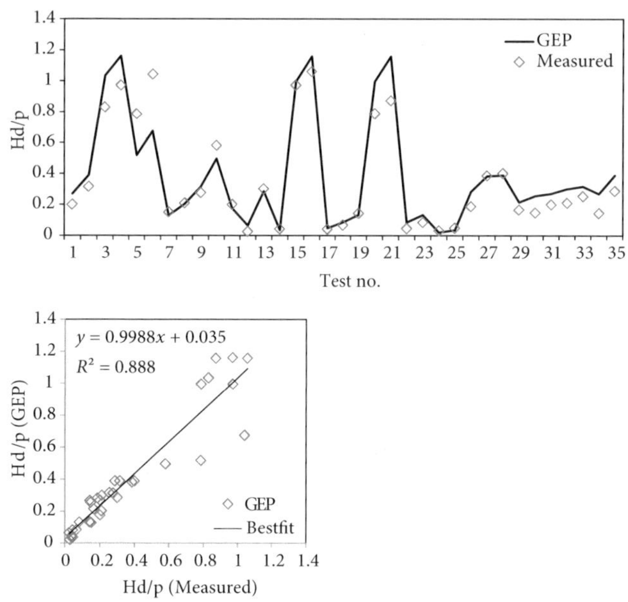

Fig. 10. GEP estimations of H_d/P versus measured ones for test set (model 2)

b/p) used for 30° of bend. Also, explicit formulations of developed GEP models are obtained (Eqns (14), (15) and (16)). The performance of GEP is found more influential when compared to multiple linear regressions (MLR) for predicting the clear-water scour of side-weir intersection along curved channel in Table 3.

Table 3. Statistical performance of GEP and regression models (MLR/MNLR)

Method	RMSE	R^2
MLR Eqn (6)	0.146	0.840
MLR Eqn (7)	0.221	0.810
MLR Eqn (8)	0.200	0.816
MNLR Eqn (10)	0.093	0.938
MNLR Eqn (11)	0.114	0.866
MNLR Eqn (12)	0.108	0.881
GEP Eqn (14)	0.105 0.117	0.882 (train) 0.893 (test)
GEP Eqn (15)	0.109 0.121	0.872 (train) 0.888 (test)
GEP Eqn (16)	0.101 0.096	0.927 (train) 0.928 (test)

9. Training and testing results of GEP modelling

The performance of GEP in training and testing sets is validated in terms of the common statistical measures R^2 (coefficient of determination), root-mean-square (RMSE):

$$R^2 = \left(\frac{\sum xy}{\sum x^2 \sum y^2} \right)^2 ; \tag{17}$$

$$RMSE = \left[\frac{\sum (X-Y)^2}{n} \right]^{\frac{1}{2}}, \tag{18}$$

where: x = observed values; y = predicted value and n = number of samples (Azamathulla *et al.* 2010).

Table 4. Sensitivity analysis for independent parameters for the testing set

	Model	RMSE	R^2
1	$H_d/p = f(V/V_c, L/r, h_1-p/h_1, b/p, \theta)$	0.117	0.893
2	$H_d/p = f(V/V_c, L/r, h_1-p/h_1, b/p)$	0.121	0.888
3	$H_d/p = f(V/V_c, L/r, b/p)$	0.125	0.878
4	$H_d/p = f(V/V_c, L/r, h_1-p/h_1)$	0.120	0.867
5	$H_d/p = f(V/V_c, h_1-p/h_1, b/p)$	0.126	0.882
6	$H_d/p = f(L/r, h_1-p/h_1, b/p)$	0.214	0.612

First, to assess the significance of each input parameter, attempt was performed on H_d/p. The GEP models are compared with one of the independent parameters removed in each case in Table 4, and deleting any independent parameter from the input set yielded larger RMSE and lower R^2 values. These five independent parameters have non-negligible influence on H_d/p and thus in this study the functional relationship given in Eqn (14) is used for GEP modelling. Also V/V_c is very effective on the scour ($R^2 = 0.612$ and RMSE = 0.214) in model 6. The testing performance of proposed GEP model showed a high generalization capacity with $R^2 = 0.893$ and RMSE = 0.117.

Conclusions

1. A new and efficient approach for the formulation of clear-water scour of a side-weir intersection along curved channel using GEP for the first time in the literature is presented in this study.
2. The proposed GEP formulation is empirical and based on experimental results. For maximum scour depth, regression formulations are proposed. The testing performance of proposed GEP model showed a high generalization capacity with $R^2 = 0.893$ and RMSE = 0.117 for maximum scour depth.
3. GEP formulations are compared with experimental results, multiple linear regression equation (MLR), multiple nonlinear regression equation (MNLR). MNLR were quite close to GEP.
4. The results of this study recommend that GEP modelling is a more sophisticated and improved than the corresponding conventional approaches for predicting clear-water scour of a side-weir intersection along curved channel.

References

Agaccioglu, H.; Onen, F. 2005. Clear water scour at a side-weir intersection along the bend, *Irrigation and Drainage* 54(5): 553–569. http://dx.doi.org/10.1002/ird.209

Agaccioglu, H.; Onen, F.; Toprak, Z. F. 2007. Scour around a side-weir at a 30° section of a 180° alluvial curved channel, *Irrigation and Drainage* 56(4): 423–438. http://dx.doi.org/10.1002/ird.304

Agaccioglu, H.; Emiroglu, M. E.; Kaya, N. 2012. Discharge coefficient of side weirs in curved channels, *Proceedings of the Institution of Civil Engineers Water Management* 165(6): 339–352.

Aytek, A.; Kisi, O. 2008. Genetic programming approach to suspended sediment modelling, *Journal of Hydraulic* 351: 288–298.

Azamathulla, H. M. D.; Ghani, A. A. B.; Zakaria, N. A.; Lai, S. H.; Chang, C. K.; Leow, C. S.; Abuhasan, Z. 2008. Genetic programming to predict ski-jump bucket spill-way scour, *Journal of Hydrodynamics* 20(4): 477–484. http://dx.doi.org/10.1016/S1001-6058(08)60083-9

Azamathulla, H. M. D.; Ghani, A. A. B.; Zakaria, N. A.; Guven, A. 2010. Genetic programming to predict bridge pier scour, *Journal of Hydraulic Engineering* 136(2): 165–169. http://dx.doi.org/10.1061/(ASCE)HY.1943-7900.0000133

Babovic, V.; Keijzer, M. 2002. Rainfall-runoff modeling based ongenetic programming, *Nordic Hydrology* 33(5): 331–346.

Bagatur, T.; Onen, F. 2013. A predictive model on air entrainment by plunging water jets using GEP and ANN, *KSCE Journal of Civil Engineering* 18(1): 304–314. http://dx.doi.org/10.1007/s12205-013-0210-7

Bilhan, O.; Emiroglu, M. E.; Kisi, O. 2011. Use of artificial neural networks for prediction of discharge coefficient of triangular labyrinth side-weir in curved channels, *Advance in Engineering Software* 42(4): 208–214. http://dx.doi.org/10.1016/j.advengsoft.2011.02.006

Chiew, Y. M. 1984. *Local scour at bridge piers:* Ph.D. Thesis, School of Engineering, University of Auckland, New Zealand, No. 355.

Cosar, A.; Agaccıoglu, H. 2004. Discharge coefficient of a triangular side-weir located on a curved channel, *Journal of Irrigation and Drainage Engineering* ASCE 130(5): 410–423. http://dx.doi.org/10.1061/(ASCE)0733-9437(2004)130:5(410)

Dorado, J.; Rabunal, J. R.; Pazos, A.; Rivero, D.; Santos, A.; Puertas, J. 2003. Prediction and modelling of the rainfall-runoff transformationof a typical urban basin using ANN and GP, *Application Artificial Intelligence* 17(4): 329–343. http://dx.doi.org/10.1080/713827142

Eldrandaly, K. 2009. Integrating gene expression programming and geographic information systems for solving a multi site land use allocation problem, *American Journal of Applied Sciences* 6(5): 1021–1027. http://dx.doi.org/10.3844/ajassp.2009.1021.1027

Emiroglu, M. E.; Kaya, N.; Agaccioglu, H. 2010. Discharge capacity of labyrinth side weir located on a straight channel, *Journal of Irrigation and Drainage Engineering* ASCE 136(1): 37–46. http://dx.doi.org/10.1061/(ASCE)IR.1943-4774.0000112

Fares, Y. R. 2000. Changes of bed topography in meandering rivers at a neck cutoff intersection, *Journal of Environmental Hydrology* 8(13): 1–18.

Ferreira, C. 2002. *Gene expression programming: mathematical modelling by an artificial intelligence*. Portugal: Angra do Heroismo. 478 p.

Ferreira, C. 2006. *Gene-expression programming: mathematical modelling by an artificial intelligence*. Berlin: Springer. 478 p.

Giustolisi, O. 2004. Using genetic programming to determine Chezyresistance coefficient in corrugated channels, *Journal of Hydroinformatic* 6(3): 157–173.

Guven, A.; Gunal, M. 2008. Genetic programming approach for prediction of local scour downstream hyraulic structures, *Journal of Irrigation and Drainage Engineering* 134(2): 241–249. http://dx.doi.org/10.1061/(ASCE)0733-9437(2008)134:2(241)

Guven, A.; Aytek, A. 2009. New approach for stage-discharge relationship: gene-expression programming, *Journal of Hydrologic Engineering* 14(8): 812–820. http://dx.doi.org/10.1061/(ASCE)HE.1943-5584.0000044

Guven, A.; Kisi, O. 2010. Estimation of suspended sediment yield in natural rivers using machine-coded linear genetic programming, *Water Resources Management* 25(2): 691–704. http://dx.doi.org/10.1007/s11269-010-9721-x

Harris, E. L.; Babovic, V.; Falconer, R. A. 2003. Velocity predictions in compound channels with vegetated floodplains using genetic programming, *International Journal of River Management* 1(2): 117–123. http://dx.doi.org/10.1080/15715124.2003.9635198

Khorchani, M.; Blanpain, O. 2005. Development of a discharge equation for side-weirs using artificial neural network, *Journal of Hydroinformatics* 7: 31–39

Kisi, O.; Emiroglu, M. E.; Bilhan, O.; Guven, A. 2012. Prediction of lateral out flow over triangular labyrinth side weirs under subcritical conditions using soft computing approaches, *Expert systems with Applications* 39(3): 3454–3460. http://dx.doi.org/10.1016/j.eswa.2011.09.035

Melville, B. W.; Chiew, Y. M. 1999. Time scale for local scour at bridge scours, *Journal of Hydraulic Engineering* ASCE 125(1): 59–65.

Onen, F.; Agaccioglu, H. 2007a. Local scours around side-weirs located on alluvial bends, *Teknik Dergi* 18(1): 4035–4053.

Onen, F.; Agaccioglu, H. 2007b. Scour at a side-weir intersection located on an alluvial river, *Nordic Hydrology* 38(2): 165–176. http://dx.doi.org/10.2166/nh.2007.005

Onen, F.; 2013. Prediction of penetration depth in a plunging water jet using soft computing approaches, *Neural Computing and Application,* September 2013. http://dx.doi.org/10.1007/s00521-013-1475-y

Taskiran, T. 2010. Prediction of California bearing ratio (CBR) of fine grained soils by AI methods, *Advances in Engineering Software* 41(9): 1115–1123.

Fevzi ONEN, Assistant Professor, Dr, Department of Civil Engineering, Faculty of Engineering, Dicle University, Diyarbakir, Turkey. Publications: author of several scientific publications. Research interests: water resources, scouring in open channel, hydrology and hydroelectric power plant.

EVALUATION OF TREATABILITY OF HIGH STRENGTH WASTEWATER IN A THREE STAGE-ROTATING BIOLOGICAL CONTACTOR

Atieh EBRAHIMI[a], Ghasem D. NAJAFPOUR[b], Manouchehr NIKZAD[c]

[a]*Department of Civil Engineering, Institute of Pardisan Fereidonkenar Iran*

[b]*Biotechnology Research Lab., Faculty of Chemical Engineering, Noshirvani University of Technology, Babol, Iran*

[c]*Faculty of civil Engineering, Noshirvani University of Technology, Babol, Iran*

Abstract. Cheese whey effluent contains biodegradable organic compounds in the range of 40 to 80 g. L^{-1}. In this study, a three–stage rotating biological contactor was fabricated as a bench scale experimental unit to remove organic matters from cheese whey. First, the treatability of cheese whey effluent in the three–stage rotating biological contactor (RBC) was evaluated. Then the effect of extended specific surface area (SSA) and recirculation rate on COD removal was investigated. The obtained results showed that the organic removal rate increased with an increase in loading rate, till other limiting parameters affect the process. Prior to application of the designated modifications to the system, maximum COD removal efficiency at HRTs of 24 and 36 h with OLR of 50 gCOD.L^{-1}.d^{-1} was 90 and 92.4%, respectively. The removal efficiency was improved as a result of increasing the SSA and recirculation rate. Also, recirculation rate may assist to increase the DO level of the wastewater, especially at high OLRs. To sum up, obtained results showed that whey effluent has been efficiently treated in a continuous operation of bench scale RBC.

Keywords: cheese whey, COD removal, Rotating Biological Contactor, Recirculation, Organic loading rate.

Introduction

The main priorities for wastewater treatment are effluent quality, cost, energy efficiency and nutrient removal/recovery (Reardon *et al.* 2013). The sustainable development of wastewater treatment technology is essential to improve process efficiency and the long term treatment plant capacity (Hoyland *et al.* 2010, Grady Jr *et al.* 2012).

Cheese whey (CW) is a by–product of the cheese and dairy effluents. Cheese processing effluents exhibit chemical oxygen demand (COD) values in the range of 0.8–102 g.L^{-1} and biological oxygen demand (BOD) values in the range of 0.6–60 g.L^{-1}which leads to a high consumption of dissolved oxygen in water bodies (Carvalho *et al.* 2013; Bylund, Pak 2003; Janczukowicz *et al.* 2008; Ebrahimi *et al.* 2009). It has been reported that cheese whey effluents have pH values in a wide range of 3.3–9.0. Typically, the aforesaid composition has a low pH. Suspended solids (SS), Total Kjeldahl Nitrogen (TKN), and total phosphorus are in the ranges of 0.1–22.0, 0.01–1.7 and 0.006–0.5 g. L^{-1}, respectively.

Whey is a source of protein and lactose; it has high content of organic compounds. Whey as raw material is used for production of lactate and polylactate which is used in packaging industry. Also, it can be used in agriculture, pharmacy and food industry as nutritional supplements and food additives. In the last decades, technological and economic advancement made it possible to recover soluble proteins from cheese whey, but still approximately half of world cheese whey production is not treated, and it is discharged as effluent (Yorgun *et al.* 2008; Smithers 2008). If for any reason such as economic, sanitary, local and lacking enough equipment, whey valorization technologies (such as protein and lactose recovery, spray drying, freeze-concentration, etc.) or direct utilization of whey for animal feed are not applicable, disposal has become an essential issue. In this case, aerobic or anaerobic treatment can be a solution to whey disposal. It is therefore necessary to process the whey even if it may be uneconomic (Kaewkannetra *et al.* 2009). In general, biological and membrane treatment processes are mainly

Corresponding author: Ghasem D. Najafpour
E-mail: najafpour@nit.ac.ir

used to treat cheese whey wastewater, but most of these processes are high energy intensive (Prazeres *et al.* 2012; Carvalho *et al.* 2013).

Rotating biological contactors (RBC) are normally called disc biofilm reactors which are an efficient alternative to wastewater treatment process such as activated sludge. The RBC technology has been widely used for the treatment of both municipal and industrial wastewaters (Mba, Bannister 2007). In this process, microorganism grows by degrading the substrate which causes microbial growth in a static biological film (Hassard *et al.* 2015; Sirianuntapiboon, Chumlaong 2013). In cases of wastewaters with high BOD/COD ratio per day such as cheese whey, some crucial operational challenges like biological clogging emerge, which are unavoidable and quite frequent in biofilters. Correspondingly, the most recognized benefit of rotating biological contactors is prevention of such phenomenon (Vasala *et al.* 2005; Carvalho *et al.* 2013; Rodgers, Zhan 2003). Moreover, high oxygen transfer efficiency in RBC is more cost-effective in the final result in comparison to other processes using diffusers or surface aerators. Other particular advantages such as high capacity to tolerate fluctuations in wastewater characteristics and to dampen shock loadings could also be achieved in a properly designed RBC (Metcalf 2003). Furthermore, it causes higher degradation efficiencies of organic loadings and resistance to toxic shocks than suspended growth process (Najafpour *et al.* 2005, 2006, 2008; Ebrahimi et al. 2010). Statistics indicate that RBC demands approximately 70–80% of the energy requirements of a trickling filter system and 40–50% of an activated sludge system (Rodgers, Zhan 2003; Metcalf 2003). In addition, activated sludge process and trickling filter system bring a substantial increase in operation complexity and insufficient efficiency. Therefore, decentralized wastewater technologies like RBC are gaining more attention in order to deal with wastewaters having high organic loading rate such as cheese whey, in terms of both the total cost and energy consumed. Najafpour and his coworkers (Najafpour *et al.* 2005) have investigated the treatability of a high strength organic load of palm oil mill industrial wastewater which contained COD of about 16 $g.L^{-1}$ through a bench scale experiment using a modular RBC reactor. After a five–day HRT, 91, 80 and 89% removal rates were achieved for COD, TKN and SS, respectively. In the most recent studies it was indicated that organic loading rate (OLR) causes a considerable impact on removal rate of biofilm developed on a polyvinyl chloride (PVC) mesh, polyester and polyurethane foam in simultaneously operated RBC reactors (Ercan, Demirci 2013). OLR, stages of RBC, recirculation, hydraulic retention time (HRT), rotation speed, media specific surface area, mass transfer rate and submergence percentage of disks are important parameters in RBC performance. There is considerable relation between some of these parameters which need to be investigated (Hassard *et al.* 2015). Staging has a positive effect on performance of RBC considering wastewater strength and its composition. However, this effect would be insignificant after four stages (Hassard *et al.* 2015). Discharge recirculation also increases the process performance by adjusting the substrate to microorganism ratio. Extensive research has been conducted on the efficiency of biofilm systems for cheese whey treatment; however, not much research was conducted on behavior of RBC systems under various conditions such as excessive daily loading rates, usage of modified extended surface discs as media support and effect of recirculation on removal efficiency in high strength wastewaters.

In the previous work (Ebrahimi *et al.* 2009), the performance of RBC in COD removal of high strength wastewater under various hydraulic loading conditions was experimented. This work is continuation of previous work with more detail studies and evaluation of additional parameters. The effect of increasing SSA and recirculation on COD removal and DO level were compared. In addition to variation of COD and HRT, other effective parameters like dissolved oxygen (DO) and TSS were evaluated. The main objective was to define the most effective process parameter which leads to the highest COD removal.

1. Materials and methods

1.1. Experimental setup

A three-stage RBC was built similar to that in our previous studies (Ebrahimi *et al.* 2009, 2010). The RBC consisted of three equal-sized compartments (75×35×30 cm), separated by fixed baffle plates with 16 light–weight PVC discs in each compartment. The total volume of RBC was 78.75 L. The discs, mounted on a galvanized metal shaft, had a diameter of 32 cm, 3 mm thickness and interspacing of 8 mm. The total surface area available for microorganism growth on discs was 7.1 m^2. A changeable speed motor (NORD motor, model SK 63/4, Germany) was rotating discs mounted on the shaft. In order to increase the SSA, some rectangular extended thin blades of PVC plates in certain dimensions were attached to the disks surface which increased the surface area by 10%. Five essential factors including HRT, OLR, recirculation, DO level and SSA were evaluated to investigate the impact of RBC operation on overall system performance. Based on steady state condition, the samples were taken for each volumetric flow rate after two times of corresponding retention time. The fresh raw wastewater was collected in a 20-liter container, then transported to the laboratory on a daily basis, refrigerated and stored at 4 °C to prevent any acidification and undesirable changes in their chemical composition. Diluted whey was fed into the reactor at pH 6.5 during the adaptation phase. Several dilutions of CW were prepared using distilled water based on requirements of the tests. In order to adjust the feed pH to 6.5, a 6M sodium hydroxide solution was used. The CW used in this study was obtained from "Gela Dairy Industrial Plant" (Amol, Iran), which was collected from ultra-filtration process for the production of cheese. The whey composition is reported in our previous work (Ebrahimi *et al.* 2009); otherwise, based on experimental conditions, the values are stated.

1.2. Analytical methods

The collected samples were settled for 30 minutes to analyze COD and biomass concentration. Standard methods for examination of water and wastewater was employed to determine TSS and DO concentrations (APHA 2005). In order to measure COD, a colorimetric method with closed reflux method was developed. Spectrophotometer (UNICO 2100, USA) at wavelength of 600 nm was used to measure the absorbance of samples. DO and pH meter (Hanna, HI9024) were used to measure pH and dissolved oxygen. The reactor temperature was basically maintained at room temperature.

2. Results and discussion

2.1. Start-up of the RBC

The RBC reactor was inoculated with 10 L of sludge with MLSS value of 4000 mg. L^{-1}, which was obtained from the mentioned industrial treatment plant. In order to develop initial biofilm, the reactor was fed with raw whey containing 0.4 gCOD.L^{-1}.d^{-1} at rotating speed of 4 rpm and HRT of 36 h. A gradual formation of biofilm on the rotating discs was observed after two weeks and fast biofilm accumulation was observed afterwards. The image of fully developed biofilm on RBC discs is shown in Figure 1.

Fig. 1. The RBC with developed biofilm after start–up

2.2. RBC performance

2.2.1. Effect of OLR and HRT on the efficiency of system

Due to daily variations of dairy effluents, the performance of RBC was investigated in 4 different groups of COD concentrations including 40, 50, 60 and 70 g.L^{-1}. The COD removals were studied at different HRTs (12, 16, 24 and 36 h). The obtained results for COD variations and COD removal in certain OLRs and HRTs are summarized in Table 1.

According to data presented in the Table 1, the maximum removal of organic load using the lowest HRT was about 82%, obtained at HRT of 12 h for COD content of 50 g.L^{-1}. This result is much higher when compared to similar reported data by other researcher which used in a similar RBC process (Hassard *et al.* 2015). The RBC obtained a COD removal efficiency of 92.4% for the COD concentration of 50 g.L^{-1}, when HRT increased to 36 h. However, it is not logical to use long HRTs in order to reach to high levels of treatment, since it would considerably raise the operating costs by decreasing the capacity of system. Also possible fluctuation of loading rate in HRTs of more than 24 h should be considered. Additionally, according to the obtained results, excessive growth of biofilm on the disks is not recommended for achieving more efficiency in the system as this may result in trapping and accumulation of inert materials that do not contribute to the degradation of the organic matters. Oxygen and nutrient passage toward interior layers can also be limited due to generation of thick biofilms on the disks (Ohl *et al.* 2004).

From the results it is conducted that at the certain COD concentration from 40 to 60 g.L^{-1}, increasing the HRT enhanced the removal efficiency. Such sensitivity was not observed in COD concentration of 70 g.L^{-1}. In fact in COD content of 70 g/l, the removal efficiency was increased from 45 to 59% due to a shift in HRT from 12 to 24 h, but when the HRT extended to 36 h, COD removal decreased to 52% in contrast with other HRTs. This can be explained by development of microbial layer which lead to reducing the dissolved oxygen required for the growth of the microorganisms. Overall COD removal with COD concentration of 70 g.L^{-1} was significantly lower compared to other groups of COD data presented in Table 1. Insufficient COD removal capacity with COD concentration of 70 g.L^{-1} can be attributed to limitation of oxygen mass transfer and specific surface area for attached microorganisms in relatively high OLR. Moreover, results indicate that the maximum COD removal efficiency of 92.4% was achieved with COD concentration of 50 g. L^{-1} and HRT of 36 h. The obtained results proved that the performance capability of RBC is mostly dependent on OLR rather than on the individual organic concentration or flow rate.

DO variation were monitored for each stage of the reactor. The DO meter probe was placed at the midpoint in each stage. Figure 2 illustrates the DO level regarding the designated HRTs in three stages of the RBC reactor. As it is indicated in this figure, the last stage of the RBC

Table 1. The COD variations and COD removal in certain OLRs and HRTs

COD in (g.L^{-1})	40				50			
HRT (h)	12	16	24	36	12	16	24	36
OLR (g COD.L^{-1}. d^{-1})	80	60	40	26.7	100	75	50	33.3
COD out (g.L^{-1})	10.8	10.13	5.5	4	9.2	6.4	5.2	3.8
COD Removal (%)	73	75	86	90	82	87	90	92.4
COD in (g.L^{-1})	60				70			
HRT (h)	12	16	24	36	12	16	24	36
OLR (g COD.L^{-1}. d^{-1})	120	90	60	40	140	105	70	46.7
COD out (g.L^{-1})	21.6	19	15.5	9.7	38.5	30.5	28.6	33.8
COD Removal (%)	64	68	74	84	45	56	59	52

had the highest content of DO. Very low DO level was observed in the first stage. The averaged DO level in the maximum COD removal condition (COD concentration of 50 g.L^{-1} and HRT of 36 h) were 2.82, 4.5 and 5.2 mg.L^{-1} in the first, second and third stage, respectively. This increase in DO level with increasing the number of stages can be due to the thick biofilm generated on the disks and effectiveness of n-reactors in series. The microorganisms attached to the disks consume oxygen to degrade organic matter which leads to decrease the remained dissolved oxygen level. In fact, staging in RBC can be considered as process enhancement and the physical dividers may separate the flow behavior within the reactor. It causes a gradual reduction in the organic matters so the reactor approaches to a plug-flow regime. This stepwise treatment allows microorganisms to be adapted to each stage condition. This procedure enhances the substrate removal rate and stability of the process. In addition staging can improve the ability of system to manage shock loads as well.

From the results it also can be observed that in a certain COD concentration, as HRT increased, the DO level gradually increased. The increasing trend for DO concentration in all stages was similar; this means for long HRTs, as the concentration of organic matter decreased, the rate of oxygen consumptions slightly declined. For each HRT, as OLR was in an increasing trend; that caused the DO levels to show a slight decrease.

Figure 3 depicts the effect of organic loading rates and number of stages on DO concentration. From the data, it is observed that the DO concentration in the first stage significantly decreased while in the following stages it showed a steady increase. In addition, by increasing the organic loading rates DO concentrations was decreased. As illustrated in Figures 2 and 3, dissolved oxygen profiles mainly followed a pattern of initial decreased in the first stage and then gradually increased in the successive stages. There is a direct correlation between the DO concentration and organic loading rate which is at high organic loading rates; it caused a considerable decrease in dissolved oxygen. It can be a result of heterotrophic uptake of dissolved oxygen in these stages. At the first stage, high utilization of substrate and heterotrophic growth were occurred; which could be related to maximum consumption of dissolved oxygen. As organic substrate is metabolized through the successive stages, heterotrophic growth decreased; that was a result of an expected reduction in oxygen demand in the last stages. The collected data are

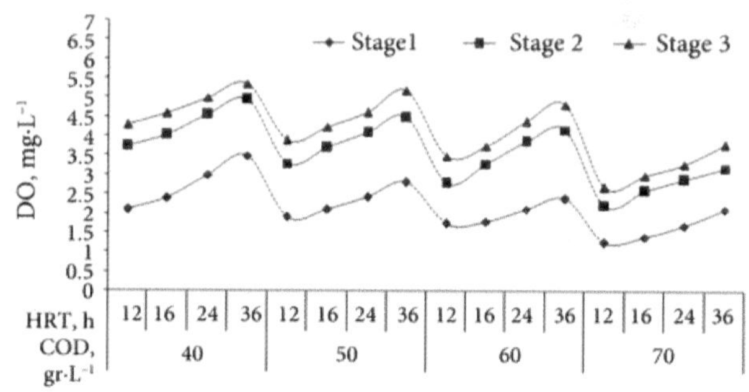

Fig. 2. DO level with respect to HRT and COD in different stages

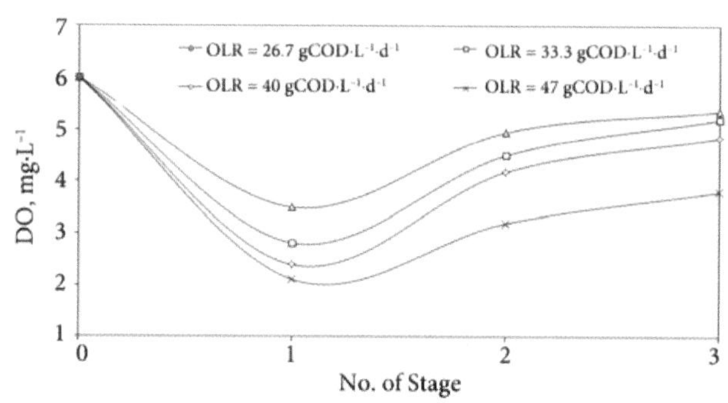

Fig. 3. DO concentration profiles in different stages of RBC with various OLR

in accordance with data reported in the literature (Chen *et al.* 2006).

The effect of organic loading rate and number of stages on TSS concentration is shown in Figure 4. In OLR of 26.7 gCOD.l^{-1}.d^{-1} TSS decreased from 8 in inlet to 3, 2 and 1.2 g.L^{-1} over first, second and third stages, respectively. In OLR of 47 gCOD.l^{-1}.d^{-1}, the TSS with the same trend dropped from 15 in inlet to 11.3, 8.9 and 7.8 g.L^{-1} in stage number of 1, 2 and 3, respectively. It showed that increasing the amount of organic loading rate; resulted in an increase in TSS concentration. Also about 66% TSS removal was achieved in the first stage in OLR of 26.7 to 40 gCOD.l^{-1}.d^{-1}; in contrary, only 25% TSS removal was attained in OLR of 47 gCOD.l^{-1}.d^{-1} (data is not shown in the figure) which is resulted in an insufficient performance of RBC in TSS removal for high organic loadings compared to low OLRs. The obtained results showed that the most TSS removal completed in the first stage. It is concluded that as the number of the stages proceeded, the TSS concentration decreased. In fact the TSS concentration was strictly dependent on both the organic loading rate and number of stages.

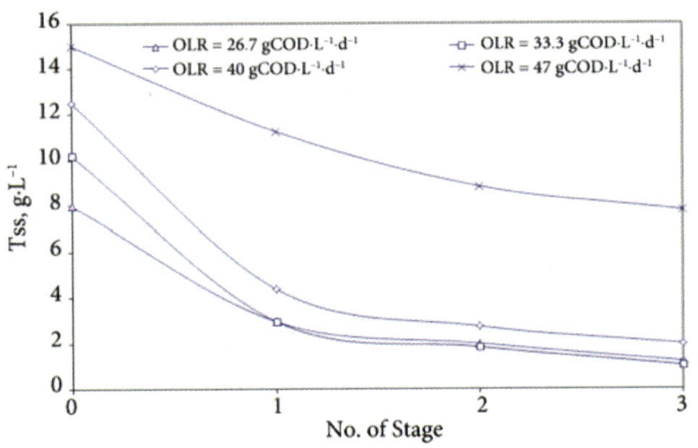

Fig. 4. TSS concentration profiles in different stages of RBC at various OLR

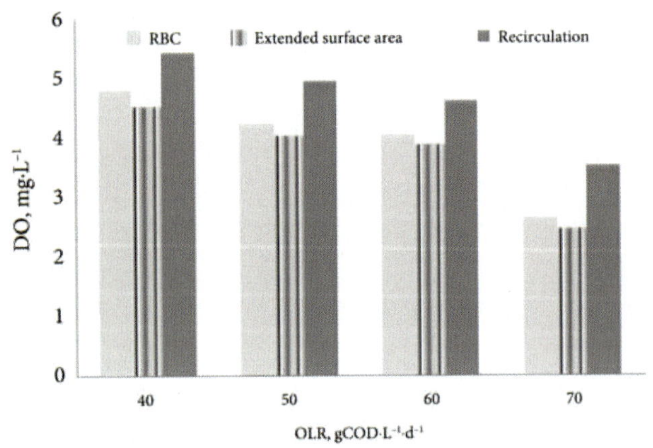

Fig. 5. DO level with respect to OLR for RBC, extended surface and recirculation

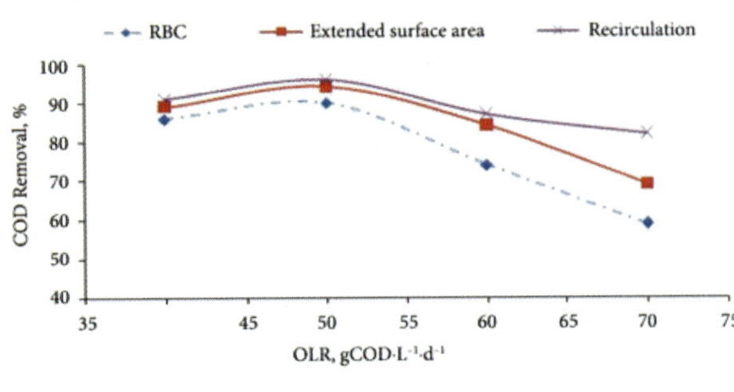

Fig. 6. COD removal with various OLR in RBC, with extended surface and recirculation at HRT of 24 h

2.2.2. Effect of the modified specific surface area and recirculation on the efficiency of the system

Figure 5 shows the effect of extended SSA and recirculation on average DO concentration for OLR values ranging from 40 to 70 gCOD.L^{-1}.d^{-1}. From the obtained results it can be observed that recirculation enhanced the average DO level. In particular, for OLR of 70 gCOD.L^{-1}.d^{-1}, recirculation seems to have a more positive impact on the DO concentration. Also it is observed that the DO level decreased slightly after increasing SSA in compare to initial condition (as shown in Fig. 5). It could be result of sufficient thickness of biofilm as to the formation of new biofilm and also more dissolved oxygen utilization.

As it was mentioned, we can conclude that increasing the SSA can enhance the capability of the treatment system to remove organic substrate. In order to increase the efficiency of the system in case of a high strength wastewater, a 50% recirculation were performed and the SSA was also extended. Figure 6 shows the COD removal efficiency for the given OLRs at HRT of 24, after increasing the surface area and applying the recirculation in compare to initial setup (see Fig. 6).

The COD removal efficiency at HRT of 24 was investigated after the SSA was increased by 10%. From the illustrated data in this figure it can be observed that for the high OLRs, the tolerance between the initial results and those obtained after modification has increased, which is an indication of the improvement of RBC performance at high OLRs. According to the presented data, the maximum COD removal efficiency at OLR of 50 gCOD.L^{-1}.d^{-1} and HRT of 24 in initial RBC, after increasing surface area and recirculation were 89, 94.2 and 96% respectively.

After increasing the surface area, in OLRs lower than 50 gCOD.l^{-1}.d^{-1}, the COD removal efficiency had a slight improvement (about 4%), but in OLR higher than 60 gCOD.L^{-1}. d^{-1}, the removal efficiency was enhanced by about 10%. Also recirculation improved COD removal rate by 12% in OLR of 70 gCOD.L^{-1}.d^{-1}. This could be attributed to the dilution of the substrate concentration applied to the system. More improved efficiency at high OLR is attributed to the presence of high concentrations of DO in the three stages of the system with recirculation compared to the system without recirculation.

Conclusions

From the results it is concluded that a properly operated RBC can effectively be employed for dealing with cheese whey effluents. The maximum COD removal efficiency in the initial configuration was 92.4% which was operated at COD concentration of 50 g.L^{-1} and HRT of 36 h. An increase in OLR (50 to 70 g COD.L^{-1}.d^{-1}), a drop (90 to 59%) in percentage of the COD removal was observed. It shows that oxygen transfer rate through biofilm would cause a limiting effect on the efficiency of substrate degradation. Hence, maximum organic removal rate is attained at the highest loading before exceeding of biofilms mass transfer restrictions. Both modification of SSA of the discs and recirculation had positively improved the COD removal efficiency, especially in higher OLRs. Although increasing the surface area slightly reduced the average DO level, recirculation had evidently enhanced the DO level. Thus, the findings of the present study well demonstrated that the recirculation may be beneficial when the inlet organic loading rate is extremely high.

Acknowledgements

The authors are gratefully acknowledged Biotechnology Research Lab. Noushirvani University of Technology (Babol, Iran) and the Institute of Pardisan Fereidonkenar for the support and facilities provided to conduct present work.

References

APHA. 2005. *Standard methods for the examination of water and wastewater.* American Public Health Association (APHA), Washington, D.C., USA.

Bylund, G.; Pak, T. 2003. *Dairy processing handbook.* Tetra Pak Processing Systems AB Lund.

Carvalho, F.; Prazeres, A. R.; Rivas, J. 2013. Cheese whey wastewater: characterization and treatment, *Science of the Total Environment* 445: 385–396. https://doi.org/10.1016/j.scitotenv.2012.12.038

Chen, Z.; Wen, Q.; Wang, J.; LI, F. 2006. Simultaneous removal of carbon and nitrogen from municipal-type synthetic wastewater using net-like rotating biological contactor (NRBC), *Process Biochemistry* 41(12): 2468–2472. https://doi.org/10.1016/j.procbio.2006.06.003

Ebrahimi, A.; Asadi, M.; Najafpour, G. 2009. Dairy wastewater treatment using three-stage rotating biological contactor (NRBC), *International Journal of Engineering* 22(2): 107–114.

Ebrahimi, A.; Najafpour, G. D.; Mohammadi, M.; Hashemiyeh, B. 2010. Biological treatment of whey in an UASFF bioreactor followed a three-stage RBC, *Chemical Industry and Chemical Engineering Quarterly* 16(2): 175–182. https://doi.org/10.2298/CICEQ100315025E

Ercan, D.; Demirci, A. 2013. Current and future trends for biofilm reactors for fermentation processes, *Critical Reviews in Biotechnology* 35(1): 1–14. https://doi.org/10.3109/07388551.2013.793170

Grady JR, C. L.; Daigger, G. T.; Love, N. G.; Filipe, C. D. 2012. *Biological wastewater treatment.* CRC Press.

Hassard, F.; Biddle, J.; Cartmell, E.; Jefferson, B.; Tyrrel, S.; Stephenson, T. 2015. Rotating biological contactors for wastewater treatment – a review, *Process Safety and Environmental Protection* 94(3): 285–306. https://doi.org/10.1016/j.psep.2014.07.003

Hoyland, G.; Vale, P.; Rogalla, F.; Jones, M. 2010. A new approach to nutrient removal using the HYBACS process, *Proceedings of the Water Environment Federation* 14(7): 81–94. https://doi.org/10.2175/193864710798208421

Janczukowicz, W.; Zieliński, M.; Dębowski, M. 2008. Biodegradability evaluation of dairy effluents originated in selected sections of dairy production, *Bioresource Technology* 99(10): 4199–4205. https://doi.org/10.1016/j.biortech.2007.08.077

Kaewkannetra, P.; Garcia-Garcia, F.; James, A.; Chiu, T. 2009. Effect of surfactants on the stability and filterability of whey suspensions, *Separation and Purification Technology* 66(2): 362–367. https://doi.org/10.1016/j.seppur.2008.12.017

Mba, D.; Bannister, R. 2007. Ensuring effluent standards by improving the design of Rotating Biological Contactors, *Desalination* 208(1): 204–215. https://doi.org/10.1016/j.desal.2006.04.079

Metcalf, E. 2003. *Inc., Wastewater Engineering, Treatment and Reuse.* New York: McGraw-Hill.

Najafpour, D. G.; Naidu, P. N.; Kamaruddin, A. H. 2008. Rotating biological contactor for biological treatment of poultry processing plant wastewater using Saccharomyces cerevisiae, *ASEAN Journal of Chemical Engineering* 2(1): 1–6.

Najafpour, G.; Yieng, H. A.; Younesi, H.; Zinatizadeh, A. 2005. Effect of organic loading on performance of rotating biological contactors using palm oil mill effluents, *Process Biochemistry* 40(8): 2879–2884. https://doi.org/10.1016/j.procbio.2005.01.002

Najafpour, G.; Zinatizadeh, A.; Lee, L. 2006. Performance of a three-stage aerobic RBC reactor in food canning wastewater treatment, *Biochemical Engineering Journal* 30(3): 297–302. https://doi.org/10.1016/j.bej.2006.05.013

Ohl, A. L.; Horn, H.; Hempel, D. 2004. Behaviour of biofilm systems under varying hydrodynamic conditions, *Water Science and Technology* 49(11–12): 345–351.

Prazeres, A. R.; Carvalho, F.; Rivas, J. 2012. Cheese whey management: a review, *Journal of Environmental Management* 110(3): 48–68. https://doi.org/10.1016/j.jenvman.2012.05.018

Reardon, R.; Davel, J.; Baune, D.; Mcdonald, S.; Appleton, R.; Gillette, R. 2013. Wastewater treatment plants of the future: current trends shape future plans, *Florida Water Resources Journal* 1(5): 8–14.

Rodgers, M.; Zhan, X.-M. 2003. Moving-medium biofilm reactors, *Reviews in Environmental Science and Biotechnology* 2(2–4): 213–224. https://doi.org/10.1023/B:RESB.0000040467.78748.1e

Sirianuntapiboon, S.; Chumlaong, S. 2013. Effect of Ni2+ and Pb2+ on the efficiency of packed cage rotating biological contactor system, *Journal of Environmental Chemical Engineering* 1(3): 233–240. https://doi.org/10.1016/j.jece.2013.04.019

Smithers, G. W. 2008. Whey and whey proteins—from "gutter-to-gold", *International Dairy Journal* 18(7): 695–704. https://doi.org/10.1016/j.idairyj.2008.03.008

Vasala, A.; Panula, J.; Neubauer, P. 2005. Efficient lactic acid production from high salt containing dairy by-products by Lactobacillus salivarius SSP. salicinius with pre-treatment by proteolytic microorganisms, *Journal of Biotechnology* 117(4): 421–431. https://doi.org/10.1016/j.jbiotec.2005.02.010

Yorgun, M.; Balcioglu, I. A.; Saygin, O. 2008. Performance comparison of ultrafiltration, nanofiltration and reverse osmosis on whey treatment, *Desalination* 229(1): 204–216.

Atieh EBRAHIMI. She is a lecturer on water and wastewater treatment engineering at Institute of Pardisan Fereidonkenar, Iran. She has completed her MSc degree and also is PhD candidate at School of Civil engineering, Babol Noshirvani University of Technology, where she furthers her research on advanced water and wastewater treatment and waste minimization. She has presented 10 papers on various biological treatment processes at conferences both home and abroad and published 3 ISI journal articles.

Ghasem D. NAJAFPOUR. Professor Ghasem D. Najafpour is a distinguished professor in Chemical Engineering and Chairman of Biotechnology Research Center, Babol Noshirvani University of Technology, Iran. He is an educated scholar with PhD Degree from University of Arkansas, USA with strong background in biological processes. He is deeply involved in research and teaching in biochemical engineering subjects and he has conducted many practical researches in enzyme technology, fermentation processes, biodiesel, biofuel, wastewater treatment and biochemical engineering. In past decades, he has supervised more than 165 master and 28 PhD students. He has published more than 360 papers in international journals and has written 10 books in the field of Chemical Engineering and Biotechnology. In year 2006, he has published his 1st edition of his book with Elsevier entitled "Biochemical Engineering & Biotechnology". The 2nd edition of "Biochemical Engineering & Biotechnology" with 20 chapters is published by Elsevier in Feb. 2015.

Manouchehr NIKZAD. He is graduated, having BSc degree in civil engineering and currently is M.Sc. student in environmental engineering at Babol Noshirvani University of Technology. He has presented 3 research papers on environment engineering subjects at conferences both home and abroad. His research interests include Geoenvironmental Engineering technologies, Environmental Modeling and Optimization, and Green Building Modeling.

EFFECTS OF ORGANIC LOADING RATE AND HYDRAULIC RETENTION TIME ON TREATMENT OF PHENOLIC WASTEWATER IN AN ANAEROBIC IMMOBILIZED FLUIDIZED BED REACTOR

Roya Pishgar[a], Ghasem D. Najafpour[b], Bahram Navayi Neya[a], Nafise Mousavi[a], Zeinab Bakhshi[a]

[a]*School of Civil Engineering, Babol Noshirvani University of Technology, Babol, Mazandaran, Iran*
[b]*Biotechnology Research Center, Faculty of Chemical Engineering, Babol Noshirvani University of Technology, Babol, Babol, Iran*

Abstract. Treatability of phenolic wastewater in an anaerobic immobilized fluidized bed reactor (AIFBR) in consequence of stepwise increment in phenolic load as well as decrease in hydraulic retention time (HRT) was investigated. The experimental data indicated that high degradation efficiencies of phenol and COD in the bioreactor at low HRTs and high organic loading rates were obtained. At constant HRT of 16 h with increase in influent phenol concentration from 98 to 630 mg/l, the average phenol and COD removals were 96 and 88%, respectively. However, further increase in phenol concentration in the feed stream to 995 mg/l resulted in decrease in phenol and COD removal efficiencies to 84 and 79%, respectively. For influent phenol concentration of 995 mg/l, the biogas production rate of 4.55 l/l.d was obtained. As HRT decreased from 3 to 0.15 day, the system showed high stability; influent phenol and COD were removed and reached to average values of 17 and 173 mg/l correspond to the removal efficiencies of about 97 and 90.5%, respectively. The bioreactor experienced a failure with further decrease in HRT to 0.1 day. Biogas production was gradually decreased from 7.04 l/l.d at HRT of 3 days to 2.23 l/l.d at HRT of 0.1 days. The value of the ratio of volatile fatty acids to total alkalinity (VFAs/TA) ranged from 0.03 to 0.24 during the entire course of operation.

Keywords: biotechnologies in environmental engineering, wastewater management.

Reference to this paper should be made as follows: Pishgar, R.; Najafpour, G. D.; Navayi Neya, B.; Mousavi, N.; Bakhshi, Z. 2014. Effects of organic loading rate and hydraulic retention time on treatment of phenolic wastewater in an anaerobic immobilized fluidized bed reactor, *Journal of Environmental Engineering and Landscape Management* 22(1): 40–49.

Introduction

One of the most important challenges of modern world is to preserve the environment against diverse pollutions. Progressive industrialization is the major reason of releasing different types of contaminants into ecosystem (Kwon *et al.* 2009; Mousavi *et al.* 2011). Phenol and its derivatives which are known as toxic carcinogenic aromatics abundantly occur in environment by disposal of phenol-containing effluents of number of industries. Some of these industries are petroleum refineries, steel plants, pulp and paper, pharmaceuticals, manufacturing of synthetic chemicals, coal conversion, coke oven, etc (Arutchelvan *et al.* 2006; Fang *et al.* 2006; Carbajo *et al.* 2010). High concentrations of phenol (up to 17,500 mg/l) have been detected in various industrial wastewaters (Carbajo *et al.* 2010). The reported range of phenol concentrations in various industrial wastewaters are summarized in Table 1 (Carlo *et al.* 2008). Specific low concentration of phenol poses significant health and environmental risks for human (Rittmann *et al.* 2001; El-Naas *et al.* 2009). Hence, phenolic compounds have been regarded as priority organic pollutant by US Environmental Protection Agency (U. S. EPA). In addition, World Health Organization (WHO) has defined permissible level of 1 μg phenol per liter of drinking water (Nuhoglu *et al.* 2005; Yan *et al.* 2006). Considering noxious affects of phenol, phenolic wastewaters must be treated before being discharged to the surrounding environment (Yan *et al.* 2005).

Different physicochemical or conventional biological methods can be employed to reduce phenol content below permitted level (Gonzalez *et al.* 2001a; Carlo *et al.* 2008). Bioprocesses are proven to be cost-effective environmentally-friendly approaches in which complete mineralization of organic contaminants may occur

Corresponding author: G. D. Najafpour
E-mail: najafpour@nit.ac.ir

Table 1. Phenol concentration in various industrial wastewaters

Wastewater	Phenol concentration (mg/l)
Refineries	6–500
Coking operations	28–3900
Coal processing	9–6800
Petrochemicals	2.8–1220
Pulp and paper	0.1–1600
Pharmaceutical	0.1–1600
Synthetic wastewater (this study)	98–995

(Moussavi *et al.* 2009; Firozjaee *et al.* 2011). Moreover, treatment of phenolic wastewater using physicochemical techniques may result in production of toxic by-products such as poisonous poly chlorinated phenols. Therefore, phenol biodegradation is preferred in comparison to chemical oxidizing or physical treatment as transferring pollution from one phase to another (Arutchelvan *et al.* 2006; Bakhshi *et al.* 2011).

Phenol removal in presence of oxygen is often preferred due to high phenol inhibition on anaerobic processes. However, complete degradation of phenol can happen in anaerobic systems by a well acclimated culture (Moussavi *et al.* 2009). Aerobic treatment of phenolic wastewater is extensively discussed in the literature (Gonzalez *et al.* 2001a; Gonzalez *et al.* 2001b; Alemzadeh *et al.* 2002; Tziotzios *et al.* 2005; Vázquez *et al.* 2006; Nair *et al.* 2007; Moussavi *et al.* 2009). Though, in late decades there is a widespread tendency to apply anaerobic biological treatment systems for treating different types of wastewaters (Şen *et al.* 2003; Bertin *et al.* 2004; Subramanyam *et al.* 2008; Asadi *et al.* 2009a; Asadi *et al.* 2009b; Carbajo *et al.* 2010; Saghafi *et al.* 2010; Wang *et al.* 2010). Anaerobic biodegradation is being favored to aerobic systems due to low energy and nutrient requirements and less sludge yield (Metcalf *et al.* 2003; Veeresh *et al.* 2005; Carbajo *et al.* 2010). Besides, methane can be produced throughout anaerobic digestion which is considered as an innovative energy source (Metcalf *et al.* 2003; Bakhshi *et al.* 2011). Table 2 listed the conducted researches on anaerobic biodegradation of phenolic wastewaters reported in the literature. Among several high rate biological processes, fluidized bed reactor (FBR) is an advanced technique with relatively low hydraulic retention times. Considering slow growth of anaerobic consortia, FBR as a novel technology provides appropriate bioreactor volume as well as high active biomass concentration. Due to

Table 2. Studies of anaerobic biodegradation of phenolic compounds

Substrate	Phenolic compound (mg/l)	COD (mg/l)	OLR (g COD/ l.d)	Phenolic compound removal (%)	COD removal (%)	HRT (day)	Reactor[e]	Ref.
p-nitrophenol	10–700	3000	2.89×10^{-4}	≥ 99	≥ 90	10.38	ABR	Kuscu *et al.* 2005
Resorcinol	212–1038	443–1892	1.91–12.34	81–97	78–94	0.125–0.25	FFFB	Latkar *et al.* 2003
Catechol	97–1018	188–1897	0.88–13.88	57–91	56–87			
Hydroquinone	158–903	426–1777	1.88–12.05	32–79	29–83			
OMW[a]	720–1300	10250–25500	4.34–17.70	66–75 5–32	32–65 4–18	0.042	UAPB-GAC UAPB-SB	Bertin *et al.* 2004
Phenol	100–1000	250–2500	0.25–2.5	94	88	1	UAPB	Bakhshi *et al.* 2011
Phenol	500–1000	5000	–	≥ 83	≥ 60	0.5	EGSB	Scully *et al.* 2006
CGW[b]	143–540	1000–2500	1.0–2.5	63 ≥	60 ≥	1	UASB	Wang *et al.* 2010
4-chlorophenol	40	1124–1738	1.7–5.3	82–90	≥ 90	0.25–0.67	UASB	Majumder *et al.* 2008
Catechol	100–1500	–	3.77–10.52	≥ 95	≥ 95	0.33	UASB	Subramanyam *et al.* 2007
Phenol	252–1176	1000–6500	2–18	84–99.9	74–91.3	0.33–1	UASB	Gali *et al.* 2006
Phenol	2080 ≥	–	15.46 ≥	≥ 99.9	–	0.43	AFBR	Carbajo *et al.* 2010
PCP[c]	1333 ≥	–	–	≥ 99.9	≥ 98	0.39–1.55	AFBR	Khodadoust *et al.* 1997
PCP + PAHs[d]	100 + 0.5–690	–	–	≥ 99.8	–	0.097–0.39	AFBR	Koran *et al.* 2001
Phenol	98–995	234–2390	0.35–12.25	84–98	79–91	0.15–3	AIFBR	This study

[a] Olive mill wastewater
[b] Coal conversion wastewater
[c] Pentachlorophenol
[d] Poly aromatic hydrocarbons (PAHs) such as naphthalene, acenaphthene, pyrene and benzo(b)fluoranthene
[e] ABR: Anaerobic baffled reactor, FFFB: Fixed film-fixed bed bioreactor, UAPB-GAC: Upflow anaerobic packed bed reactor filled with granular activated carbon, UAPB-SB: Upflow anaerobic packed bed reactor filled with silica beads, EGSB: Expanded granular sludge bed-based bioreactor, UASB: Upflow anaerobic sludge blanket, AFBR: Anaerobic fluidized bed bioreactor; AIFBR: Anaerobic immobilized fluidized bed bioreactor

adhesion of microorganisms to a solid support, the sludge retention time and the hydraulic residence time are uncoupled in this process. Thus, application of high organic loading rates (OLR) is feasible in such system (Şen *et al.* 2003; Carbajo *et al.* 2010; Bakhshi *et al.* 2011). FBR has been successfully used to treat a broad spectrum of either readily or hardly biodegradable wastes in addition to hazardous pollutants (Borja *et al.* 1995; Perez *et al.* 2007; Lohi *et al.* 2008; Sowmeyan *et al.* 2008; Kuyukina *et al.* 2009).

The literature survey reveals that treatment of phenolic wastewater in anaerobic immobilized fluidized bed reactor (AIFBR) is rarely carried out. In this study performance of a laboratory-scale AIFBR on treatment of synthetic wastewater containing high concentrations of phenol was evaluated. The present research was aimed at investigating the effect of increasing phenolic and organic loading rate and decreasing HRT on phenol and COD removal as well as biogas production. Simultaneously variation of pH, VFAs and alkalinity of wastewater were monitored in the entire operation period. The results may be customized for optimizing the operation of a full-scale wastewater treatment plant treating phenolic effluents.

1. Materials and methods

1.1. Fluidized bed bioreactor

Continuous biodegradation of phenol was accomplished using an anaerobic immobilized fluidized bed reactor (AIFBR). A laboratory-scale configuration was used as experimental setup (Fig. 1). The system consisted of a 7.4 cm internal diameter and 114.5 cm long cylindrical Plexiglas column to form the 5.24 l reactor. The reactor was filled with mature immobilized microorganisms entrapped in calcium alginate gel beads up to 30% of its total volume. The working volume of the reactor was about 3.5 l. The flow was continuously fed into the reactor from a conical section at the bottom that could provide a homogenous fluidization. Recycle flow was drawn from the top of the reactor by a peristaltic pump (BT600, Prefluid, China). The synthetic phenol solution along with recycle stream was fed upward into the reactor. Synthetic phenolic wastewater was transferred into reactor using a peristaltic pump (BT100, Prefluid, China). Bed expansion in the bioreactor was kept around 100% with the aid of the recycle flow. The system was embedded in a transparent talc encasement to provide constant temperature at about 25 °C. Produced biogas was released from effluent by aid of a 1000 ml separation funnel which was filled with polyethylene supports up to half of its volume. The liberated biogas was then led to a glass cylindrical collecting tube where the volume of gas was defined by displacement of inside water. The collecting tube was fabricated with an internal diameter of 15 cm and height of 150 cm. Two sampling ports were installed on AIFBR for monitoring influent and effluent characteristics. The startup and operation periods of AIFBR are illustrated below.

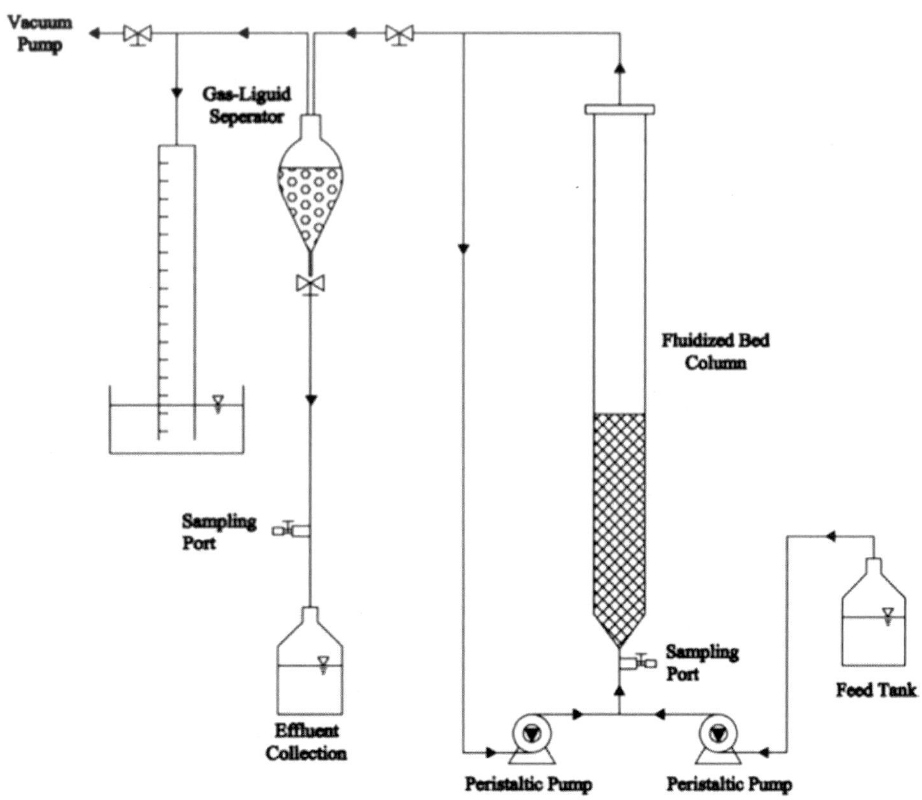

Fig. 1. Schematic diagram of a lab-scale AIFBR used as experimental setup in continuous operations

1.2. Inoculums

The seed inoculums were isolated from the effluents of two industries including coke oven, Isfahan, Iran and pulp and paper, Sari, Iran as well as activated sludge taken from wastewater treatment facility of the latter. An acclimation period of 5 months was implemented to adapt the mixed culture to phenol as sole carbon/ energy source. Adaption steps were performed in 250 ml Erlenmeyer flasks under anaerobic condition in desiccators. During this period, the microorganisms were exposed to increasing phenol concentrations. At the beginning, low concentration of phenol i. e. 100 mg/l was introduced to the culture in presence of 100 mg/l glucose as co-substrate. Upon complete removal of 100 mg/l phenol within 20 days, exposure of the culture to solely phenolic media was started. As phenol removal efficiency of over 95% was observed, microorganisms were inoculated to a fresh medium containing higher phenol concentration. Inoculums concentration was about 10% (v/v). Finally, in duration of 12 days, the adapted culture was capable to degrade 96 and 10% of 700 and 1000 mg/l phenol, respectively.

1.3. Immobilization protocol

The adapted microorganisms were harvested at the stationary phase, and then entrapped and immobilized in calcium-alginate gel beads. After adding sterilized sodium alginate solution (2% w/v) into cell suspension, the mixture was introduced drop wise into sterilized calcium chloride solution (6% w/v); thus, immobilization procedure was accomplished. Sodium alginate (SIGMA) solution was prepared by adding 10 g of the alginate powder was added to 500 ml of distilled water. 60 g of anhydrous granulated form of calcium chloride (Sharlau, Spain) was separately added to 1 l of distilled water to prepare calcium chloride bath. A peristaltic pump (BT100, Prefluid, China) and a silicon tube with internal diameter of 1.1 mm were used to instill the mixture of sodium alginate solution and cell suspension in the calcium chloride bath. Spherical bio-particle beads with average diameter of 3–4 mm were ultimately obtained.

1.4. Wastewater characteristics

The synthetic phenolic wastewater was composed of phenol, yeast extract as well as a basal medium which contained all the necessary micro- and macro-elements. The composition of basal medium was (in mg/l): K_2HPO_4, 522.54; KH_2PO_4, 408.27; and NH_4Cl, 200; NaCl, 200; KCl, 200; $CaCl_2.2H_2O$, 150; $MgCl_2$, 100; $MgSO_4.7H_2O$, 5; $FeSO_4.7H_2O$, 10; $CoCl_2.2H_2O$, 0.2; $NiCl_2.2H_2O$, 0.2; $ZnCl_2$, 0.2; $CuCl_2.2H_2O$, 0.2; $MnCl_2.4H_2O$, 0.2; $NaMoO_4.2H_2O$, 0.5; H_3BO_3, 0.2; $NaHCO_3$, 400. Concentration of yeast extract was kept constant in the feed at 800 mg/l. By addition of 400 mg/l $NaHCO_3$, pH of influent was maintained at 6.9 ± 1. To prepare synthetic wastewater, phenol and nutrients stock solutions were diluted with distilled water to appropriate concentrations. All of the chemicals were supplied by Merck (Darmstadt, Germany).

1.5. Startup and operation strategies

The startup period was carried out to acquaint the immobilized culture to continuous condition in phenol removal plus evaluating the effect of increasing organic load on phenol biodegradation. During startup period, the bioreactor was run at constant hydraulic retention time of 16 h, and phenol influent concentration was gradually increased from 98 to 995 mg/l in stepwise manner (equivalent to phenol loading rate of 0.15 to 1.48 g phenol/l.d). Concentration of corresponding COD varied in the range of 234 to 2390 mg/l. The startup period was implemented in 5 operation phases for duration of 65 days. Initially, phase 1 (0–5 days) was performed, where phenol feed and OLR were 98 mg/l and 0.35 g COD/l.d, respectively. Second phase started on 6th day and lasted 8 days with influent phenol concentration of 236 mg/l and OLR of 0.85 g COD/l.d. At phase 3 (day 15 to day 27), initial phenol concentration was increased to 440 mg/l (OLR of 1.58 g COD/l.d). In the next 16 days, influent phenol concentration was kept constant at 628 mg/l which corresponds to OLR of 2.24 g COD/l.d. In the final phase between day 44 to day 65 (for the last 21 days) phenol concentration increased to 995 mg/l (OLR of 3.56 g COD/l.d). Under steady-state condition, phenol in feed stream was gradually increased while more than 96% of phenol was removed.

In the next stage, the effect of hydraulic retention time (HRT) on the process performance was evaluated. Eight experimental runs were carried out with various HRTs including 3, 2, 1, 0.5, 0.3, 0.2, 0.15, 0.1 day. This stage of experiment lasted 77 days; the values of phenol and COD in the influent were maintained at 720 and 1830 mg/l, respectively. The bioreactor was operated at OLRs of 0.61, 0.92, 1.83, 3.72, 6.15, 9.20, 12.25, 18.34 g COD/l.d, which corresponded to the above HRT values. The phenol loading rate was 0.24–7.22 g phenol/l.d. Descending order of HRT was applied in this stage of experiment. Steady-state condition was obtained when phenol concentration in the effluent remained practically constant.

1.6. Sampling and Analytical methods

Samples of influent and effluent of the bioreactor were daily analyzed in duplicate. Parameters including chemical oxygen demand (COD), residual phenol, volatile fatty acids (VFAs), total alkalinity (TA) and pH were determined. COD was measured by closed reflux method. Residual concentrations of phenol were defined by direct photometric method using 4-aminoantipyrene and alkaline potassium ferricyanide. Direct titration was applied for determination of VFAs and TA. The above procedures were performed in accordance with Standard Methods for Examination of Water and Wastewater (Eaton *et al.* 2005). Produced gas was determined by water displacement. pH measurements were carried out with a pH meter (pH 212, HANNA, Germany).

2. Results and discussion

2.1. Startup period; effect of phenolic load

Response of the AIFBR to incremental increase of phenolic load during the startup period was evaluated. The influent and effluent phenol and corresponding COD concentrations, phenol and COD removal, biogas production, VFAs, total alkalinity, and pH are presented in Figure 2. Startup period was completed for duration of 65 days. This entire operation period was divided into five phases, where stepwise increase in phenol feed from 98 to 995 mg/l at a constant HRT of

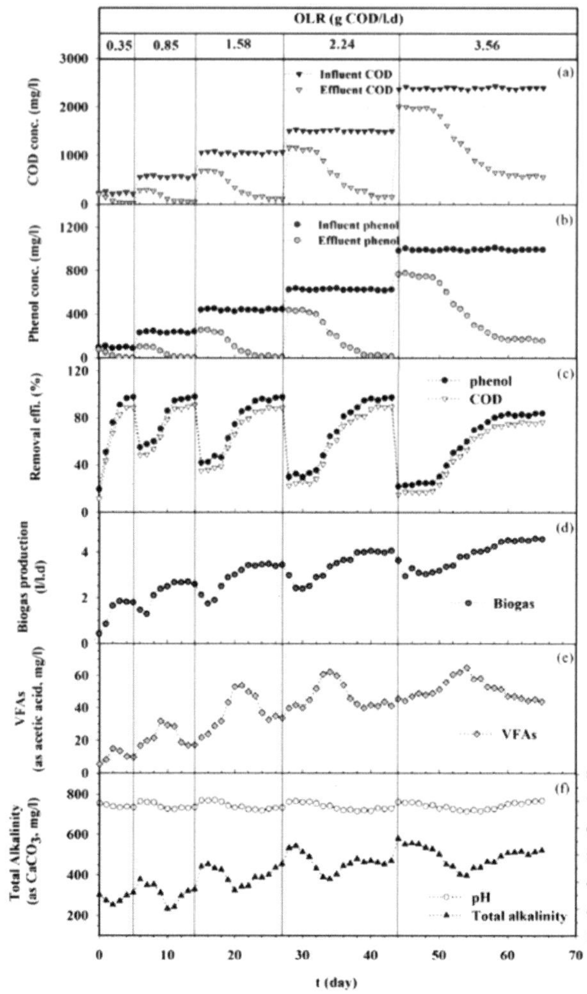

Fig. 2. Response of the AIFBR to increment of phenolic load during startup

16 h was investigated. Figure 2(a–d) shows phenol and COD removal and biogas production during the startup period.

In Phase 1 (days 1–5), the AIFBR was fed with phenol and corresponding COD of 98 and 234 mg/l, respectively. The value of OLR was 0.35 g COD/l.d. Phenol degradation was immediately started. The bioreactor could remove about 44 mg/l of input phenol at the end of the first day of the experiment. Finally, 98% of influent phenol and 89% of influent COD were removed in this phase of experiment. Maximum production of biogas was observed at the final days of the phase when phenol concentration in the effluent remained constant at about 2.6 mg/l. At day 4, the volumetric biogas rate was 1.81 l/l.d.

Phase 2 was performed within day 6 to day 14 with influent phenol concentration of 236 mg/l, correspondent to COD concentration and OLR of 556 mg/l and 0.85 g COD/l.d, respectively. Although the microorganisms were strictly adapted to phenol concentration of 700 mg/l, a lag period in degradation plus a sudden reduction in removal efficiencies of phenol and COD were diagnosed at the beginning of every single phase.

With increase in phenol concentration up to about 230 mg/l, phenol and COD removal efficiencies reduced to 55 and 48%, respectively. During initial stage of second phase (day 6 to day 9), the phenol concentration in the effluent only decreased to about 102 mg/l. Also biogas production rate decreased from 1.79 l/l.d at day 5 to 1.29 l/l.d at day 7 which is equivalent to 28% reduction.

At day 11 of experiment, phenol concentration in the effluent degraded to 12.5 mg/l with removal efficiency of 95%. At the end of this phase, COD removal of 89% and volumetric biogas rate of 2.67 l/l.d were achieved. In third phase (days 15–27), the OLR was maintained at 0.66 g COD/l.d with input phenol concentration of 441 mg/l and COD of 1050 mg/l for duration of 12 days. As the influent concentration of phenol increased, the initial lag period in phenol and COD degradation was slightly prolonged. At the initial day of this phase, phenol and COD removal efficiencies drastically decreased to 42 and 35%, respectively. During the first 4 days of 3rd phase, only 46% of influent phenol was removed; thence the phenol and COD removal efficiencies increased. Moreover, volumetric biogas rate diminished and reached 1.74 l/l.d at day 16. At day 24 of the experiment, phenol and COD removal efficiencies enhanced to 96 and 87%, respectively; accordingly the effluent concentrations of 32 mg/l phenol and 196 mg/l COD were determined in this day. In subsequence, the system experienced steady-state condition untill the day 27th of operation. At the end of this phase, low concentrations of phenol (15 mg/l) and COD (141 mg/l) in the effluent were obtained. Recovery in biogas production from 1.74 to 3.42 l/l.d occurred within days of 16 to 22 of the experiment. Biogas production achieved the value of 3.45 l/l.d at the end of this phase. The increment in organic load obviously resulted in further growth of the microorganisms. It seemed that consequent increase in density of the alginate beads had happened. This phase required high recycle flow rate for 100% fluidization of the support beads could prove the assumption of beads weight increased.

Phase 4 of startup operation was carried out within days 28 to 44. This phase lasted 17 days; influent phenol concentration was increased to 628 mg/l. Increment in phenol feed led to corresponding COD concentration of 1520 mg/l and OLR of 2.24 g COD/l.d. In this phase, phenol degradation rate decreased in compare to previous phases. It indicated the substrate inhibition which was imposed by high phenol concentration on the metabolism of the culture. At day 29, the effluent concentrations of 439 mg/l phenol and 1165 mg/l COD were observed which were equivalent to removal efficiencies of 30 and 22%, respectively. Removal efficiencies slightly increased to 36% for phenol and 28% for COD at day 32. With further increase in removal efficiencies, 64% of influent phenol concentration was removed and reached 224 mg/l phenol in the effluent at the day 34th of experiment. Maximum phenol and COD removal of 95 and 87% were accomplished at the end of this phase. Biogas production rate represented descending trend throughout two initial days of 4th phase. However, it recovered and obtained the steady value of 4.08 l/l.d within days 37 to 43.

Final phase of startup period was started at day 44 and lasted till day 65. In this phase, the influent phenol and COD concentration were 995 and 2390 mg/l, respectively. Therefore, the value of OLR was kept constant at 3.56 g COD/l.d. Due to intense inhibition of high phenol concentration, its removal efficiencies showed a dramatic decrease to 22% at first day of this phase. At the same day, COD removal of 15% was obtained. For 7 days at the beginning of this phase, no significant reduction in phenol and COD concentration was observed. At day 56, phenol removal efficiency increased to 72% which corresponded to effluent phenol and COD concentrations of 277 and 839 mg/l, respectively. The AIFBR completed final phase in stable condition. Steady-state condition was obtained at the time interval between the days 58 to 65 with a phenol effluent concentration of 140 mg/l correspond to removal efficiencies of 84% for phenol and 79% for COD. The volumetric biogas rate of 4.55 l/l.d was achieved at the end of 5th phase which was the highest value for the startup period. Carbajo and coworkers (Carbajo *et al.* 2010) achieved the maximum volumetric biogas rate of 4.4 l/l.d with average yield of 0.28 l CH_4/g COD removed in an AFBR while the OLR was progressively increased to 15.46 g COD/l.d. However, they observed that with increase in OLR biogas composition was changed; the percentage of CO_2 increased as the CH_4 percentage decreased.

2.2. Variation of VFAs during startup period

The concentrations of effluent volatile fatty acids measured as acetic acid during phenolic waste treatment are presented in Figure 2e. Stepwise increase in influent phenol led to ascending attitude in VFAs concentration at the beginning of every operation phase; subsequently after obtaining a peak value, a gradual decrease was followed in VFAs value. From one phase to another, the peak concentration of VFAs was increased as phenolic load was added. Decrease in VFAs concentration may be correlated to the methanogenic stage when VFAs and organic compounds were utilized and stable end product of methane and carbon dioxide were formed. During startup operation, VFAs concentration varied in the range of 5.5 to 65 mg/l. At the end phase of startup, the highest value of effluent VFAs i.e. 65 mg/l was observed. Bakhshi *et al.* (Bakhshi *et al.* 2011) reported the range of 34–64.7 mg/l for VFAs concentration as the value of OLR increased from 0.25 to 2.5 g COD/l.d during phenol treatment in UAPB. Subramanyam and Mishara (Subramanyam *et al.* 2007) found out that the value of VFAs which was produced during catechol degradation in a UASB varied from 15 to 65 mg/l. However, in presence of resorcinol and catechol, VFAs concentration of 30 to 63.5 mg/l was observed in the UASB (Subramanyam *et al.* 2008); that may be due to intense toxicity of two inhibitory substrates and retardation in methanogenic activity which resulted in consequent accumulation of VFAs. In this period of the operation, the ratio of VFAs/TA varied between 0.03–0.16. It is according to the reported data in literature (Subramanyam *et al.* 2007; Subramanyam *et al.* 2008; Bakhshi *et al.* 2011).

2.3. pH and alkalinity during startup period

Figure 2f shows the variation of pH and total alkalinity (TA) in the effluent of AIFBR during startup operation period. Anaerobic systems are sensitive to pH variation. Production of organic acids and VFAs either in hydrolysis or acidogenesis periods of anaerobic digestion can lead to acidic condition (Rittmann *et al.* 2001; Bakhshi *et al.* 2011). The pH values of below or above the range of 6.5–7.1 may adversely affects the biological process by disturbance and interruption in bacterial activity, especially in methanogenesis (Subramanyam *et al.* 2008). Thus, the value of pH in AIFBR was adjusted between 6.8 ± 0.2 by adding $NaHCO_3$ with the concentration of 400 mg/l. Subramanyam and Mishara (Subramanyam et al. 2007) observed instability in the pH of a UASB degrading phenolic waste in absence of alkalinity. The effluent alkalinity as $CaCO_3$ in the AIFBR was 233 to 579 mg/l. As VFAs concentration increased at the beginning of every phase of startup operation, the effluent alkalinity contributed a diminution, though, it was enhanced later.

2.4. Effect of HRT on the AIFBR performance

In this period of operation, the AIFBR was continuously fed with a solution of 720 mg/l phenol and corresponding COD of 1830 mg/l for 77 days. The feed with phenol concentration of 700 mg/l was well treated due to high performance of the AIFBR at the similar phenolic load during the startup period. The HRT initially was fixed at 3 days with OLR of 0.61 g COD/l.d; it was progressively declined from 3 to 0.1 days in eight experimental runs between the days 65 to 142 of the experiment. As a result, the OLR ranged between 0.61–18.34 g COD/l.d. For each HRT, removal capacity of phenol and COD in the bioreactor were monitored periodically. Once phenol and COD concentrations in the effluent remained constant, HRT was decreased to a lower value.

Figure 3 shows the performance of the AIFBR at different HRTs, in terms of the influent and effluent concentrations and removal efficiencies of phenol and COD, biogas production, the effluent VFAs, pH and alkalinity. The trend of organic load reduction with respect to time in term of decreasing HRT is shown in Figure 3. Reduction of HRT means the bioreactor operates faster. In fact, this figure represents the actual data obtained with respect to HRT.

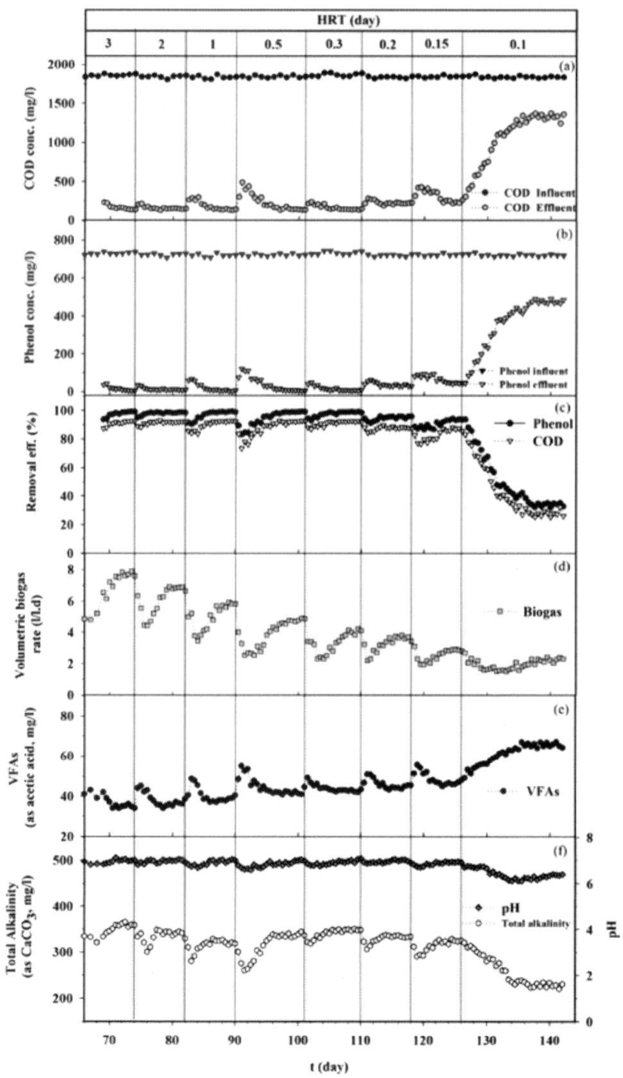

Fig. 3. Effect of decreasing retention time on the AIFBR performance

In three initial runs of the experiments when HRT varied between 1–3 days, the AIFBR presented an appropriate performance; the bioreactor immediately reached a very stable condition due to long HRTs. Finally, the effluent phenol and COD concentrations found to be less than 7 and 145 mg/l, respectively. In these three sets of experimental runs, the removal efficiencies of over 98% for phenol and above 91% for COD were achieved.

Nevertheless, the system required a prolonged time interval for removing phenol from the effluent at HRT of 0.5 day. With decrease in the value of HRT to 0.5 day, the bioreactor showed instability in 6 initial days of the phase (days 90–96). The concentration of phenol in the effluent increased from 8 mg/l at day 90 to 121 mg/l at day 91. At the same day (91st day), the effluent concentration of 488 mg/l for COD was determined. However, the AIFBR finally reached steady-state condition between the days of 96 to 101. At HRT of 0.5 day, the influent phenol and COD were removed with efficiencies of 97 and 90%, respectively. During the experimental runs of 5 to 7 (HRT of 0.3 to 0.15 day), the bioreactor was able to obtain high removal efficiencies for phenol and COD with average values of 95 and 89%, respectively. However, the stable removal efficiency of phenol and COD at the end of each run were progressively increased with further decrease in the value of HRT. For the HRT of 0.15 days (OLR of 12.25 g COD/l.d), the performance of the bioreactor was initially disturbed for 4 days; the phenol and COD concentrations of effluent were 96 and 420 mg/l, respectively. Although the system showed an appropriate behavior in phenol elimination at day 123; but a decrease in final removal efficiencies was observed for phenol (93%) and COD (86%) rather than former runs. Residual phenol of 48 mg/l and COD of 235 mg/l were determined at the HRT of 0.15 days. However, as the value of HRT was declined to 0.1 day, the bioreactor intensely experienced an unstable condition. At the end run, progressive increase in undegraded phenol and COD concentrations in the effluent was observed and biomass washout rapidly occurred. The average concentration of 17 mg/l for phenol and 173 mg/l for COD in the effluent of the AIFBR was detected for the HRTs of 3 to 0.15 days. Biogas production was continuously monitored when the HRT was decreased during 77 days of the experiment (Figure 3d). As the HRT decreased and correspondent increase in OLR happened, biogas production rate was diminished. When HRT was fixed to a lower value, a sharp drop in biogas production occurred and volumetric biogas rate went through a minimum value. However, it subsequently followed an increasing trend to a stable condition. The maximum biogas production e.g. 7.04 l/l.d was observed in the first run with the HRT of 3 days. At the end of run, no significant biogas production (1.72–2.38 l/l.d) was observed due to severe instability in the performance of the AIFBR.

The AIFBR responded successfully to phenol biodegradation at high OLRs and low HRTs in compare to other studies in the literature (see Table 2). The long HRT of 10.38 days was applied in an ABR for p-nitrophenol degradation where removal efficiency of 99 was achieved at loading rate as high as 33.9 g p-nitrophenol /m^3.d. Khodadoust *et al.* (Khodadoust *et al.* 1997) achieved 99.9% removal of pentachlorophenol in anaerobic granulated activate carbon fluidized bed reactor (GAC-AFBR) at an empty bed contact time as low as 9.3 h. In presence of poly aromatic hydrocarbons, high degradation efficiency of pentachlorophenol (i.e. 99.8%) also occurred in the GAC-AFBR with 46.5% average conversion to its dechlorination intermediates. Also low concentrations of influent poly aromatic hydrocarbons were indentified in the effluent (Koran *et al.* 2001). Adequate treatment of phenolic wastewater was reported in an AFBR at constant HRT of 0.43 days and OLR of up to 15.46 g COD/l.d by Carbajo *et al.* (Carbajo *et al.* 2010).

2.5. Effect of HRT on the Variation of VFAs

The variation of VFAs in the effluent between the days of 65 and 142 are depicted in Figure 3e. VFAs concentration was progressively increased as the HRT decreased. Increase in the concentration of VFAs at the beginning of each run could be explained by accumulation of organic acids due to facing with a higher value of HRT. But, the system was later adapted to the new situation when HRTs ranged between 0.15–3 days and reached to a very stable condition at the end of any individual run. The effluent VFAs varied in the range of 36–66 mg/l. The maximum value of VFAs as acetic acid, i.e. 66 mg/l, was obtained at HRT of 0.1 days; that may be due to the failure in the AIFBR. The parameter VFAs/TA was between 0.09–0.2 as the HRT decreased from 3 to 0.15 days, though it increased to 0.24 at the final run which indicated high concentrations of VFAs as well as low activity of methanogens.

2.6. Effect of HRT on the pH and alkalinity

The value of pH and alkalinity relevant to decrease in HRT are shown in Figure 3f. Within the runs of 1 to 7 with HRTs of 0.15–3 days, when the system presented a favorable behavior in phenol removal, pH and alkalinity of the effluent was in the range of 6.59–7.11 and 261–362 mg/l, respectively. At the end run with HRT of 0.1 days, the value of pH dropped to 6.11 at day 135. Alkalinity concentration at day 126 decreased from 325.5 to 227 mg/l at day 140. Addition of 400 mg/l $NaHCO_3$ was accomplished for the pH stability, though, addition of this content of alkalinity could not counterbalance the acidic condition at HRT of 0.1 days because of the high accumulation of VFAs.

Conclusion

1. Phenol containing wastewater was effectively treated in as anaerobic immobilized fluidized bed reactor (AIFBR) at high organic loading rates (i.e. 0.35–12.25 g COD/l.d) with a culture entrapped in calcium alginate beads as biocatalyst.
2. At startup period in which phenolic load was stepwise increased, over 95% of phenol and 87% of COD was removed at phenol loading of 98–630 mg/l, OLR of 0.35–2.44 g COD/l.d and phenol loading rate of 0.15–0.94 g phenol/l.d. However, a decrease in phenol and COD removal to 84% and 79% at phenol feed of 995 mg/l was observed due to intense substrate inhibition.
3. The highest biogas production of 4.55 l/l.d occurred at 995 mg/l phenol during startup.
4. In this period, VFAs, pH and alkalinity in the effluent of the AIFBR varied between 5.5–65 mg/l, 6.5–7.1 and 233–579 mg/l, respectively. The value of VFAs/TA ratio was 0.03 to 0.16.
5. The AIFBR was favorably tolerated the decrease in hydraulic retention time from 3 days to 0.15 day. In this operation period, removal efficiency of 93–98% for phenol and 86–91% for COD was obtained for constant phenol loading of 720 mg/l and OLRs of 0.61–12.25 g COD/l.d. However, further decrease of HRT to 0.1 days (OLR of 18.34 g COD/l.d) led to instability and failure in reactor performance.
6. For HRT of 0.1 days, VFAs/TA was 0.24 which showed a disruption in anaerobic process due to low HRT.

Acknowledgment

The authors are grateful for the facilities provided to carry out present research in Biotechnology Research center, Babol Noshirvani University of Technology.

References

Alemzadeh, I.; Vossoughi, F.; Houshmandi, M. 2002. Phenol biodegradation by rotating biological contactor, *Biochemical Engineering Journal* 11(1): 19–23. http://dx.doi.org/10.1016/S1369-703X(02)00011-6

Arutchelvan, V.; Kanakasabai, V.; Elangovan, R.; Nagarajan, S.; Muralikrishnan, V. 2006. Kinetics of high strength phenol degradation using *Bacillus brevis*, *Journal of Hazardous Materials* 129(1–3): 216–222. http://dx.doi.org/10.1016/j.jhazmat.2005.08.040

Asadi, M.; Ebrahimi, A.; Najafpour, G. H. D. 2009a. Dairy wastewater treatment using three-stage rotating biological contactor (NRBC), *International Journal of Engineering* 22(2): 107–114.

Asadi, M.; Najafpour, G. D.; Hashemiyeh, B. A.; Mohammadi, M. 2009b. Removal of Acetone from Contaminated Air in Biofilter using *Pseudomonas putida*, *American–Eurasian J. Agric. & Environ. Sci.* 5(5): 712–719.

Bakhshi, Z.; Najafpour, G.; Navayi, N. B.; Kariminezhad, E.; Pishgar; Moosavi, N. 2011. Recovery of UAPB from high organic load during startup for phenolic wastewater treatment, *Chemical Industry and Chemical Engineering Quarterly* 17(4): 517–524. http://dx.doi.org/10.2298/CICEQ110428037B

Bertin, L.; Berselli, S.; Fava, F.; Petrangeli-Papini, M.; Marchetti, L. 2004. Anaerobic digestion of olive mill wastewaters in biofilm reactors packed with granular activated carbon and "Manville" silica beads, *Water Research* 38(14–15): 3167–3178. http://dx.doi.org/10.1016/j.watres.2004.05.004

Borja, R.; Banks, C. J. 1995. Response of an anaerobic fluidized bed reactor treating ice-cream wastewater to organic, hydraulic, temperature and pH shocks, *Journal of Biotechnology* 39(3): 251–259. http://dx.doi.org/10.1016/0168-1656(95)00021-H

Carbajo, J.; Boltes, K.; Leton, P. 2010. Treatment of phenol in an anaerobic fluidized bed reactor (AFBR): continuous and batch regime, *Biodegradation* 21(4): 603–613. http://dx.doi.org/10.1007/s10532-010-9328-1

Carlo, R.; Laura, A. 2008. Technologies for the removal of phenol from fluid streams: a short review of recent developments, *Journal of Hazardous Materials* 160(2–3): 265–288.

Eaton, A.; Franson, M. 2005. *Standard methods for the examination of water & wastewater*, Amer. Public Health Assn.

El-Naas, M.; Al-Muhtaseb, S.; Makhlouf, S. 2009. Biodegradation of phenol by *Pseudomonas putida* immobilized in polyvinyl alcohol (PVA) gel, *Journal of Hazardous Materials* 164(2–3): 720–725. http://dx.doi.org/10.1016/j.jhazmat.2008.08.059

Fang, H.; Liang, D.; Zhang, T.; Liu, Y. 2006. Anaerobic treatment of phenol in wastewater under thermophilic condition, *Water Research* 40(3): 427–434. http://dx.doi.org/10.1016/j.watres.2005.11.025

Firozjaee, T. T.; Najafpour, G. D.; Khavarpour, M.; Bakhshi, Z.; Pishgar, R.; Mousavi, N. 2011. Phenol Biodegradation Kinetics in an Anaerobic Batch Reactor. World Environmental and Water Resources Congress, California, American Society of Civil Engineers (ASCE).

Gali, V. S.; Kumar, P.; Mehrotra, I. 2006. Biodegradation of phenol with wastewater as a cosubstrate in upflow anaerobic sludge blanket, *Journal of Environmental Engineering* 132(11): 1539–1542. http://dx.doi.org/10.1061/(ASCE)0733-9372(2006)132:11(1539)

Gonzalez, G.; Herrera, G.; García, M.; Pena, M. 2001a. Biodegradation of phenolic industrial wastewater in a fluidized bed bioreactor with immobilized cells of *Pseudomonas putida*, *Bioresource Technology* 80(2): 137–142. http://dx.doi.org/10.1016/S0960-8524(01)00076-1

Gonzalez, G.; Herrera, M.; García, M.; Pena, M. 2001b. Biodegradation of phenol in a continuous process: comparative study of stirred tank and fluidized-bed bioreactors, *Bioresource Technology* 76(3): 245–251. http://dx.doi.org/10.1016/S0960-8524(00)00092-4

Khodadoust, A. P.; Wagner, J. A.; Suidan, M. T.; Brenner, R. C. 1997. Anaerobic treatment of PCP in fluidized-bed GAC bioreactors, *Water Research* 31(7): 1776–1786. http://dx.doi.org/10.1016/S0043-1354(97)00005-5

Koran, K.; Suidan, M.; Khodadoust, A.; Sorial, G.; Brenner, R. 2001. Effectiveness of an anaerobic granular activated carbon fluidized-bed bioreactor to treat soil wash fluids: a proposed strategy for remediating PCP/PAH contaminated soils, *Water Research* 35(10): 2363–2370. http://dx.doi.org/10.1016/S0043-1354(00)00475-9

Kuscu, O. S.; Sponza, D. T. 2005. Performance of anaerobic baffled reactor (ABR) treating synthetic wastewater containing p-nitrophenol, *Enzyme and Microbial Technology* 36(7): 888–895. http://dx.doi.org/10.1016/j.enzmictec.2005.01.001

Kuyukina, M. S.; Ivshina, I. B.; Serebrennikova, M. K.; Krivorutchko, A. B.; Podorozhko, E. A.; Ivanov, R. V.; Lozinsky, V. I. 2009. Petroleum-contaminated water treatment in a fluidized-bed bioreactor with immobilized *Rhodococcus* cells, *International Biodeterioration & Biodegradation* 63(4): 427–432. http://dx.doi.org/10.1016/j.ibiod.2008.12.001

Kwon, K. H.; Yeom, S. H. 2009. Optimal microbial adaptation routes for the rapid degradation of high concentration of phenol, *Bioprocess and Biosystems Engineering* 32(4): 435–442. http://dx.doi.org/10.1007/s00449-008-0263-z

Latkar, M.; Swaminathan, K.; Chakrabarti, T. 2003. Kinetics of anaerobic biodegradation of resorcinol catechol and hydroquinone in upflow fixed film-fixed bed reactors, *Bioresource Technology* 88(1): 69–74. http://dx.doi.org/10.1016/S0960-8524(02)00261-4

Lohi, A.; Alvarez Cuenca, M.; Anania, G.; Upreti, S.; Wan, L. 2008. Biodegradation of diesel fuel-contaminated wastewater using a three-phase fluidized bed reactor, *Journal of Hazardous Materials* 154(1–3): 105–111. http://dx.doi.org/10.1016/j.jhazmat.2007.10.001

Majumder, P. S.; Gupta, S. 2008. Degradation of 4-chlorophenol in UASB reactor under methanogenic conditions, *Bioresource Technology* 99(10): 4169–4177. http://dx.doi.org/10.1016/j.biortech.2007.08.062

Metcalf, E.; Eddy, H. 2003. *Wastewater engineering: treatment and reuse*. McGraw-Hill Book Company.

Mousavi, N.; Najafpour, G. D.; Bakhshi, Z.; Pishgar, R. 2011. Performance of Anaerobic Baffled Reactor for Biodegradation of Phenol, *Iranica Journal of Energy and Environment* (IJEE) 2(3): 229–234.

Moussavi, G.; Mahmoudi, M.; Barikbin, B. 2009. Biological removal of phenol from strong wastewaters using a novel MSBR, *Water Research* 43(5): 1295–1302. http://dx.doi.org/10.1016/j.watres.2008.12.026

Nair, I. C.; Jayachandran, K.; Shashidhar, S. 2007. Treatment of paper factory effluent using a phenol degrading *Alcaligenes sp.* under free and immobilized conditions, *Bioresource Technology* 98(3): 714–716. http://dx.doi.org/10.1016/j.biortech.2006.02.034

Nuhoglu, A.; Yalcin, B. 2005. Modelling of phenol removal in a batch reactor, *Process Biochemistry* 40(3–4): 1233–1239. http://dx.doi.org/10.1016/j.procbio.2004.04.003

Perez, M.; Rodriguez-Cano, R.; Romero, L.; Sales, D. 2007. Performance of anaerobic thermophilic fluidized bed in the treatment of cutting-oil wastewater, *Bioresource Technology* 98(18): 3456–3463. http://dx.doi.org/10.1016/j.biortech.2006.11.005

Rittmann, B.; McCarty, P. 2001. *Environmental biotechnology: principles and applications*. New York, McGraw-Hill.

Saghafi, S.; Bakhshi, Z.; Najafpour, G. D.; Kariminezhad, E.; Rad, H. A. 2010. Biodegradation of Toluene and Xylene in an UAPB Bioreactor with Fixed Film of *Pseudomonas putida*, *American-Eurasian Journal of Agricultural & Environment Sciences* 9(1): 01–07.

Scully, C.; Collins, G.; O'Flaherty, V. 2006. Anaerobic biological treatment of phenol at 9.5–15 C in an expanded granular sludge bed (EGSB)-based bioreactor, *Water Research* 40(20): 3737–3744. http://dx.doi.org/10.1016/j.watres.2006.08.023

Şen, S.; Demirer, G. 2003. Anaerobic treatment of real textile wastewater with a fluidized bed reactor, *Water research* 37(8): 1868–1878. http://dx.doi.org/10.1016/S0043-1354(02)00577-8

Sowmeyan, R.; Swaminathan, G. 2008. Performance of inverse anaerobic fluidized bed reactor for treating high strength organic wastewater during start-up phase, *Bioresource Technology* 99(14): 6280–6284. http://dx.doi.org/10.1016/j.biortech.2007.12.001

Subramanyam, R.; Mishra, I. 2007. Biodegradation of catechol (2-hydroxy phenol) bearing wastewater in an UASB reactor, *Chemosphere* 69(5): 816–824. http://dx.doi.org/10.1016/j.chemosphere.2007.04.064

Subramanyam, R.; Mishra, I. 2008. Co-degradation of resorcinol and catechol in an UASB reactor, *Bioresource Technology* 99(10): 4147–4157. http://dx.doi.org/10.1016/j.biortech.2007.08.060

Tziotzios, G.; Teliou, M.; Kaltsouni, V.; Lyberatos, G.; Vayenas, D. 2005. Biological phenol removal using suspended growth and packed bed reactors, *Biochemical Engineering Journal* 26(1): 65–71. http://dx.doi.org/10.1016/j.bej.2005.06.006

Vázquez, I.; Rodriguez, J.; Maranon, E.; Castrillón, L.; Fernandez, Y. 2006. Study of the aerobic biodegradation of coke wastewater in a two and three-step activated sludge process, *Journal of Hazardous Materials* 137(3): 1681–1688. http://dx.doi.org/10.1016/j.jhazmat.2006.05.007

Veeresh, G. S.; Kumar, P.; Mehrotra, I. 2005. Treatment of phenol and cresols in upflow anaerobic sludge blanket (UASB) process: a review, *Water Research* 39(1): 154–170. http://dx.doi.org/10.1016/j.watres.2004.07.028

Wang, W.; Han, H.; Yuan, M.; Li, H.; Fang, F.; Wang, K. 2010. Treatment of coal gasification wastewater by a two-continuous UASB system with step-feed for COD and phenols removal, *Bioresource Technology* 102(9): 5454–5460. http://dx.doi.org/10.1016/j.biortech.2010.10.019

Yan, J.; Jianping, W.; Hongmei, L.; Suliang, Y.; Zongding, H. 2005. The biodegradation of phenol at high initial concentration by the yeast *Candida tropicalis*, *Biochemical Engineering Journal* 24(3): 243–247. http://dx.doi.org/10.1016/j.bej.2005.02.016

Yan, J.; Jianping, W.; Jing, B.; Daoquan, W.; Zongding, H. 2006. Phenol biodegradation by the yeast *Candida tropicalis* in the presence of m-cresol, *Biochemical Engineering Journal* 29(3): 227–234. http://dx.doi.org/10.1016/j.bej.2005.12.002

Roya PISHGAR has a Master of Science of Environmental Engineering and a Bachelor of Science of Civil Engineering at the School of Civil Engineering, Babol Noshirvani University of Technology. Has published 6 journal papers and 5 conference papers and has attended the 1st Environmental Remediation Technologies Conference. Research interests include: water and wastewater treatment, waste management, air pollution.

Ghasem D. NAJAFPOUR is a Professor of Chemical Engineering at the Babol Noshirvani University of Technology, Iran. An Associate Professor of Chemical Engineering at the Universiti of Sains Malaysia (1998–2005). An Associate Professor at the Chemical Engineering Department, University of Mazandaran, Iran (1995–1998). An Assistant Professor at the Department of Chemical Engineering, University of Mazandaran, Iran (1983–1995). Visiting Professor, Sabbatical Leave & Research Scientist at the Department of Chemical Engineering, University of Arkansas, USA (1990–1993). Teaching Assistant at the University of Arkansas, USA (1976–1980, 1981–1983). Head of Biotechnology Research Center and Editor in Chief of IJE, IJEE, MEJSR, JAES and WASJ. Has published 312 National, International ISI indexed Journals and Proceedings and 9 books.

Bahram NAVAYI NEYA is an Associate Professor of Civil Engineering at the Babol Noshirvani University of Technology, Iran. He has a PhD of Civil Engineering-Hydraulic Structures at the Moscow Power Engineering Institute, Russia; Master of Science of Civil Engineering-Hydraulic Structures; Bachelor of Science of Civil Engineering at the Amir Kabir university of Technology (Tehran Polytechnic), Iran. Has published 19 English and Persian papers, 53 national and international conference papers and 3 books. Research interests include: design and analysis of water network, elasticity and plasticity theories, plates and shells, concrete behavioral models, design and analysis of concrete dams.

Nafise MOUSAVI. Has a Master of Science of Environmental engineering and a Bachelor of Science of Civil engineering at the School of Civil Engineering, Babol Noshirvani University of Technology, Babol, Mazandaran, Iran. Has published 6 journal papers and 5 conference papers and has attended the 1st Environmental Remediation Technologies Conference). Research interests include: water and wastewater treatment, waste management, air pollution.

Zeinab BAKHSHI. Has a Master's degree of Science of Environmental engineering and a Bachelor's degree of Civil engineering at the School of Civil Engineering, Babol Noshirvani University of Technology, Babol, Mazandaran, Iran. Has published 8 journal papers and 5 conference papers. Research interests include: water and wastewater treatment, membrane technology.

15

WASTEWATER AND LANDFILL LEACHATE TESTING: ACUTE TOXICITY BIOTEST RESULTS EVALUATION

Anolda ČETKAUSKAITĖ[a], Milda Zita VOSYLIENĖ[b], Nijolė KAZLAUSKIENĖ[b], Virginija KALCIENĖ[c]

[a]*Department of Biochemistry and Molecular Biology, Faculty of Natural Sciences, Vilnius University, Čiurlionio g. 21/27, LT-03101 Vilnius, Lithuania*

[b]*Nature Research Centre, Akademijos g. 2, LT-08412 Vilnius, Lithuania*

[c]*Environmental Studies Centre, Faculty of Natural Sciences, Vilnius University, Čiurlionio g., 21/27, LT-03101 Vilnius, Lithuania*

Abstract. This diverse review discusses biotest species and results scoring systems, which were applied to aquatic toxicity assessment of effluents/wastewater (WW) and landfill leachate (LL). European and American aquatic toxicity testing is reviewed. An example of Lithuanian research data on LL biotesting with aquatic organisms of different phylogenetic and ontogenetic levels is presented. Acute toxicity WW and LL is assessed on the basis of (L(E)C50, acute Toxic Units (TUa), pT values, and, by applying different simple result scoring systems or toxicity thresholds. The differences in legislation and recommendations for biotest application in WW and LL aquatic toxicity testing are compared. It is concluded that WW and LL lowest acute toxicity data (TUa value 0.3) should be considered equally as risk to aquatic environment, and technical management decisions should be made. The universal features of toxicity scoring systems, the problems of inventory of old small landfills and cost effective approach are discussed.

Keywords: water pollution, aquatic toxicity, landfills, environmental management.

Introduction

Classification systems for the toxicity of wastewaters, surface waters: publications in Europe, USA and Canada. Bulich (1982) proposed one of the first aquatic bioassay scoring systems. Thereafter, a variety of standard ecotoxicity tests and test batteries were used in Europe and America. These included variety of different systems of classification and toxicological data ranking. Certain recommendations for classification were presented in US EPA, SETAC, HELCOM, and other publications (Tonkes *et al.* 2005; Huschek, Hansen 2006; Hansen 2007; HELCOM LAND 2007). Other systems were reported in collective publications of researchers (Blaise, Férard 2005) or research articles containing substantial reviews (Ronco *et al.* 2005; Libralato *et al.* 2010; Zhang *et al.* 2012). Short term tests and different systems of classification were applied both for Whole Effluent Toxicity (WET) testing and Whole Effluent Assessment (WEA), (Blaise, Férard 2005); and also, for ambient and *in situ* surface water quality (SWQ) testing (Den Besten, Munawar 2005; Palma *et al.* 2008).

Use of biotest data for WEA and pollution permit limit in Europe, USA and Canada. Historically, chemical analysis of effluents or wastewaters (WW) was more developed than biotesting in many European countries. Consequently, permit limits for WW were based on individual constituent chemical permit limit, i.e. concentrations measured by analytical methods (AA or GC), and recognized as toxic from individual chemical toxicity testing. As the EU Water Framework Directive specified good biological and chemical status of receiving waters in EU states, both types of analyses were used in the EU in WW monitoring (Directive 2008/105/EC; European Union 2010a). 2007 OSPAR Report (and guidance document of OSPAR Commission) on WEA presented acute test protocols, which were used in different EU countries. Additionally, nine tests were recommended for freshwaters and seven tests for marine or brackish waters (OSPAR Commission 2007).

In different EU countries, organizations and companies, e.g. Concawe (CONservation of Clean Air and

Corresponding author: Milda Zita Vosylienė
E-mail: zmvosyliene@email.lt

Water in Europe) and e.g. Arkema (a French multinational chemical producer), the biological species used in most protocols are: i) bacteria *Vibrio fischeri,* marine spp. (Belgium, Southern Ireland, Portugal, Germany, Netherlands, Sweden, UK, Arkema, Concawe), ii) algae *Pseudokirchneriella subcapitata* (Belgium, S. Ireland, Portugal, Netherlands, Sweden, UK, Arkema, Concawe), iii) algae *Scenedesmus subspicatus,* freshwater spp. (Germany), iv) algae *Skeletonema costatum,* marine spp. (S. Ireland, UK), v) crustacean *Daphnia magna,* freshwater spp. (Belgium, S. Ireland, Portugal, Germany, Netherlands, UK, Arkema, Concawe), vi) crustacean *Nitocra spinipes* (Sweden, Concawe), vii) crustaceans *Tisbe battagliai* and *Acartia tonsa,* marine spp. (S. Ireland, UK), viii) fish *Oncorhynchus mykiss* (Belgium, S. Ireland, Germany), ix) *Danio rerio* fish egg test (Germany), x) water plants - *Lemna minor* (Portugal), (OSPAR Commission 2007). In various countries, different systems of WET or WEA biotest data were applied as an alternative to a chemical approach (and/or permit limits). The MATC - Maximum Acceptable Toxicant Concentration [MATC = geometric mean of NOEC × LOEC] is used for WW accepting water bodies in USA; Toxic Units - in USA and widely in European countries. In 2005, Germany started to use Lowest Ineffective Dilution (LID), (Federal Ministry for... 2004).

Attempts towards the harmonization of WET and WEA biotesting, and ranking systems. What changes appeared in the recommendations of HELCOM for WET evaluation and WEA after 2007? Based on COHIBA (Control of Hazardous Substances in Baltic Sea Region) Project results, HELCOM MONAS (Helsinki Commission Monitoring and Assessment Group) in 2011 proposed recommendations for WET and WEA, (monitoring of WW) in Baltic Sea region. Different single species biotests (for acute toxicity determination), and evaluation of their toxicity data (by settling toxicity limits) were recommended (HELCOM MONAS 2011). For acute toxicity estimation, the following freshwater tests were recommended according to EN ISO standards: 1) *Daphnia magna* immobility test (24 and 48 h), 2) algae growth inhibition test (short 48 h version), and 3) luminescent bacteria (15 and 30 min light inhibition) test (HELCOM MONAS 2011). For WEA, additionally, the "test battery" (TB) was recommended, which included at least 2–3 different trophic levels organisms (Fig. 1). The use of only one test species was not recommended for unclear sensitivity reasons (COHIBA 2012).

Currently, no single ISO standard method provides guidance for the scoring of ecotoxicity test results. Usually ISO methods describe how to calculate L/EC50 using a Graphical and Probit Method. The exception is guidance on statistical interpretation of ecotoxicity data of tests applied to characterize soils; it recommends test strategies for the protection of ground and surface waters (ISO 17616:2008). As these guidance methods for ecotoxicity scoring were under development, different countries developed the WET and WEA control systems individually. Thus, variation of ranking scales and toxicity thresholds is substantial [q.v. S. Ireland (Hernan, O'Rourke 2011); Germany (Gartiser *et al.* 2010); Poland (Mankiewicz-Boczek *et al.* 2008); Portugal (Mendonça *et al.* 2009); Italy (Tigini *et al.* 2011)].

EU Regulations and recommendations for landfill leachates biotesting. Contrary to EU guidelines for WET testing, recent recommendations for landfill leachates (LLs) and waste testing originated from another - Directive on waste (2008/98/EC; European Union 2010b). This Directive lists hazard criteria for waste classification. Waste inherent hazardous property H14 - "ecotoxicity" is defined as a "mirror entry" in this EU document (2008/98/EC). Furthermore, such classification should be based on EU legislation on chemicals, i.e. Council Regulation (EC) No 440/2008, in which for individual chemicals and chemical mixtures hazard for aquatic environment is defined by phrases H410 - H413. Thus, from an initial examination, it appears that the same biotesting principles are valid for wastes as for chemical substances (Thomas *et al.* 2009). Two types of acute toxicity test batteries (aquatic and soil) are proposed for solid waste and LL in Germany after 2005–2007 European inter-laboratory comparison ring test (Wilke *et al.* 2008; Moser *et al.* 2011; Pandard, Römbke 2013). Despite the requirement to cover the traditional three trophic levels with biotest species, the aquatic tests battery for waste testing and hazard classification does not include fish (bioluminescent bacteria, algae and crustacean tests are proposed). This is contrary to Council Regulation (EC) No 440/2008 recommendations where fish, but not fast toxicity screening bioluminescent bacteria (*V. fischeri*) test can be applied for individual chemicals and their mixtures testing. Figure 1 indicates that for waste ecotoxicity (H 14) characterization, the minimum number of tests needed is six (three tests per LL (or waste eluate) and three - per solid waste), as the eluate, as well as, solid waste tests should be represented (Wuttke, Walther 2013). This eliminates the value of rapid rather than time consuming characterization of waste LL only, especially, when such LL, equally as WW, is discharged to aquatic environment. More detail examination of these Recommendations (Moser, Römbke 2009; Wuttke, Walther 2013) gives evidence of large number recommended tests (≤14, including significant number of chronic toxicity) and not cost-effective approach for LL urgent characterization. This raises the open question - which test species and data evaluation systems should be used for LLs?

Ecotoxicological reseach of toxicants, WW and LL in Lithuania. Ecotoxicity investigations of various individual toxicants or their mixtures, WW, LLs were undertaken with standard aquatic organisms of different phylogenetic levels (bacterium, crustacean, algae, fish (Četkauskaitė, Braženaitė 2004; Kalcienė, Četkauskaitė 2007); and ontogenetic stages (fish embryos, larvae, adults) (Kazlauskienė, Vosylienė 2008; Svecevičius 2006; Vosylienė *et al.* 2010; Svecevičius, Kazlauskienė 2011; Montvydienė *et al.* 2014). Chemical analysis and biotest battery data based ecotoxicological evaluations of municipal WW from regional WW treatment plants of Lithuania cities were implemented in 2007 and 2010 (Manusadžianas *et al.* 2010; COHIBA 2012). At present, the centralised ecotoxicity and chemical analyses of surface waters, industrial and municipal WW, variably treated LL are performed in the laboratories of Environmental Research Department of the Lithuanian EPA.

In this work, aquatic toxicity test data scoring systems are compared and reviewed. Main purposes of this review are: i) to present an easily usable unified scheme for acute toxicity result evaluation as a summary of

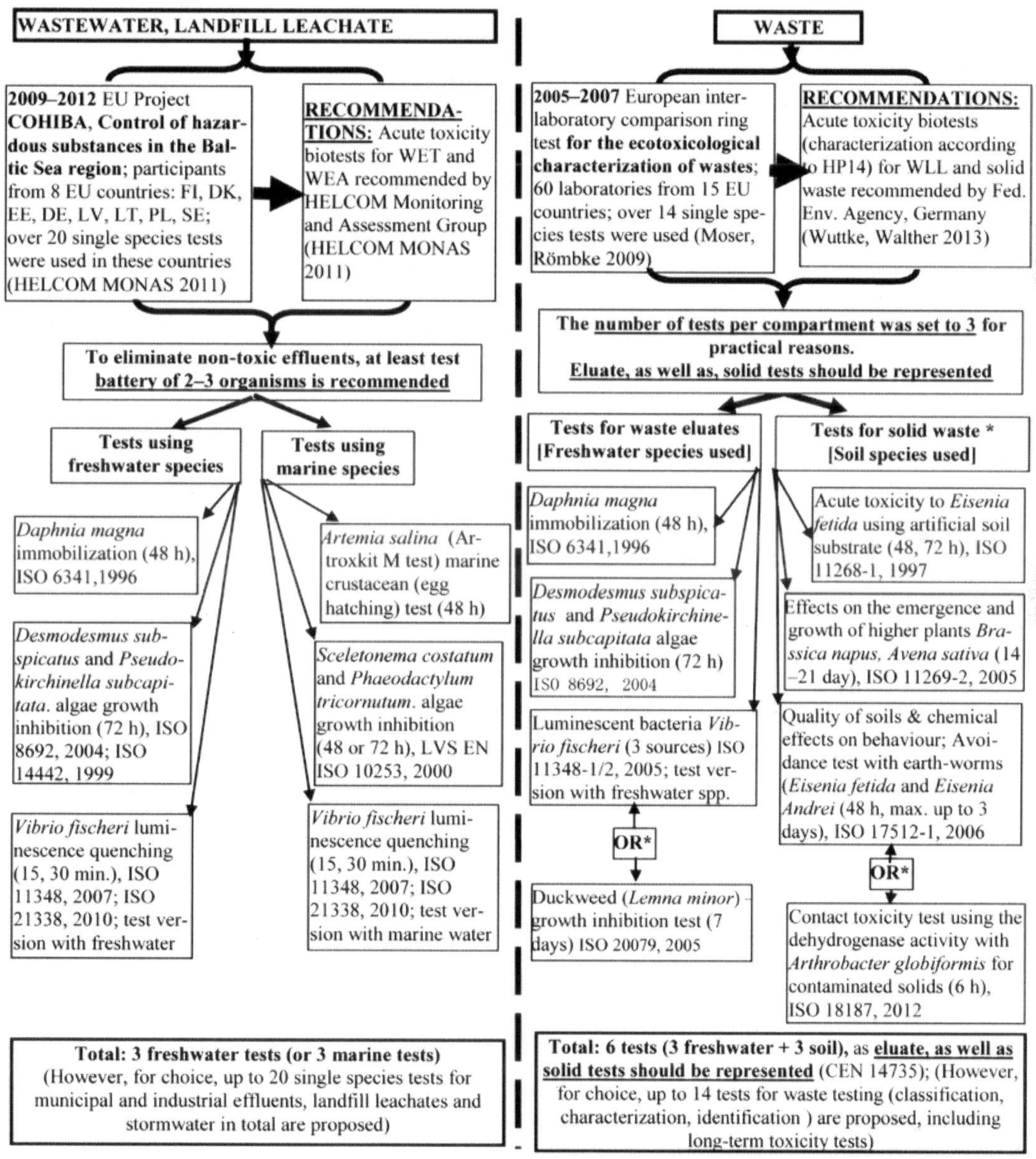

Note: * a different set of basic test battery is represented in Moser, Römbke (2009) and in 2013 Recommendations of Federal Environmental Agency, Germany (Wuttke, Walther 2013), despite the fact that both were based on the same European ring test for the ecotoxicological characterization of wastes.

Fig. 1. Differences in the number of recommended biotests for toxicity testing of WW, LL, and waste (including waste eluate). (Source: Scheme was compiled on the basis of comparative analysis of the existing literature on 2009–2012 EU Project COHIBA (HELCOM MONAS 2011) and 2005–2007 European inter-laboratory comparison ring test (Moser, Römbke 2009; Wuttke, Walther 2013). References to these ISO standard methods q.v. (HELCOM MONAS 2011; Moser, Römbke 2009; Wuttke, Walther 2013) intervals of reference TUa values from WW research; ii) to demonstrate toxicity ranges and lowest values, when samples should be evaluated as toxic; and, iii) to define, which acute toxicity values are appropriate to recommend for LL or WW treatment.

1. Materials and methods

Series and monographs related to WET and aquatic toxicity testing (Blaise, Férard 2005; Norberg-King 2005) were assessed. Science Direct and Medline searches "on TU (or LC50) based ecotoxicity classes (categories)", "wastewater permit limits" and others were performed.

ISO standard documents on biotesting and HELCOM guidelines were examined and compared. Experimental articles in the field of WW and LLs toxicity testing were analysed.

The acute toxicity of the Kairiai LL (North Lithuania) after an incident bund wall overflow was studied at the Laboratory of Ecology and Physiology of Hydrobionts (Ecological Physiology and Toxicology Sector) of the Institute of Ecology, Nature Research Centre, Vilnius. The acute toxicity bioassays up to 96 h duration were carried out using different biotests: higher plant – *Lepidium sativum* (seed germination), crustacean – *Daphnia magna,* schrimp *Thamnocephalus platyurus,* fry and fingerling (juvenile) fish *Oncorhynchus mykiss* (Kazlauskienė *et al.* 2004; Šaulienė *et al.* 2004) (Table 1). Experimental fish acclimatization, handling, physico-chemical water properties, fish care and killing were performed in accordance with recommendations of EU Commission of 18 June 2007 (notified under document number C (2007) 2525) on guidelines for the accommodation and care of animals used for experimental and other scientific purposes (2007/526/EC).

Statistical details for the estimation of L/EC50 and TUa values can be found in (Kazlauskienė *et al.* 2004; Svecevičius, Kazlauskienė 2011).

2. Review, results and discussion

2.1. Principles of aquatic toxicity classification systems

Over 40 different schemes for the classification of aquatic toxicity of chemical mixtures, WW, LLs, sediment porewater, elutriates were proposed and used in European States, USA, and Canada, in the period 1982–2012. For example, the principal acute toxicity parameters recommended for WET classification are: standard lethality (LC) – LC50, LC25, immobilization (IC) – IC25 and even NOEC (Not Observed Effect Concentration) of WW, i.e. non-toxic effect, for LID or pT values (Gartiser *et al.* 2010; Krebs 2005) or other end-points for acute bioassays of 24–96 h duration (HELCOM MONAS 2011). In general, systems of WET classification are based on several approaches: i) comparison of WW samples with clean water controls using dilution methods and simple indices, i.e. constant dilution factor or per cent of WW volume producing an effect; these are determined in diluted samples; ii) using threshold(s) or WET Permit limit(s), expressed in acute Toxic Units, TUa, [TUa = 100/L(C)50], or chronic Toxic Units, TUc, [TUc = 100/NOEC]; and, iii) use of integration of toxicity (amount or strength), by using WW flow rate or toxicity spread area, or by integrating type of toxicity etc. (COHIBA 2010).

First group. Simple scoring systems that use individual indices or scores of toxicity ranking of WW, sediment elutriates, LL fall within following types of data:

a) **Dilution factor type data,** i.e. **LID** – defined as the reciprocal value of the lowest ineffective concentration (NOEC) (HELCOM MONAS 2011); LID values were defined as limits for toxicity or dilution factor; e.g. for algae – when "no inhibition >20% is ascertained for higher dilution factors"; for sediment elutriates, porewater, **pT** value, i.e. negative binary log of the first non-toxic dilution of WW or sediment elutriate (dilutions are performed by factor two) (Krebs 2005); **G** value – lowest dilution factor, when such dilution of WW causes no effect on test-organisms in TB (German regulatory values) (Latif, Licek 2004).
b) **Percentage of WW volume, giving certain effect,** i.e. **PE** – percentage effect caused by WW volume (**% WW/v**) generating IC50 value in Microtox® test (Bulich 1982); or, EC50 value in TB (Tonkes *et al.* 1999); it can also represent percentage of natural waters, stormwaters, soil percolates volume generating IC50 in Toxkit based biotests (Ronco *et al.* 2005).
c) **L(E)C50 based classification** for waste (Wilke *et al.* 2008; Pandard, Römbke 2013).
d) **TUa** (acute toxic units) **as an initial, background scoring value for indices or scales** – for WW (HELCOM LAND 2007) and for waste (Lapa *et al.* 2006). **TUa** based more sophisticated indices: **AvTx** – average of toxicity index, i.e. the sum of the endpoint values in TUs divided by the maximum number of possible/used tests (Manusadžianas *et al.* 2010). **Equitox parameter** – $\log_{10}(100/EC50)$ (Vindimian 2005); **WET** – Whole effluent toxicity index, WET = $\log_{10}[1 + (n/N) ' \Sigma T]$, (where, n = number of positive toxic responses, N = number of bioassays used, T – toxic units); **MST** – most sensitive test index is an integrated evaluation of WW toxicity by TB; represents the highest TU-score obtained among the tests used in the TB (Manusadžianas *et al.* 2010).

Second group. Simple scoring systems are based on the thresholds of indices or indices based scales, or numerical guidelines for the same data types:

a) **Effluent Permit limit (EPA emission limit value)** was assigned by the S. Ireland EPA: <10 TUa, i.e. "slightly toxic" or only ≤50% toxic effect can be reached after 10 dilutions of WW (if more, further action needed) (Hernan, O'Rourke 2011).

b) **WW acute toxicity categories (five)** classified by Irish Shannon Aquatic Toxicity Laboratory (SATL) were in more detail, than the permit limit of Ireland EPA: i) non-toxic (TU < 3); ii) slightly toxic (3–10 TUs); iii) toxic (10–50 TUs); iv) very toxic (50–100 TUs) (Hernan, O'Rourke 2011).

c) **WET Index Threshold** (or Permit limit) is expressed as follows: a) 0.3 acute toxic unit (TUa), and b) 1.0 chronic toxic unit (TUc) (Ronco *et al.* 2005).

d) **LID** – lowest ineffective dilution as a **threshold,** also named **T** – Toxicity limit value. These values for industrial WW at the point of discharge toxicity to fish eggs were usually 2 (sometimes up to 6), umu-test 1.5–3, algae test 16, *Daphnia* 2–8, bacteria 4–32 at 10 (or 20) % effect (HELCOM MONAS 2011).

e) **TUa values based five toxicity classes** were proposed for WW and natural waters (TB use of data) evaluation: a) no-acute toxicity (TU < 0.4); b) slight acute toxicity (0.4 < TU < 1.0); c) acute toxicity (1.0 < TU < 10); and, d) high acute toxicity (10 < TU > 100), very high acute toxicity (TU > 100), (Ronco *et al.* 2005); more simple version was used for waste leachates (Lapa *et al.* 2006).

f) **Threshold toxicity** (acute and chronic) **value for Toxicity identification evaluation (TIE)** of effluent (WW) toxicant is >1.2 TU (giving 95–100% success for toxicant identification); <1.2 TU – minimal WW toxicity strength (for WW, which are typically toxic only at full strength) (Goodfellow, McCelloh 2005).

g) **EC50 and WW dilution based threshold**: If EC50 is ≤ the concentration in a 10% dilution of waste eluate (leachate), then a distinct negative effect is observed (potentially harmful substance in waste) (Pandard, Römbke 2013).

Third group. More complex scoring systems have more sophisticated, derived and complex indices, based on:

a) **Integration of toxicity amount using WW flow rate ETI** – effluent toxicity index, derived applying PLS (Partial Least Square) regression for TB data and effluent flow rate (Vindimian 2005).

b) **Integration of toxicity strength, persistence and specificity in trophic chain** – **TP** – toxic print is WW (effluent) characteristic; it is calculated by multiplying average toxicity AvTx and coefficient n (real number of tests which indicated (geno) toxic results). **PEEP** index – potentially ecotoxic effect probe, calculated value reflecting ecotoxicity of WW, their persistence and flow rate (Manusadžianas *et al.* 2010) or WastePEEP in the WASTOXHAZ procedure (Férard, Ferrari 2005).

c) **EDAR** – effect dilution average ration index (Ronco *et al.* 2005).

d) **Methods integrating types of toxicity** (or comparing dependence of relations) – **WTI** – WW toxicity index based on MSD (Minimum Significant Difference value for deriving Toxicity threshold) (Libralato *et al.* 2010).

Comparison of preliminary experimental data within the first group of indices of simple scoring systems can assist researchers or monitoring institutions laboratories in their further choice of testing, TB or TIE steps.

Examples presented in the second group show that Permit limits for WW, equally as definitions of what is "very toxic", vary substantially among different systems of evaluation in different countries. Such differences must be resolved.

Third group extended analytical methods and mathematical apparatus bearing methodologies are hardly suitable for routine every day work with substantial numbers of samples in department laboratories of monitoring institutions; they also create difficulties for understanding by municipality decision makers.

For pollution evaluation purposes, simple toxicity scoring systems and thresholds are valuable for following pragmatic reasons: a) less sampling/labour intensity, when there is no need to cover exposure time and/or hazard area evaluation; b) less special mathematical processing in obtaining results; c) easier application for monitoring institutions or department laboratories of routine testing (samples will be sent to large regional laboratories); d) easier understandable for non-specialists members of society or municipality decision makers.

2.2. WET testing and ranking in Lithuania

2.2.1. WET classification systems in Lithuania

During the period 1996–2011, the systems of the evaluation of effluent acute toxicity were based on following factors:

- WET evaluation performed on the basis of comparison of non-toxic clean water controls with diluted WW; acute toxicity (LC50, EC50, IC50, LT50) values were calculated on the basis of different dilutions of WW and LLs (Kazlauskienė *et al.* 2004; Vosylienė *et al.* 2010).
- Ranking of acute toxicity of WW and LLs was achieved using TUs approach (Šaulienė *et al.* 2004; Manusadžianas *et al.* 2010), or applying other scoring systems (MST, AvTx) (Manusadžianas *et al.* 2010).

It is indicated, that the majority of studies, performed throughout this period, had no wider comparison of experimental data with different toxicity indices and/or thresholds (Fig. 2).

2.2.2. Toxicity evaluation according individual toxicity scoring schemes and thresholds: an example of LL of Kairiai landfill (Lithuania)

Comparison of the Kairiai LL data (TUa units) with different TU-based aquatic toxicity scoring systems is

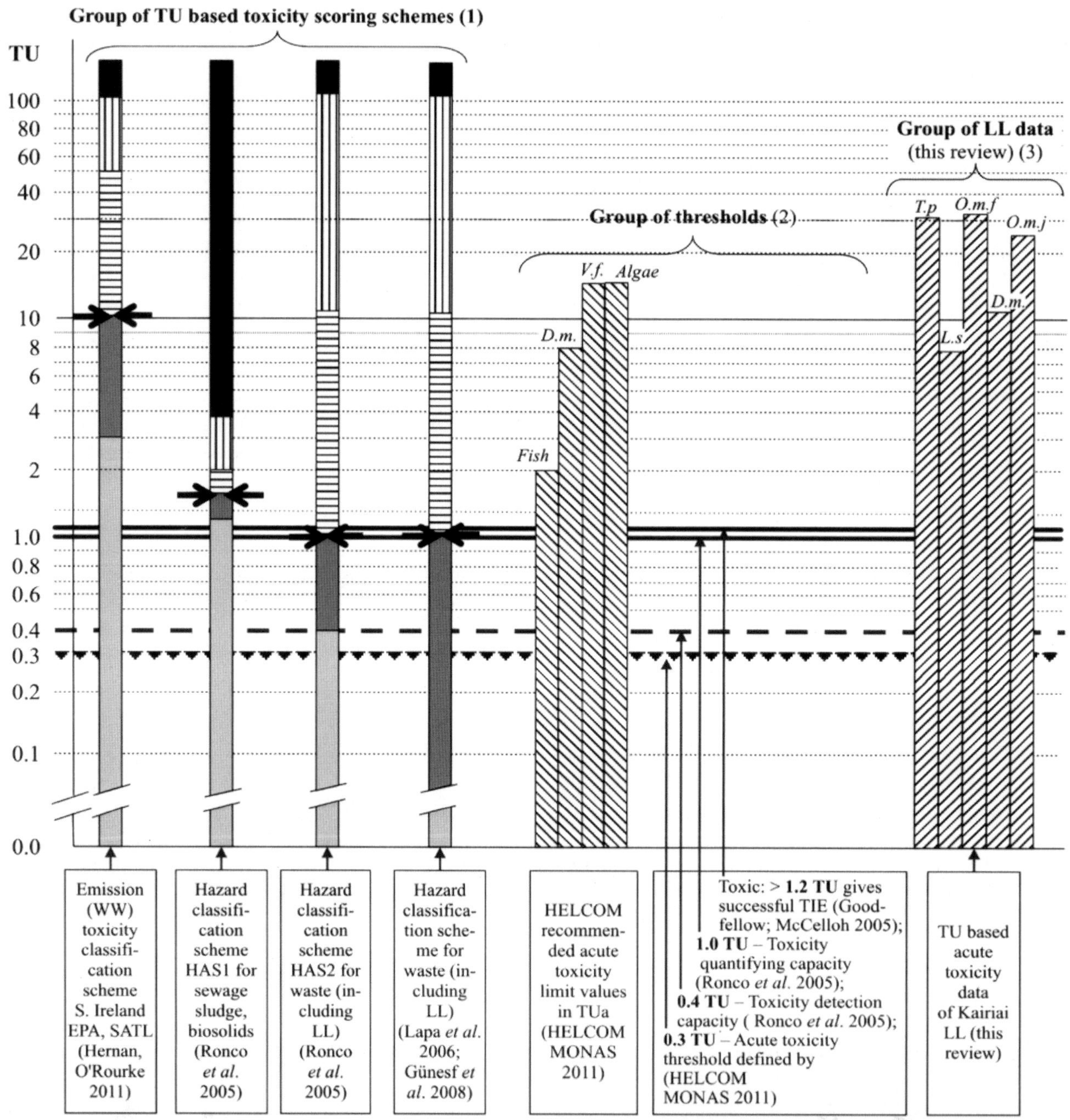

Fig. 2. The scheme of aquatic toxicity of emissions (WW, sewage sludge, LL, etc.) evaluation: the group of hazard classification schemes (1), the group of toxicity thresholds (2) and an example of Kairiai LL acute toxicity data (3), (see Table 1).

Notes:

Non Toxic; Slightly Toxic; Toxic; Very Toxic; Extremely toxic

HELCOM limit values; Kairiai LL data

Black horizontal opposite arrows crossing bars means the limit between "slightly toxic" and "toxic"

Abbreviations: *V. f.* – luminescent bacterium *Vibrio fischeri; L.s. Lepidium sativum; D.m.* – *Daphnia magna; T.p.* – *Thamnocephalus platyurus; O.m.*f. – *Oncorhynchus mykiss* fry; *O.m.* j. – *Oncorhynchus mykiss* juveniles (fingerling).

presented in Table 1. According to L(E)C50 data, the sensitivity of test organisms to leachate is indicated in following sequence: *Oncorhynchus mykiss* fry > *Thamnocephalus platyurus* > *Oncorhynchus mykiss* fingerling > *Daphnia magna* > *Lepidium sativum* (roots). LC50 of LL to fish fry comprised 3.0% and was comparable to *T. platyurus* LC50 – 3.22%. These data confirmed the sensitivity of fish to toxicants at early stages of development and low sensitivity of plant biotests. When the toxicity of the medium is high, all test organisms exposed demonstrated similar high toxicity levels. *O. mykiss* fry was much more sensitive (TUa = 33.3) than fingerling (TUa = 23) and these values significantly exceeded HELCOM MONAS (2011) recommended acute toxicity limit value for fish (TU = 2). Data obtained with *D. magna* (TUa = 11.3) were closer to the COHIBA (2012) recommended limit value for *Daphnia* test is TUa = 8. The significant sensitivity of shrimp *T. platyurus* biotest that corresponded to data obtained for tested effluents in COHIBA (2012) is noted.

2.2.3. LL data evaluation using groups of toxicity scoring systems, toxicity limit values and toxicity threshold groups

In the TUs based scale, it is possible to compare the ranges of WET, waste (including LL) by using acute toxicity (or hazard) classification schemes and groups of the different toxicity thresholds defined toxicity limit values (Fig. 2). The resulting picture gives evidence that ranges indicating what is toxic (0.3–10 TUs) are overlaping in different schemes, thresholds and toxicity limit values. Additionally, it clearly indicates that the Kairiai LL test data: a) exceed different acute toxicity thresholds (0.3, 0.4, 1.0, 1.2 TU); b) significantly exceed HELCOM recommended acute

Table 1. An example of the sensitivity of test organisms of different phylogenetic level and ontogenesis to Kairiai LL* in acute toxicity tests

Test organism / Toxicity	*Lepidium sativum* (roots, 48 h)	*Daphnia magna*	*Thamnocephalus platyurus*	*Oncorhynchus mykiss* fry	*Oncorhynchus mykiss* fingerling
Primary data L(E)C50, % WW/v (Literature source)	12.7 (Kazlauskienė *et al.* 2004)	8.83 (Šaulienė *et al.* 2004)**	3.22 (Šaulienė *et al.* 2004)**	3.0 (Kazlauskienė *et al.* 2004)	4.35 (Kazlauskienė *et al.* 2004)
L(E)C50, % of WW/v based toxicity category (Tonkes *et al.* 1999)	Minor acutely toxic (10–100% WW/v)	Moderately acutely toxic (1–10% WW/v)	Moderately acutely toxic (1–10% WW/v)	Moderately acutely toxic (1–10% WW/v)	Moderately acutely toxic (1–10% WW/v)
Toxicity class (Bulich 1982)	Highly toxic sample (<25%)	Highly toxic sample (<25%)	Highly toxic sample (<25%)	Highly toxic sample (<25%)	Highly toxic sample (<25%)
TUa	7.8	11.32	31.05	33.3	23
Toxicity according to five categories of SATL (S. Ireland) (Hernan, O'Rourke 2011)	Slightly toxic (3–10 TUs)	Toxic (10–50 TUs)	Toxic (10–50 TUs)	Toxic (10–50 TUs)	Toxic (1–50 TUs)
Toxicity according to toxicity emission limit of S. Ireland EPA (Hernan, O'Rourke 2011)	No real toxicity (<10 TUs)	Toxic (>10 TUs) Further testing and action needed	Toxic (>10 TUs) Further testing and action needed	Toxic (>10 TUs) Further testing and action needed	Toxic (>10 TUs) Further testing and action needed
Toxicity above WET acute TU threshold (Ronco *et al.* 2005)	2.6 times	37.7 times	103.5 times	111 times	76.7 times
pTv (pT value) or no adverse effect dilution (Krebs 2005)	6	n.d.	n.d.	8	7
pT value based Toxicity category (Krebs 2005)	Extremely toxic	n.d.	n.d.	Extremely toxic	Extremely toxic
Toxicity limit values for industrial WW(HELCOM MONAS 2011)	n.a.	Toxic (TU = 8)	n.a.	Toxic (TU = 2)	Toxic (TU = 2)

Notes: n.a. – not applicable; n.d. – not determined.

* Kairiai landfill is located 5 km east of Šiauliai city, Lithuania (55° 55′ 42.7″, 23° 23′ 42.81″, WGS). Incident bund wall overflow through surrounding channels reached local creek and Ginkūnai pond (2004).

** These L(E)C50, % primary data are presented with permission from Authors and Publishers from following source: Šaulienė *et al.* (2004).

toxicity limit values; and, c) represent "toxic" and "extremely toxic" ranges in scoring schemes. This comparison of different toxicity scores systems together with data of interest in TU range can assist for preliminary definition of sample toxicity when such WW or LL is discharged to aquatic environment. Moreover, analysis of **scoring/threshold** data indicated a requirement to implement strict toxicity scoring system with low level of acute TU values for category „toxic" (i.e. mainly 0.3–2 TUa). Thus, it is proposed that samples of WW and LL should be considered equally as toxic starting from a TUa value 0.3.

2.3. Different universal features of certain toxicity scoring systems

More recently, there appeared another important feature of aquatic toxicity data scoring systems – their universality. The same values based scoring systems are used for different origin aquatic samples:

1) pT-values based classification, applied for WWs (Krebs 2005), was used also for sediment materials (porewater, elutriates, extracts, and solid phases);
2) TU based acute toxicity classes initially were proposed for waste dumps and WW; later these were used for aquatic toxicity evaluation (Sauco *et al.* 2013), with microbiotests in many countries for different environmental samples: natural waters (Greeley *et al.* 2011), WWs (Kontana *et al.* 2008; Carbonell *et al.* 2010); soils, sediments, wastes, biotoxins, etc. Alternatively, TU based classes were used in municipal solid waste (MSW) incineration bottom ash toxicity evaluation (Lapa *et al.* 2006; Tsiridis *et al.* 2012);
3) Potential Ecotoxic Effect Probe, PEEP index was used for WET, integrating TB data (TU form), persistence and WW flow rate (Q), was later modified (without Q, etc.) and also applied as Waste-PEEP for LLs ecotoxicity scoring (Férard, Ferrari 2005).

Universality of aquatic toxicity data scoring systems is also reflected in their application to tests performed with different species. pT-values based scoring system was initially (1987/1988) proposed for the *V. fischeri* bioluminescent test use only. Later, it was expanded to scoring data of the tests with algae, daphnids, fish (Zha, Wang 2006) and genotoxicity test results (Krebs 2005).

2.4. Use of fish test for WW and LL toxicity testing

Different positions exist regarding fish tests used in waste and LLs toxicity testing. Fish tests were not included into 2013 recommendations (ethical considerations) from a 2007 EU ring test of solid waste (including waste leachate) classification, identification and basic characterization because of "animal protection reasons and the large quantity of eluate needed" (Pandard, Römbke 2013). Contrary to this position, the review on species used in the period 1983–2003 showed that fish tests were applied for evaluation of waste LLs (Férard, Ferrari 2005). Aquatic biotests were recommended for testing of substances and mixtures in the Council Regulation (EC) No 440/2008 where fish tests were included. Fish embryo (non-protected life stages) toxicity tests were recognized as suitable in International Workshop 2008 on fish protection in scientific testing (Embry *et al.* 2010). Fish tests were used in 2009–2012 EU Project COHIBA (HELCOM MONAS 2011). Fish tests were also used for MSW and MSW incineration ash toxicity evaluation (Stiernström *et al.* 2011).

2.5. Problems of environmental inventory of old small landfills

The significant amount of scientific, regulatory, technological, and environmental problems exist today for waste landfills hazard inventory and LLs aquatic toxicity control (Fig. 3). It is realistic to note, that, in the former Soviet Block EU States (fSvBEUS), there exist old MSW landfills from years 1960–1990 (without MSW incineration ash), where the large volume of naturally occurring LLs remains a problem. As there are at least 126 (small) landfill sites in Northern Lithuania, it is not unreasonable to deduce that there are >1000 such sites in the fSvBEUS. For example, according to implementation of waste management projects plans, supported by EU, 116 out of 126 small local landfills were closed in North Lithuania regions (Decision of Šiauliai Regional Development Council (Šiaulių Regiono Plėtros Taryba 2012)). However, contrary to new MSW big landfills, small regional landfills which were closed with minimal technical facilities (drainage, cover, etc.) should be evaluated and, perhaps, will be waiting best available technologies (BAT): high density polyethylene bottom lining, reverse osmosis (RO) filtration of LL, gas collection systems, etc. As it is not economically viable to transfer the content of small landfills to modern new landfills with BAT, this serious problem should be incorporated into the Agenda for EU Environment Directorate.

Studies on the effect of leachate from a landfill to the underground water (often used for extraction for potable water; or, discharged to rivers) of the surrounding area demonstrated that underground water quality has been compromised by such leachate (Vasarevičius *et al.* 2005; Hossain *et al.* 2014). This potential public health hazard to potable water resources cannot be ignored. The landfills could remain a permanent pollution sources for long periods of time after closing (Pablos *et al.* 2011; Klauck *et al.* 2013). For example, physico-chemical investigation, phytotoxicity testing and use of battery of biotests (leech, crustacean and fish) showed that the Kairiai landfill closed in 2007 remains a serious source of long-lasting pollution, which adversely affects the neighboring water bodies

(Svecevičius *et al.* 2013; Montvydienė *et al.* 2014). Such problems can be potentially even worse when leachates are percolating into trans-boundary Rivers (i.e. Nemunas River rises in Belarus and the final reaches are shared with Russia).

Various methods (biological, physical and chemical) in fSvBEUS, especially Romania and Poland are described, applied and proposed for treatment of leachate for reducing environmental risk (Białowiec *et al.* 2007; Kulikowska, Klimiuk 2008; Şchiopu, Gavrilescu 2010; Şchiopu *et al.* 2012) and even cost-effective methods using anthropogenic soils rehabilitation through phytoremediation (Cretescu *et al.* 2013). Treatment of the leachates is costly. Şchiopu *et al.* (2012), (Romania) proposed the use of RO. Whilst it is efficient, these Authors does not advice how to dispose of the RO concentrate (solute) pragmatically. Concentrate treatment in WW treatment plants could be not possible because of heavy metals or thiourea (causing nitrification and other sewage works problems). It should be noted, that new recycling technologies (heavy

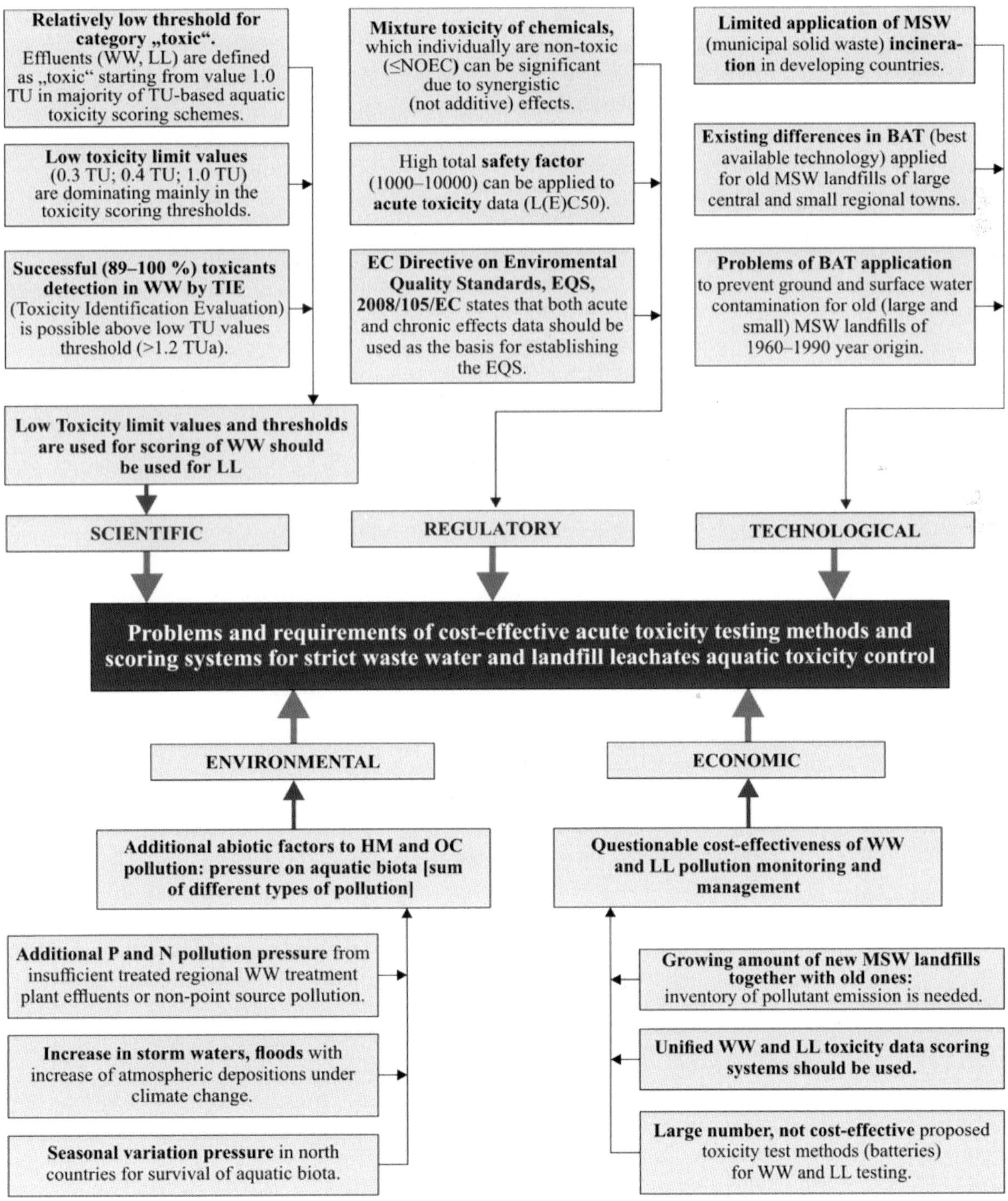

Fig. 3. Pollution of aquatic environment caused by wastewaters and landfill leachates: related problems and multiple evidence of requirement for strict aquatic toxicity scoring and control of emissions (more widely in Review, Results and Discussion section)

metal bio-mining, combustion gas collection, ethanol production from bio-waste, etc.) are waiting to be applied to waste landfill sites. Otherwise, this growing amount of waste will exacerbate environmental, social, and economic problems (Fig. 3).

2.6. Waste and LL biotesting: ecotoxicity criteria, recommendations, and problems of cost effective approach

The characterization of hazard of small regional landfills falls into criterion H15 apart of H14 of WFD 2008/98/EC: "H15. Waste capable by any means, after disposal, of yielding another substance, e.g. a leachate, which possesses any of characteristics listed above". Thus, despite the size and closure status, regional municipal solid waste landfills hazard criteria (primary H14, H15) should be resolved with appropriate test methods. To re-iterate, it is obvious that different EU Directives are applied to water, WW and waste (including LLs), (European Union 2010a, 2010b). Consequently, different recommendations for water, WW and waste (including LLs) acute toxicity testing and data scoring were elaborated and accepted (q.v. Introduction). As mentioned in Introduction, two types of acute toxicity test batteries (aquatic and soil) were proposed for waste and LL in Germany after 2005–2007 European inter-laboratory comparison ring test (Moser *et al.* 2011; Wuttke, Walther 2013). The aforesaid collectively creates a situation that for aquatic toxicity testing of WW, waste, LL is proposed costly, multiple biotesting, including: multiple species, acute and chronic tests (for water and soil environment) and different scoring systems. Finally, if legally enforceable in EU, these should be accepted for practical use in and will place a financial burden on local municipalities or EPA branch laboratories of EU states, including developing countries with limited resources. Alternatively, the experience of >30 years of test use, species selection, data scoring systems for toxicity assessment of water environment (WW, surface water, aquatic life) is much more abundant than waste (including leachates) testing. This experience should be used, and recommendations (for waste testing, classification, water environment testing, etc.) require harmonization together on the lowest cost possible basis. Examples presented in this review show that Permit limits for WW also vary substantially among different systems of evaluation in different countries. For pragmatic reasons the Authors deliberately limited this Review to EU, USA and Canada, but are aware that there are a plethora of other test methods used in Pacific Rim Countries, South America, Africa, Middle East and not least the Commonwealth of Independent States including Russia. The questions arise – which scoring system or thresholds are suitable for correct choice, when toxicity data (L(E)C50 or TUa, or pTv, LID, etc.) are available from test with two or three species – each from different trophic level and which samples of WW or LL should be indicated as toxic (e.g. see second concluding remark)?

Thus, the globally unified, low cost, acute toxicity testing methods and scoring systems for strict WW and LL toxicity control are of primary need.

Concluding remarks

Multiple test methods, models and acute toxicity scoring schemes (>50) for WW testing were created in the period 1982–2014. A significant number of acute toxicity classification schemes, thresholds, toxicity limit values (TLVs) were created in the TUa based scale. Additionally, acute toxicity L(E)C50 data can be easily converted to TUa. The analysis of these groups gives multi-evidence that ranges indicating what is slightly toxic – toxic (0.3–10 TUs) are overlapping. This permits comparison and preliminary definition of sample (WW, LL) toxicity within many scoring systems.

On the basis of comparison of the multiple TUa based scoring systems and thresholds, it is proposed that samples of WW and LL should be considered equally as toxic starting from TUa value 0.3 and subsequent decisions for technical management should be undertaken for WW and LL equally. Long-term and not cost effective soil toxicity testing for LL might be omitted. With experience of >30 years of biotesting and subsequent data scoring, the aquatic toxicity scoring systems are more elaborated and relevant for toxicity evaluation of effluents (including LL) discharged to aquatic environment.

The legacy of >1000 old landfill sites in the fSvBEUS is a major issue requiring very urgent resolution. Concern is related to variably (if any) treated LL, especially of those waste landfills, which are small (and/or medium) size, old (i.e. created in 1970–90's), non-insulated (with high density PE) and have no proper leachate treatment (i.e. reverse osmosis) resulting in slow discharges to aquatic environment.

Authors choice from results (Table 1) is the use of the most sensitive early life stages juvenile fish (i.e. fish fry) for acute toxicity evaluation of LL and WW. However, the reader can choose otherwise. Some of the sensitive biotest methods, when algae, invertebrate and fish are used, are slow and expensive for routine monitoring as compared to analytical chemistry. Hence, having so determined that a discharge is toxic, then a target chemical(s) can be rapidly and cheaply monitored by analytical chemistry, and thus ensure effective environment management.

In view of the gross disparities of biotests and their associated scoring systems usage, the open question remains: which species to use in aquatic samples testing and data evaluation of landfill leachates in different countries? This requires urgent global attention. Hence, the only way forward is for the United Nations Specialist Agencies to recommend a global, small set of tests and ideally one only scoring system.

Acknowledgements

This work was funded by Research Council of Lithuania, Project No. MIP-038/2012. The authors would like to thank M. Richardson, BASIC, formally of Crawley, England, for reading, esteemed comments and improving the English of this article.

References

Białowiec, A.; Wojnowska-Baryła, I.; Agopsowicz, M. 2007. The efficiency of evapotranspiration of landfill leachate in the soil–plant system with willow *Salix amygdalina* L., *Ecological Engineering* 30(4): 356–361. http://dx.doi.org/10.1016/j.ecoleng.2007.04.006

Blaise, C.; Férard, J. F. 2005. Effluent assessment with the PEEP (Potential Ecotoxic Effect Probe), in C. Blaise, J. F. Férard (Eds). *Small-scale freshwater toxicity investigations.* Vol. 2. Hazard assessment schemes. Dordrecht, the Netherlands: Springer, 69–88.

Bulich, A. 1982. A practical and reliable method for monitoring the toxicity of aquatic samples, *Journal Process Biochemistry* 17: 1–3.

Carbonell, G.; Fernandez, C.; Tarazona, J. V. 2010. A cost/effective screening method for assessing the toxicity of nutrient rich effluents to algae, *Bulletin Environmental Contamination and Toxicology* 85(1): 72–78. http://dx.doi.org/10.1007/s00128-010-0035-8

COHIBA. 2010. *Control of hazardous substances in the Baltic Sea region. Whole effluent assessment (WEA). Proposed recommendations for the use of toxicity limits.* Finish Environment Institute SYKE. 23 p.

COHIBA. 2012. How to control and manage hazardous substances in the Baltic Sea region. Final summary report of the COHIBA Project, in A. Pilke, T. Nakari, E. Schultz, P. Munne, E. Brorström-Lundén, H. Andersson, C. Mathan, F. Marscheider-Weidemann, M. Durkin (Eds.). *COHIBA project consortium* [online], [cited 20 April 2014]. Available from Internet: www.cohiba-project.net/publications

Council Regulation (EC) No 440/2008 of 30 May 2008 laying down test methods pursuant to Regulation (EC) No 1907/2006 of the European Parliament and of the Council on the Registration, Evaluation, Authorisation and Restriction of Chemicals (REACH), *Official Journal of the European Union* L 142, 31.5.2008. 739 p.

Cretescu, I.; Pohontu, C.; Iticescu, C.; Cioroi, M.; Ciocinta, R. C.; Bucur, D. 2013. Treatment of landfill leachate using *Zea mays* and *Triticum* sp. on antropogenic soils, *Journal of Food Agriculture and Environment* 11(3,4): 1507–1512.

Četkauskaitė, A.; Braženaitė, J. 2004. Effects of mixtures of oleic acid with chlorinated herbicides on *Vibrio fischeri* bacteria, *Biologija* 4: 36–39.

Den Besten, P. J.; Munawar, M. 2005. *Ecotoxicological testing of marine and freshwater ecosystems: emerging techniques, trends and strategies.* Boca Raton, London, New York, Singapore: CRC Press, Taylor and Francis Group.

Directive 2008/105/EC of the European Parliament and of the Council on environmental quality standards in the field of water policy, *Official Journal of the European Communities* L 348/84, 24.12.2008, 84–97.

Directive 2008/98/EC of the European Parliament and of the Council of 19 November 2008 on waste and repealing certain Directives, *Official Journal of the European Union* OJ L 312/3, 22.11.2008, 3–26.

Embry, M. R.; Belanger, S. E.; Braunbeck, Th. A.; Burgos, M. G.; Halder, M.; Hinton, D. E.; Léonard, M. A.; Lillicrap, A.; Norberg-King, T.; Whale, G. 2010. The fish embryo toxicity test as an animal alternative method in hazard and risk assessment and scientific research, *Aquatic Toxicology* 97(2): 79–87. http://dx.doi.org/10.1016/j.aquatox.2009.12.008

EU Commission recommendation of 18 June 2007 (2007/526/EC) on guidelines for the accommodation and care of animals used for experimental and other scientific purposes (notified under document number C (2007) 2525), (text with EEA relevance), *Official Journal of the European Union* L 197/1, 30.7.2007, 1–89.

European Union. 2010a. *Water Environment. Water is for life: How the Water Framework Directive helps safeguard Europe's resources.* European Commission Luxembourg: Publications Office of the European Union. Printed in Belgium. 28 p.

European Union. 2010b. *Environment. Being wise with waste: the EU's approach to waste management.* European Commission Luxembourg: Publications Office of the European Union. Printed in Belgium. 20 p.

Federal Ministry for the Environment, Nature Conservation and Nuclear Safety. 2004. *Waste Water AbwV (Ordinance on Requirements for the Discharge of Waste Water into Waters).* 121 p. [Promulgation of the New Version of the Waste Water Ordinance of 17 June 2004, coming into force on 1 January 2005]. Germany. Federal Law Gazette BGBl. 2004, Teil I. 1108–1184.

Férard, J. F.; Ferrari, B. 2005. WSATOXHAZ: a bioanalytical strategy for solid wastes assessment, in C. Blaise, J. F. Férard (Eds.). *Small-scale Freshwater Toxicity Investigations,* Vol. 2. Hazard assessment schemes. Dordrecht, the Netherlands: Springer, 331–375.

Gartiser, S.; Hafner, C.; Hercher, C.; Kronenberger-Schäfer, K.; Paschke, A. 2010. Whole effluent assessment of industrial wastewater for determination of BAT compliance. Part 1: paper manufacturing industry, *Environmental Science and Pollution Research* 17(4): 856–865. http://dx.doi.org/10.1007/s11356-009-0289-z

Günesf, E. H.; Günesf, Y.; Talenle, I. 2008. Toxicity evaluation of industrial and land base sources in a river basin, *Desalination* 226: 348–356. http://dx.doi.org/10.1016/j.desal.2007.02.116

Goodfellow, W.; McCelloh, W. 2005. Minimal toxicity necessary for an effective toxicity identification evaluation. Case Study 6.31, in T. Norberg-King (Ed.). *Toxicity reduction and toxicity identification evaluations for effluents, ambient waters, and other aqueous media.* Pensacola, USA: SETAC Press, 338–344.

Greeley, M. S. Jr.; Kszos, L. A.; Morris, G. W.; Smith, J. G.; Stewart, A. J. 2011. Role of a comprehensive toxicity assessment and monitoring program in the management and ecological recovery of a wastewater receiving stream, *Environmental Management* 47(6): 1033–1046. http://dx.doi.org/10.1007/s00267-011-9679-3

Hansen, P. D. 2007. Risk assessment of emerging contaminants in aquatic systems, *Trends Analytical Chemistry* 26(11): 1095–1099. http://dx.doi.org/10.1016/j.trac.2007.10.001

HELCOM LAND 12/2007. Practical guidance on whole effluent assessment (WEA). Doc. code 4.2/2, in *Land-based Pollution Group 12th Meeting*, 22–24 May 2007, Espoo, Finland. 31 p.

HELCOM MONAS 15/2011. Helsinki commission, in *HELCOM Monitoring and Assessment Group 15th Meeting*, 4–7 October 2011, Vilnius, Lithuania. 18 p.

Hernan, R.; O'Rourke K. 2011. *Aquatic toxicity testing in Ireland*, 2010. Ireland: Shannon. 27 p.

Hossain, Md. L.; Das S. R.; Hossain M. K. 2014. Impact of landfill leachate on surface and ground water quality, *Journal of Environmental Science and Technology* 7(6): 337–346. http://dx.doi.org/10.3923/jest.2014.337.346

Huschek, G.; Hansen, P. D. 2006. Ecotoxicological classification of the Berlin River system using bioassays in respect to the European Water Framework Directive, *Environment Monitoring Assessment* 121(1–3): 15–31. http://dx.doi.org/10.1007/s10661-005-9061-0

ISO 17616:2008. *Soil quality – Guidance on the choice and evaluation of bioassays for ecotoxicological characterization of soils and soil materials.* 11 p.

Kalcienė, V.; Četkauskaitė, A. 2007. Environmental and synthetic sulfhydryl group inhibitors: effects on bioluminescence and respiration of *Vibrio Fischeri, Alternatives to Laboratory Animals* 35: 93–100.

Kazlauskienė, N.; Vosylienė, M. Z.; Marčiulionienė, D.; Montvydienė, D.; Beržinskienė, J. 2004. The assessment of toxicity of Kairiai landfill leachate and water of drainage ditches by use of biological methods, in *Proceedings of the International Scientific Conference of 80 years Anniversary of Lithuanian University of Agriculture (Vandens telkinių apsauga ir valdymas [Protection and management of water bodies])*, 23–24 October 2004, Kaunas, Lithuania, 100–109 (in Lithuanian).

Kazlauskienė, N.; Vosylienė, M. Z. 2008. Characteristic features of the effect of Cu and Zn mixtures on rainbow trout *Oncorhynchus mykiss* in ontogenesis, *Polish Journal of Environmental Studies* 17(2): 291–293.

Klauck, C. R.; Rodrigues, M. A. S.; da Silva, L. B. 2013. Toxicological evaluation of landfill leachate using plant (*Allium cepa*) and fish (*Leporinus obtusidens*) bioassays, *Waste Management Research* 31(11): 1148–1153. http://dx.doi.org/10.1177/0734242X13502388

Kontana, A.; Papadimitriou, C. A.; Samaras, P.; Zdragas, A.; Yiangou, M. 2008. Bioassays and biomarkers for ecotoxicological assessment of reclaimed municipal wastewater, *Water Science and Technology* 57(4): 947–953. http://dx.doi.org/10.2166/wst.2008.196

Krebs, F. 2005. The pT-method as a hazard assessment scheme for wastewaters, in C. Blaise, J. F. Férard (Eds.). *Small-scale freshwater toxicity investigations.* Vol. 2. Hazard Assessment Schemes. Dordrecht, the Netherlands: Springer, 115–137.

Kulikowska, D.; Klimiuk, E. 2008. The effect of landfill age on municipal leachate composition, *Bioresource Technology* 99(13): 5981–5985. http://dx.doi.org/10.1016/j.biortech.2007.10.015

Lapa, N.; Barbosa, R.; Camacho, S.; Monteiro, R.; Fernandes, M. H. F. V.; Oliveira, J. F. S. 2006. Leaching behaviour of a glass produced from a MSWI bottom ash, in P. M. Vilarinho (Ed.). *Materials Science Forum* 514–516: 1736–1741. http://dx.doi.org/10.4028/www.scientific.net/msf.514-516.1736

Latif, M.; Licek, E. 2004. Toxicity assessment of wastewaters, river waters, and sediments in Austria using cost-effective microbiotests, *Environmental Toxicology* 19(4): 302–309. http://dx.doi.org/10.1002/tox.20027

Libralato, G.; Volpi, G. A.; Francesco, A. 2010. How toxic is toxic? A proposal for wastewater toxicity hazard assessment, *Ecotoxicology and Environmental Safety* 73(7): 1602–1611. http://dx.doi.org/10.1016/j.ecoenv.2010.03.007

Mankiewicz-Boczek, J; Nałęcz-Jawecki, G.; Drobniewska, A. M.; Sumorok, B.; Izydorczyk, K.; Zalewski, M.; Sawicki, J. 2008. Application of a microbiotests battery for complete toxicity assessment of rivers, *Ecotoxicology and Environmental Safety* 71(3): 830–836. http://dx.doi.org/10.1016/j.ecoenv.2008.02.023

Manusadžianas, L.; Sadauskas, K.; Vitkus, R. 2010. Comparative study of indices used in toxicity evaluation of effluents, *Desalination* 250: 383–389. http://dx.doi.org/10.1016/j.desal.2009.09.061

Mendonça, E.; Picado, A.; Paixão, S. M.; Silva, L.; Cunha, M. A.; Leitão, S.; Moura, I.; Cortez, C.; Brito, F. 2009. Ecotoxicity tests in the environmental analysis of wastewater treatment plants: case study in Portugal, *Journal of Hazardous Materials* 163(2–3): 665–670. http://dx.doi.org/10.1016/j.jhazmat.2008.07.012

Montvydienė, D.; Marčiulionienė, E. D.; Kazlauskienė, N.; Svecevičius, G.; Jefanova, O.; Žukauskaitė, Z. 2014. Impact of closed Kairiai landfill on the Ginkunai Pond, Lithuania, in *The 9th International Conference "Environmental engineering"*, 22–23 May 2014, Vilnius, Lithuania.

Moser, H.; Römbke, J. 2009. *Ecotoxicological characterization of waste – Results and experiences of an European ring test.* New York: Springer Ltd. 308 p. http://dx.doi.org/10.1007/978-0-387-88959-7

Moser, H.; Roembke, J.; Donnevert, G.; Becker, R. 2011. Evaluation of biological methods for a future methodological implementation of the Hazard criterion H14 "ecotoxic" in the European waste list (2000/532/EC), *Waste Management Research* 29(2): 180–187. http://dx.doi.org/10.1177/0734242X10367844

Norberg-King, T. 2005. *Toxicity reduction and toxicity identification evaluations for effluents, ambient waters, and other aqueous media.* Pensacola, USA: SETAC Press. 496 p.

OSPAR Commission. 2007. *Practical Guidance Document on Whole Effluent Assessment.* OSPAR Hazardous Substances Series. Publication Number: 316/2007. 33 p.

Palma, P.; Alvarenga, P.; Palma, V.; Matos, C.; Fernandes, R. M.; Soares, A.; Barbosa, A. R. 2008. Evaluation of surface water quality using an ecotoxicological approach: a case study of the Alqueva Reservoir (Portugal), *Environmental Science and Pollution Research* 17(3): 703–716. http://dx.doi.org/10.1007/s11356-009-0143-3

Pablos, M. V.; Martini, F.; Fernández, C.; Babín, M. M.; Herraez, I.; Miranda, J.; Martínez, J.; Carbonell, G; San-Segundo, L.; García-Hortigüela, P.; Tarazona, J. V. 2011. Correlation between physicochemical and ecotoxicological approaches to estimate landfill leachates toxicity, *Waste Management* 31(8): 1841–1847. http://dx.doi.org/10.1016/j.wasman.2011.03.022

Pandard, P.; Römbke, J. 2013. Proposal for a "harmonized" strategy for the assessment of the HP 14 Property, *Integrated Environmental Assessment and Management* 9(4): 665–672. http://dx.doi.org/10.1002/ieam.1447

Ronco, A.; Rossini, G. B.; Sobrero, C.; Apartin, C.; Castillo, G.; Diaz-Baez, M. C.; Ramirez, A. E.; Ahumada, I.; Mendoza, J. 2005. The application of hazard assessment schemes using the water toxicity testing battery, in C. Blaise, J. F. Férard (Eds.). *Small-scale freshwater toxicity investigations.* Vol. 2. Hazard Assessment Schemes. Dordrecht, the Netherlands: Springer, 233–255.

Sauco, S.; Gómez, J.; Barboza, F. R.; Lercari, D.; Defeo, O. 2013. Modified whole effluent toxicity test to assess and decouple wastewater effects from environmental gradients, *PLOS ONE* 8(6): e66285. http://dx.doi.org/10.1371/journal.pone.0066285

Şchiopu, A.-M.; Gavrilescu, M. 2010. Municipal solid waste landfilling and treatment of resulting liquid effluents, *Environmental Engineering and Management Journal* 9(7): 993–1019.

Şchiopu, A.-M.; Piuleac, C.; Cojocaru, C.; Apostol, I.; Mămălygă, I.; Gavrilescu, M. 2012. Reducing environmental risk of landfills: leachate treatment by reversal osmosis, *Environmental Engineering and Management Journal* 11(12): 2319–2331.

Stiernström, S.; Hemström; K; Wik, O.; Carlsson, G.; Bengtsson, B-E.; Breitholtz, M. 2011. An ecotoxicological approach for hazard identification of energy ash, *Waste Management* 31(2): 342–352.
http://dx.doi.org/10.1016/j.wasman.2010.05.019

Svecevičius, G. 2006. Acute toxicity of hexavalent chromium to European freshwater fish, *Bulletin of Environmental Contamination and Toxicology* 77(5): 741–747.
http://dx.doi.org/10.1007/s00128-006-1126-4

Svecevičius, G.; Kazlauskienė, N. 2011. Behavioral responses in rainbow trout *Oncorhynchus mykiss* as indicators of sublethal exposure to heavy metals, in D. Cygas, K. D. Froehner (Eds.) "*Environmental Engineering*", *VIII International Conference, Selected Papers*, 19–20 May 2011, Vilnius, Lithuania, 374–377.

Svecevičius, G.; Kazlauskienė, N.; Slučkaitė, A.; Makaras, T. 2013. Toxicological assessment of the effects of closed landfill on neighboring hydroecosystem, in *Proceedings of the 4th International Conference on Environmental Management, Engineering, Planning and Economics (CEMEPE 2013) & SECOTOX Conference,* 24–28 June 2013, Mykonos island, Greece, 799–804.

Šaulienė, I.; Motiekaitytė, V.; Marčiulionienė, D.; Montvydienė, D. 2004. Application of biotests for toxicity evaluation of leachate and toxic tolerant vegetation processes in exploited landfill, in *Proceedings of the International Scientific Conference of 80 years Anniversary of Lithuanian University of Agricultture (Vandens telkinių apsauga ir valdymas [Protection and management of water bodies]).* Kaunas, Lithuania. 194–198.

Šiaulių Regiono Plėtros Taryba. 2012. *Sprendimas Nr. 51/5S-25. Dėl Šiaulių regiono komunalinių atliekų tvarkymo plano 2012–2020 m. patvirtinimo. 2012.05.03, Šiauliai* [Decision Regarding Confirmation of the Šiauliai Region Municipal Waste Management Plan for years 2012–2020. Šiauliai, 03.05.2012]. 82 p. (in Lithuanian).

Thomas, D. J. L.; Tyrrel, S. F.; Smith, R.; Farrow, S. 2009. Bioassays for the evaluation of landfill leachate toxicity, *Journal of Toxicology and Environmental Health* Part B, 12: 83–105.
http://dx.doi.org/10.1080/10937400802545292

Tigini, V.; Giansanti P.; Mangiavillano, A.; Pannocchia, A.; Varese, G. C. 2011. Evaluation of toxicity, genotoxicity and environmental risk of simulated textile and tannery wastewaters with a battery of biotests, *Ecotoxicology and Environmental Safety* 74(4): 866–873.
http://dx.doi.org/10.1016/j.ecoenv.2010.12.001

Tonkes, M.; de Graaf, P. J. F.; Graansma, J. 1999. Assessment of complex industrial effluents in the Netherlands using a whole effluent toxicity (or WET) approach, *Water Science Technology* 39(1): 55–61.
http://dx.doi.org/10.1016/S0273-1223(99)00253-X

Tonkes, M.; den Besten, P. J.; Leverett, D. 2005. Bioassays and tiered approaches for monitoring surface water quality and effluents, in P. J. Den Besten, M. Munawar (Eds.). *Ecotoxicological testing of marine and freshwater ecosystems: emerging techniques, trends and strategies.* Boca Raton, London, New York: CRC Press, Taylor and Francis Group, 43–86.

Tsiridis, V.; Petala, M.; Samaras, P.; Kungolos, A.; Sakellaropoulos, G. P. 2012. Environmental hazard assessment of coal fly ashes using leaching and ecotoxicity tests, *Ecotoxicology and Environmental Safety* 75(2): 212–220.
http://dx.doi.org/10.1016/j.ecoenv.2012.07.011

Vasarevičius, S.; Čegariova, J.; Sližytė, D. 2005. Investigation and evaluation of landfill leachate permeability in the soil, *Journal of Environmental Engineering and Landscape Management* 13(3): 108–115.

Vindimian, E. 2005. A multitest index of effluent toxicity by PLS regression, in C. Blaise, J. F. Férard (Eds.). *Small-scale freshwater toxicity investigations.* Vol. 2. Hazard assessment schemes. Dordrecht, the Netherlands: Springer, 89–114.

Vosylienė, M. Z.; Kazlauskienė, N.; Baršienė, J.; Andreikėnaitė, L.; Milukaitė, A.; Taujanskis, E. 2010. Ecotoxicity and genotoxicity relating to fish in wastewaters discharged from the Vilnius treatment plant, *Water Science and Technology* 62(4): 859–865. http://dx.doi.org/10.2166/wst.2010.916

Wilke, B. M.; Riepert, F.; Koch C.; Kühne T. 2008. Ecotoxicological characterization of hazardous wastes, *Ecotoxicology and Environmental Safety* 70(2): 283–293.
http://dx.doi.org/10.1016/j.ecoenv.2007.10.003

Wuttke, J.; Walther, M. 2013. *Recommendations for the ecotoxicological characterization of wastes.* Section 1.5 Municipal waste management, hazardous waste, focal point to the Basel Convention. Federal Environment Agency, Germany. 27 p.

Zha, J.; Wang, Z. 2006. Acute and early life stage toxicity of industrial effluent on Japanese medaka (*Oryzias latipes*), *Science of the Total Environment* 357(1–3): 112–119.
http://dx.doi.org/10.1016/j.scitotenv.2005.04.038

Zhang, X.; Wiseman, S.; Giesy, J. 2012. Toxicology of water, in A. Luch (Ed.). *Molecular, clinical and environmental toxicology.* Experientia Supplementum 101, Basel AG Springer, 1–27.
http://dx.doi.org/10.1007/978-3-7643-8340-4_2

Anolda ČETKAUSKAITĖ. Dr, retiree. PhD in Biology (Biochemistry, bioenergetics), Institute of Biochemistry of Republic of Lithuania, 1984. Senior research biochemist, 1988. Assoc. Prof., Department of Biochemistry and Biophysics, Faculty of Natural Sciences, Vilnius University, 1999–2012; affiliated Assoc Prof., Dept. Biochem. Mol. Biol., 2013. Scientific publications: 54. Presentations in over 20 international conferences, workshops, symposiums. Member of SETAC 2000–2010. Author of textbooks in Lithuanian language for bachelor and master degree university students: 1) Četkauskaitė A. Ecotoxicology: Mechanisms of Action of Chemical Compounds. Textbook, ISBN 9986-19-335-4, Vilnius: Vilnius University Publishing House, 1999. 452 p.; 2) Četkauskaitė A. Biochemical Toxicology. Textbook ISBN 978-9955-25-457-7, Kaunas: Kaunas University of Technology Publishing House "Technologija", 2008, 163 p.; 3) Četkauskaitė A. Ecological Biochemistry (Part I). Textbook in CD format, Kaunas: Kaunas University of Technology Publishing House "Technologija", 2008, 56 p. Research interests: ecotoxicology, aquatic toxicology, mechanisms of action of individual xenobiotics and their mixtures, QSAR, molecular biomarkers of xenobiotic toxicity, fate of xenobiotics in the environment.

Milda Zita VOSYLIENĖ. Dr, retiree. Laboratory of Ecology and Physiology of Hydrobionts. Institute of Ecology Nature Research Centre. PhD of Biomedical sciences, (physiology), Institute of experimental and clinical medicine, 1968. Publications during the last 10 years: 34 scientific publications in journals and books involved into the ISI list and 12 articles published in journals involved into "ISI Web of Science". During the last 12 y participated with the presentations in over 20 international scientific seminars, workshops, and conferences. Research interests: ecological physiology, fish physiology, ichthyohematology, ecotoxicology. 2007 – Lithuanian Science Award for the cycle "The use of biological tests to assess aquatic toxicity" (1986–2006), with co-authors.

Nijolė KAZLAUSKIENĖ. Dr, Laboratory of Ecology and Physiology of Hydrobionts. The head of Ecological physiology and toxicology sector Institute of Ecology Nature Research Centre. Dr of Biomedical sciences (physiology), Sechenov name Institute of Evolutionary biochemistry and physiology (Russia), 1987. Senior scientific worker, 2003. Publication: 97 scientific publications. Participated with the presentations in over 40 international scientific seminars, workshops, conferences, congresses and symposiums. Research interests: ecological physiology, fish physiology, ecotoxicology, ecology. 2007 – Lithuanian Science Award for the cycle "The use biological tests to assess aquatic toxicity" (1986–2006), with co-authors.

Virginija KALCIENĖ. Dr, lecturer at the Centre for Ecology and Environment Research, Vilnius University. PhD of Physical sciences (biochemistry), Vilnius University, 2011. Junior scientific worker at Department Biochemistry and Biophysics of Vilnius University, since 2009 up to 2011. Publications: 2 scientific publications. Presentations in 7 scientific conferences. Research interests: ecotoxicology, environmental risk assessment of chemicals, environmental chemistry, ecology.

16

EFFICIENCY EVALUATION OF AN ALGISTATIC TREATMENT BASED ON BARLEY STRAW IN A HYPERTROPHIC POND

Emilie Prygiel[a], Adeline Charriau[a,b], Romain Descamps[a], Jean Prygiel[a,c], Baghdad Ouddane[a], Gabriel Billon[a]

[a]*Géosystèmes, UMR 8217 CNRS – University Lille 1, Villeneuve d'Ascq, France*
[b]*Research Group on Water, Soil and the Environment EA 4330, Faculty of Sciences and Technology, University of Limoges, France*
[c]*Artois-Picardie Water Agency, Douai, France*

Abstract. Eutrophication processes were closely examined in a hypertrophic pond that was subjected to a barley straw treatment to inhibit cyanobacteria blooms. A phytoplankton activity was present from February to November 2010 with Chlorophyll *a* concentrations that ranged between 8 and 83 μg L^{-1}. The good oxygenation of the pond provided by primary production all along the year seems to maintain an oxic layer at the top of the sedimentary column. A high-frequency monitoring of algal development pointed out the end of the diatom bloom at the end of April and the growth of green algae from beginning of May. One month later, the development of cyanobacteria supplanted other algae several times. The implementation of barley straw does not seem to hinder the growth of cyanobacteria in June, August and September and do not promote the development of other algae potentially favoured by such a treatment. As a consequence, a barley straw treatment does not seem to be really appropriate for this kind of ponds in Northern France and more tests should still be undertaken to prove its real efficiency.

Keywords: eutrophic pond, sediment, nutrients, phytoplankton, environment monitoring, barley straw.

Reference to this paper should be made as follows: Prygiel, E.; Charriau, A.; Descamps, R.; Prygiel, J.; Ouddane, B.; Billon, G. 2014. Efficiency evaluation of an algistatic treatment based on barley straw in a hypertrophic pond, *Journal of Environmental Engineering and Landscape Management* 22(1): 1–13. http://dx.doi.org/10.3846/16486897.2013.801847

Introduction

Input of an excess of nutrients, especially phosphorus, in relative closed systems, where water flow is low and water residence time is high, can result in algal proliferations, and particularly cyanobacteria blooms (Bai *et al.* 2009; Kisand 2005). In Europe, the most common cyanobacteria gena are *Microcystis*, *Anabaena*, *Aphanizomenon*, *Oscillatoria*, *Nodularia* and *Nostoc*. Although primary production necessitates at least carbon dioxide, nitrate, phosphate and silicate, it has been generally observed in freshwater ponds where nitrate and/or phosphate are the main limiting factors of algal development. More precisely, in hypertrophic lakes/ponds, where phosphate can reach high concentrations, phosphorus cannot be efficiently removed from water bodies with gaseous species (H_3P) and is progressively accumulated in sediments, and to a lesser extent in water (Cao *et al.* 2011; Xiang, Zhou 2011). Nitrate can, therefore, be the limiting nutrient permitting the proliferation of heterocystous cyanobacteria (O'Neil *et al.* 2012, and references therein), even if other interacting hydrodynamic and climatic parameters also play a key role in such excessive algal development (mixing conditions, sunlight, food web structure, water temperature, transparency and residence time) (Khan, Ansari 2005). In freshwater environments, when nitrogen becomes the limiting factor (that is generally characterised by the molar ratio N/P lower than 16) and nitrate concentrations drop down to 50–100 μg L^{-1}, another source of nitrogen, like atmospheric N_2, becomes necessary for algal development, especially in summer when light and temperature are optimal. In this case, only heterocystous cyanobacteria such as *Anabaena*, *Aphanizomenon*, *Cylindrospermopsis*, *Nodularia* and *Nostoc* are able to assimilate N_2 (Paerl *et al.* 2006).

Several preventive actions and curative treatments, including the reduction of nutrient loadings from the catchment, the spreading of copper sulphate and limestone, and sediment dredging, may be applied to limit cyanobacteria blooms (Le Jeune *et al.* 2006). In the 1990s, a promising technique based on barley straw

Corresponding author: Gabriel Billon
E-mail: gabriel.billon@univ-lille1.fr

decomposition was developed and applied successfully in several water bodies (Brownlee *et al.* 2003; Ó hUallacháin, Fenton 2010; Purcell *et al.* 2013). Both the way of action and the long-term efficiency of such a treatment are controversial (Hagström *et al.* 2010): the decomposition of the lignin present in straw produces polyphenolic compounds that have an algistatic effect on several cyanobacteria and other algae as well (Ferrier *et al.* 2005; Pęczuła 2013), whereas several phytoplankton species seem to be stimulated (Ferrier *et al.* 2005; Grover *et al.* 2007). Another hypothesis proposed by Everall and Lees (1997) and Iredale *et al.* (2012) is the production of hydrogen peroxide during the photo-oxidation of a particular type of lignin present in barley straw, and which is highly toxic towards phytoplankton.

The present work aims at studying the response of a hypertrophic lake (Pont-Rouge Pond, located in Northern France in the Artois-Picardie watershed) subjected to an algistatic treatment using barley straw. In this area, numerous ponds and lakes have experienced an increase in the frequencies of these blooms, in particular with the development of *Microcystis* and *Planktothrix* (Willame, Hoffmann 2005). As a consequence, harmful algal blooms threaten economic activities and appear as a crucial issue in terms of water quality degradation and human health (Newcombe *et al.* 2012). A chemical and biological monitoring was performed in 2010, before and during the treatment, to highlight its potential impact on algal development, and particularly on cyanobacteria proliferations. The fate of phosphorus, and to a lesser extent nitrogen, was in parallel closely examined in the water column and surface sediments; chemical redox indicators were additionally taken into account to predict the possible mobilisation of phosphate at the water-sediment interface during anoxic periods. Finally, this biogeochemical approach permits us to better understand the phytoplankton cycle in a hypertrophic pond and to point out the difficulties to fight against eutrophication processes.

1. Materials and methods

1.1. Location and presentation of the site

The Pont-Rouge Pond is a 13 ha water area located in the city of Le Quesnoy (Northern France, see Fig. 1). This pond was used in the past as a defence structure, with still existing ramparts, and is nowadays dedicated to leisure activities, including fishing and swimming. The pond is not a fully closed system, considering that it receives water from a small stream (L'Ecaillon stream, in the south-eastern part of the pond) during the rains, and that a water lock (north-western part of the pond) allows maintaining a constant water level. The average depth of the pond is 1.92 m with a maximal depth of 3 m. The volume of the pond has

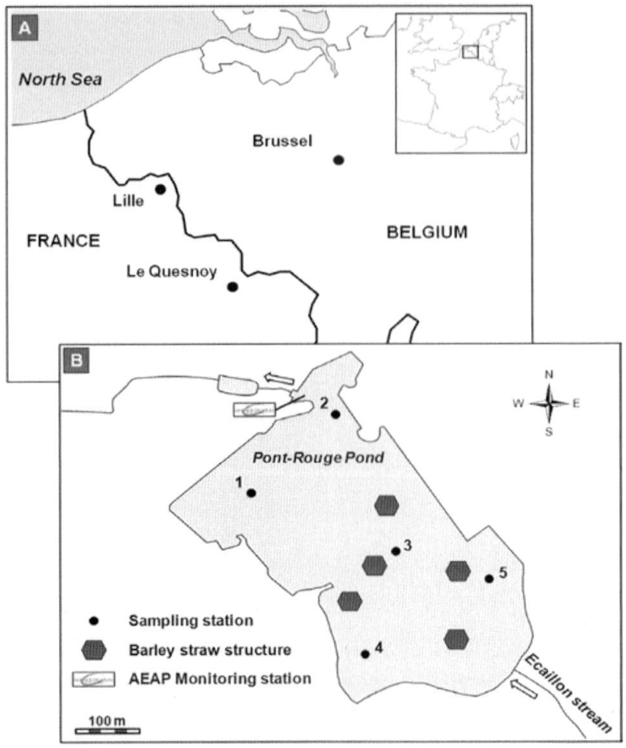

Fig. 1. Location of the city of Le Quesnoy (A) and map of the Pont-Rouge Pond including the location of sampling stations, barley straw structures and the monitoring station (B)

been estimated to be 240,000 m^3 with an average flowing-through time approximately equal to 39 days. Cyanobacteria blooms were noticed for the first time in the 1990s. To fight against these proliferations, several preventive actions or treatments have been tested, such as chemical treatments with copper sulphate (1990), alumina sulphate (1991) and calcium salts (2006), reduction of phosphorus inputs (1994–1995), sediment dredging (2000) and implementation of oxygenation structures. However, an increase in the frequency of these blooms was registered, often with the presence of *Anabaena sp*, *Aphanizomenon sp*, *Aphanothece sp* and *Planktothrix agardhii*.

1.2. Biogeochemical sampling and analysis

Five sampling stations were defined in the pond to monitor physico-chemical, chemical and biological parameters in the overlying water and sediments from February to November 2010 (Fig. 2).

Dissolved oxygen, conductivity, pH and turbidity were measured *in situ* by using a WTW multimeter (Oxi 340) connected to appropriate electrodes and a turbidimeter (Hach Lange, TSS portable). A Secchi disk was also used to evaluate water transparency.

For each station, water samples, collected in acid pre-cleaned bottles, were filtered on cellulose acetate membranes (Alltech, 0.45 μm) and immediately frozen until nutrient analyses. For the determination of metal concentrations, 100 μL of ultrapure nitric acid (Merck)

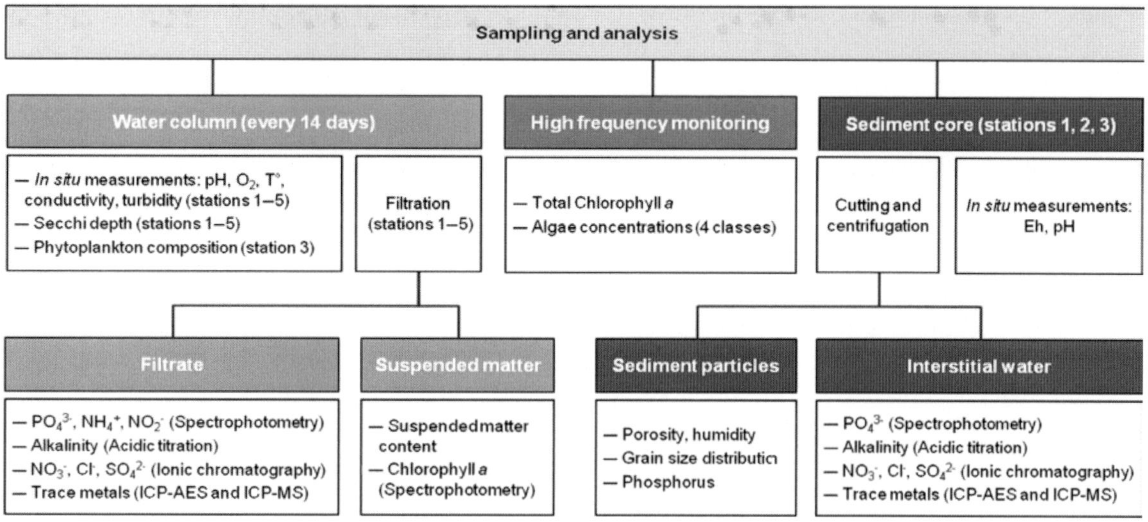

Fig. 2. Sampling and analysis design for water, sediment and phytoplankton

were immediately added to 10 mL of filtered samples that were kept at 4 °C until elemental analyses. Dissolved orthophosphate (also called DRP: Dissolved Reactive Phosphorus), ammonium and nitrite concentrations were determined according to standard spectrophotometric methods proposed by the International Organisation for Standardisation (ISO).

Measurements were performed using a spectrophotometer (Varian Cary 300 Scan UV-Visible) with a 5-cm length sample cuvet. The limits of detection (LOD) were 10 μg L^{-1} for PO_4^{3-} and 2 μg L^{-1} for NH_4^+ and NO_2^-. Nitrate contents were determined using an ionic chromatography (Dionex Ion Pac AS 18) with an LOD of 0.1 mg L^{-1}. Alkalinity was estimated by titration (Metrohm 736 GP Titrino) with a 0.02 M hydrochloric acid solution (LOD = 0.2 mmol L^{-1}). Dissolved iron and manganese concentrations were measured in acidified water samples by ICP-AES (Inductively Coupled Plasma-Atomic Emission Spectroscopy) (Varian Vista Pro, axial view), with an LOD of 0.01 and 0.05 mg L^{-1}, respectively. Suspended matters collected on glass fibre membranes (Whatman GF/F, 0.7 μm) were immediately frozen until the determination of Chlorophyll *a* (Chl-*a*) concentrations. Briefly, filters were crushed into centrifugation tubes in the presence of 10 mL of a mixture acetone-Milli-Q water (90–10, v/v), stored overnight at 4 °C, and then centrifuged at 3000 rpm for 15 min. Chl-*a* was measured in the supernatant using the previously mentioned spectrophotometer at the wavelength of 665 nm.

Sediment samplings were carried out in stations 1, 2 and 3 in spring 2010 (04 March, 01 April and 20 April, respectively). At each station, 2 sediment cores were carefully collected using 35-cm-long Perspex tubes. The first one was used to assess directly in the field the profiles *versus* depth of pH and redox potential by using a glass electrode (Toledo Mettler, penetration type) and a platinum electrode (Toledo Mettler), respectively, both combined with a Ag/AgCl, [KCl] = 3 M, reference electrode. In the laboratory, the second core was dedicated to the analysis of iron, manganese, nitrate and phosphate concentrations in pore waters, as well as the water percentage, porosity, granulometric distribution and phosphorus contents in bulk sediment. For that purpose, the second core was sliced in 2-cm thick slices under a nitrogen atmosphere inside a glove box. A part of these sliced sediments was kept in plastic bags previously purged with N_2 for solid analyses and the remaining part was centrifuged (2500 rpm for 20 min) in order to recover supernatant waters. Centrifuge tubes were returned to the N_2 flushed glove box, where pore waters were filtered with cellulose acetate membranes (Alltech, 0.45 μm) and immediately acidified with ultra-pure nitric acid (100 μL in 10 mL of pore water). Dissolved iron and manganese concentrations were measured as previously described. Total phosphorus concentrations in sedimentary particles were measured by spectrophotometry after the following extraction procedure. Briefly, dried sediments were first pyrolysed at 450 °C during 24 hours to mineralise phosphorus bound to organic matter. 200 mg of sediment particles were then attacked with 3.5 M hydrochloric acid during 24 hours at room temperature. The leaching solution was thereafter filtrated at 0.45 μm and analysed by spectrophotometry after the formation of an antimony–phosphomolybdate complex, subsequently reduced in the presence of ascorbic acid. A standard sediment (CRM 684) was used to validate the protocol and the recovery was found to be better than 90%. Granulometric distributions were evaluated with a laser granulometer (laser Malvern Mastersizer 2000).

1.3. High-frequency phytoplankton monitoring

An Algae Online Analyser (AOA, BBE Moldaenke) was deployed from May to June 2010 to record at a high frequency (1 measurement every 10 min) the evolution of phytoplankton. The AOA was embedded in the Artois-Picardie Water Agency (AEAP) monitoring station located in the northern part of the pond just upstream from the water lock (Fig. 1). The pump allowing a continuous water sampling was immersed to a depth of 50 cm. This autonomous fluorometer measures the concentrations of total Chl-*a* and four classes of algae (*Chlorophyceae, Bacillariophyceae, Cyanophyceae* and *Cryptophyceae*) thanks to five light emitting diodes (450, 252, 570, 590 and 610 nm) exciting the characteristic pigments of phytoplankton groups. Additionally, in order to get a more accurate view of the phytoplankton composition, 1 L of water was sampled twice a month, at 30-cm depth in the middle of the pond (Fig. 1, station 3). 5 mL of lugol were immediately added to preserve algae before qualitative and quantitative analyses.

1.4. Thermodynamic equilibrium calculations

The visual Minteq software version 2.61 (Gustafsson 2012) was used to calculate the saturation index of calcite in the overlying water (Eqn 1):

$$SI = \log \frac{\{Ca^{2+}\}\{CO_3^{2-}\}}{K_S},$$

where $\{Ca^{2+}\}$ and $\{CO_3^{2-}\}$ are the activities of Ca^{2+} and CO_3^{2-} species at the thermodynamic equilibrium and K_s the solubility product of $CaCO_3$. Calculations were undertaken from the following input data: pH, alkalinity, temperature values and the concentrations of Ca^{2+}, NO_3^-, Mg^{2+}, Na^+, Cl^-, SO_4^{2-} and PO_4^{3-} components.

1.5. Barley straw treatment procedure

The amount of barley straw (*Hordeum vulgare*) has been calculated from the surface of the pond according to CEH (2004), 3 tons of straw being necessary to achieve the recommended value of 50 g of straw per m^2. Barley straw was introduced in 150 loosely packed rolls that were prepared on site using a tree-wrapping machine and 3 km of tubular netting. Rolls were then attached on five anchorage floating structures. Additionally, 50 buoys were added to insure the flotation of the rolls and to permit the degradation of barley straw under aerobic conditions during the monitoring experiment (Waybright *et al.* 2009). Barley straw was introduced in the pond in the beginning of May 2010.

2. Results and discussion

2.1. Physico-chemical characterisation of the overlying water

The physico-chemical parameters recorded in the overlying water show a good homogeneity between the five stations during the monitoring experiment (Table 1). Temperature values increase from 5.8 °C in late February to a maximum of 24.8 °C in July, then decreasing back to 9.7 °C in November. Conductivity values range from 400 to 480 μS cm^{-1} during the experiment with maxima occurring in July, because of higher temperatures and lower rainfalls compared to winter time. The whole water body of the Pont-Rouge Pond is generally well oxygenated with a supersaturation during daytime throughout the year, with the exceptions of July and August when saturation dropped down to 80% (Table 1). However, these measurements were done twice a month and one cannot exclude short anoxic events during a few days, particularly close to the water-sediment interface.

pH values ranged between 7.7 and 8.8, in accordance with the carbonate-rich geological background. pH evolution during the monitoring period was probably mostly related to the primary production that consumed dissolved CO_2 in the daytime and released it during the night. However, hydrogen carbonate concentrations were generally constant with alkalinity values in the range of 1.6–2.8 mmol L^{-1} (Table 1). The lowest alkalinity values on 20 May, 4 June and 18 August corresponded to higher levels of ammonium, nitrate and nitrite, revealing organic matter degradation processes.

Finally, the values of turbidity (4.9–29.6 FNU) and water transparency (0.4–1.3 m) (Table 1) indicated that light did not penetrate deep into the water column. Therefore, this shallow euphotic zone would tend to enhance the development of cyanobacterial species adapted to low luminosity, for instance, *Planktothrix sp.* (Schwaderer *et al.* 2011; von Sperling *et al.* 2008).

2.2. Geochemical properties of sediments

Porosity and water contents in surface sediments were relatively high with values equal to 80% and 90%, respectively. In deeper depths, sediments were settled with significant differences according to the sampling sites: at 10-cm depth, the water percentage was 72% in station 2 and only 27% at station 3. Granulometric distributions were also estimated and pointed out a heterogeneity of sediments, dominated by clay particles (size < 4 μm), followed by silty particles ($4 <$ size < 63 μm). The sandy fraction was negligible in all sampling sites with less than 1% of the total fraction.

Redox potential values (given against the reference Ag/AgCl, [KCl] = 3 M electrode) were positive above the water-sediment interface and dropped drastically in the first 4 cm of the sedimentary column with an amplitude of 400–700 mV (Fig. 3). These profiles indicate clearly that early diagenetic processes mediated by bacterial activity occurred in sediments. For the aquatic ecosystem of the Pont-Rouge Pond, the sediment pool acts as a

Table 1. Physico-chemical and chemical parameters monitored in the overlying water from February to November 2010: average values for the 5 stations and relative standard deviations for these 5 results (RSD, in%)

	26-Feb		25-Mar		15-Apr		28-Apr		20-May		4-Jun		18-Jun		1-Jul	
	Avge	RSD	Avge	RSD	Avge	RSD	Avge	RSD	Avge	RSD	Avge	RSD	Avge	RSD	Avge	RSD
Water temperature (°C)	5.8	1.2	11.3	1.2	11.4	2.1	15.7	1.6	17.2	3.4	20.1	1.5	19.0	0.6	24.8	0.2
Conductivity (μS cm^{-1})	417	1.3	438	0.3	406	5.9	423.0	0.3	450.0	0.2	444.0	0.3	449.0	0.6	465.0	0.3
Dissolved oxygen (mg L^{-1})	14.9	2.3	14.0	3.0	11.7	5.0	9.8	3.9	10.5	1.6	8.9	5.0	9.3	9.4	8.8	17.4
Saturation (%)	125	6.1	130	2.7	133	3.5	121	1.3	130	2.5	120	5.0	123	8.7	128	17.5
pH	8.1	0.4	8.0	0.4	8.6	1.8	8.8	1.0	8.2	0.9	8.3	1.2	8.6	0.3	8.4	0.8
Turbidity (FNU)	*	*	*	*	14.2	17.8	9.3	27.0	5.0	20.4	5.6	39.3	*	*	7.1	18.6
Transparency (m)	*	*	*	*	0.5	20.0	0.7	11.3	1.2	10.3	1.3	16.5	0.6	0.0	1.0	9.3
Alkalinity (mmol L^{-1})	2.4	4.8	2.3	5.1	2.6	4.9	2.5	7.3	1.7	7.6	1.9	4.9	2.8	4.6	2.6	17.2

	13-Jul		22-Jul		18-Aug		26-Aug		9-Sep		22-Sep		11-Oct		15-Nov	
	Avge	RSD	Avge	RSD	Avge	RSD	Avge	RSD	Avge	RSD	Avge	RSD	Avge	RSD	Avge	RSD
Water temperature (°C)	24.4	0.4	22.6	1	18.6	0.5	19.9	0.6	17.9	0.6	16.6	1.1	15.1	1.7	9.7	2.9
Conductivity (μS cm^{-1})	482	0.3	483	0.2	415	14.3	438	0.2	433	0.3	427	0.7	427	0.2	401	1.7
Dissolved oxygen (mg L^{-1})	5.5	9.7	5.4	3.5	4.8	7.9	8.1	5.2	8.3	15.8	9.0	5.7	10.1	7.3	12.0	5.6
Saturation (%)	80	8.2	78	2.6	78	7.7	107	5.8	107	15.9	113	6.1	125	10	104	4.9
pH	8.0	0.7	8.1	0.6	8.2	2.0	8.2	1	8.1	0.3	8.5	1.1	8.3	0.3	7.7	0.7
Turbidity (FNU)	15.3	16.3	18.2	4.3	*	*	16.1	16.1	11.6	6.6	11.3	34.6	9.7	4.7	27.2	13.7
Transparency (m)	0.7	16.9	0.6	4.7	0.5	0	0.6	10.6	0.7	0	0.8	15.3	0.8	3.5	0.4	10.6
Alkalinity (mmol L^{-1})	2.7	22.9	2.4	16.9	1.6	12.1	2.0	10.6	2.1	7.2	2.4	3.1	2.3	2.3	2.4	3.8

*data not available

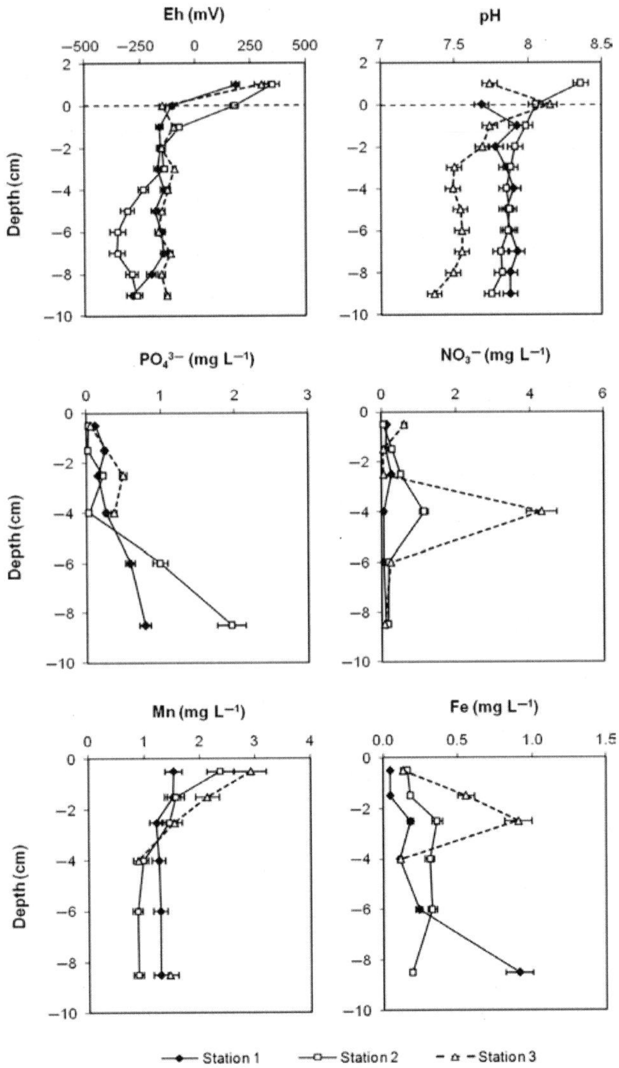

Fig. 3. Depth profiles of Eh (in mV *vs.* Ag/AgCl), pH and concentrations (in mg L^{-1}) of nitrate, phosphate, iron and manganese in the dissolved phase of sedimentary pore waters (sampling stations 1, 2 and 3). Negative values in the *y*-axis correspond to depths in the sediment core

bioreactor consuming quantitatively dissolved oxygen present close to the water-sediment interface. Just below, the consumption of oxygen by bacterial biomass, i.e. the biologically mediated reduction of nitrate, manganese and iron (hydr)oxides, occurred in sediment. The production of Mn^{2+} and Fe^{2+} in pore waters was clearly evidenced with concentrations much higher than in overlying waters: 880–2900 mg L^{-1} for Mn compared to 2–60 μg L^{-1} in overlying waters, and 45–4900 mg L^{-1} for Fe compared to 10–98 μg L^{-1} in overlying waters (Fig. 3).

As predicted by thermodynamic data, the Mn (hydr)oxides reduction occurs at higher redox potential than that of Fe (hydr)oxides, explaining the high concentrations of Mn^{2+} observed at 0.5 cm depth, whereas Fe^{2+} concentrations started to increase significantly from 1- to 2-cm depth. These profiles also indicated that the oxic and hypoxic conditions, related to biodegradable organic matter mineralisation processes, were probably restricted to the very upper layers of sediments and that phosphate release may occur quickly in case of anoxic processes at the water-sediment interface (see Section 2.3).

2.3. Nitrogen and phosphorus behaviour in relation with algal development

In the Pont-Rouge Pond, Chl-*a* concentrations ranged from 8 to 83 μg L^{-1} during the monitoring experiment (Table 1). In February and November, values remained relatively high of around 35 μg L^{-1}, explaining at least partly why phosphate was consumed all along the year without any regeneration during winter time (see thereafter). Significant decrease in Chl-*a* values took place after maxima observed in mid-April and in mid-June, indicating a quick degradation of primary production. However, phaeopigments (Phaeo) concentrations did not increase significantly during these periods and the ratio Chl-*a* / Phaeo remained higher than 1. Generally, it seems that primary production is active all along the year with a continuous recycling of dead organisms. In these conditions, sudden input of detrital material that would consume quantitatively and quickly dissolved oxygen was rather limited. As a consequence, no anoxia bound to the mineralisation of freshly died algal cells has been noticed.

The evolution of phytoplankton composition was determined from March to October 2010 (Fig. 4). In March 2010, phytoplankton was dominated by flagellated *Chrysophyceae* (class of *Chromophyceae*), and particularly by the species *Erkena sp.* In spring, centric diatoms (mostly represented by *Stephanodiscus sp.*) colonised the pond. At the same time, green algae started developing, with a predominance of *Ankyra lanceolata* in May. From June, cyanobacteria were detected, sometimes at high density: *Aphanizomenon sp.* dominated three times the algal population (04 June 2010, 17 July 2010 and 22 July 2010) and *A. flos-aquae* development resulted in a scum on 4 June at the surface of the pond. At the end of summer, *Aphanocapsa sp*, *Coelomoron* sp and *Anabaena sp* were punctually the main algal species. Finally, a minimal diversity (12 taxons) was measured in May–June, whereas a maximal diversity of 56 taxons was observed at the end of August.

Although a regular manual sampling was carried out monthly, or even every two weeks, allows the observation of some changes in algal species and their concentrations, the dynamic of the phytoplankton behaviour can only be evidenced by an increasing number of measurements. The high-frequency monitoring using the Algae Online Analyser (AOA, BBE moldaenke) confirmed the end of the diatom blooms in late April (Fig. 5) followed by a clear water period with low Chl-*a* values.

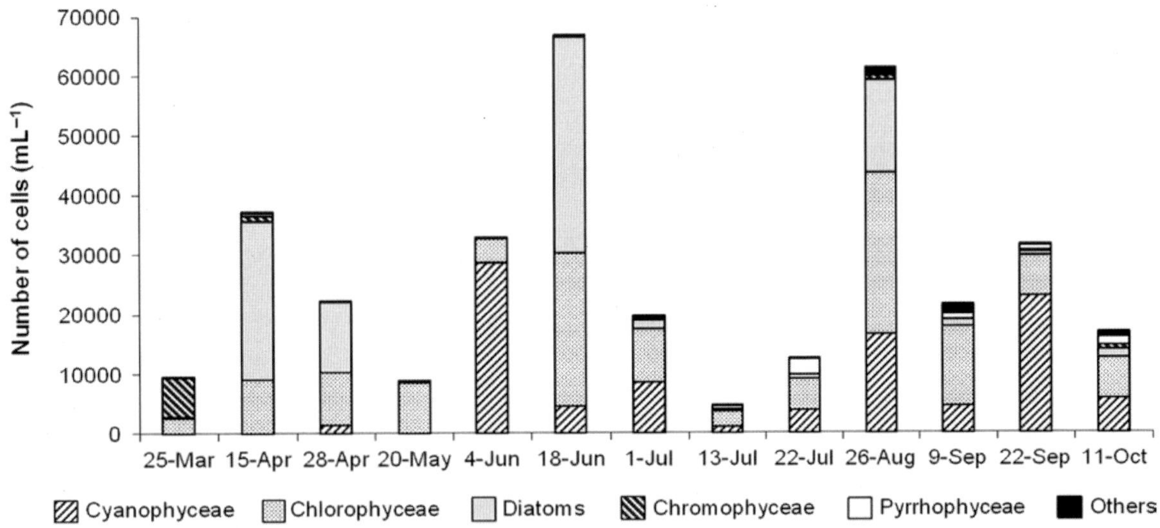

Fig. 4. Evolution of phytoplankton main groups in the overlying water of the Pont-Rouge Pond: *Cyanophyceae*, *Chlorophyceae*, *Diatoms*, *Chromophyceae* (without *Diatoms*), *Pyrrhophyceae* and others (included *Euglenophyceae* and undetermined species)

From the beginning of May, an increase of green algae was recorded (with higher Chl-*a* concentrations). On 24 May, cyanobacteria started to develop and from 4 June, their concentrations (10 µg L^{-1}) were higher than those of green algae (approximately 3 µg L^{-1}) leading to a visible bloom of *Aphanizomenon flos-aquae.* This sharp view of algal succession clearly evidenced that the growth rate of each algal species took place at the timescale of the day. For instance, diatoms and cyanobacteria species were only significantly detected over approximately 1–2 weeks in May and June 2010 and their amounts were multiplied by 10–20 within 2–3 days. These quick evolutions can be linked to the grazing pressure by zooplankton and other invertebrates, the change of physico-chemical conditions and/or weather, although no direct correlation has been established between these primary production events and luminosity, rainfall and/or temperature during the high-frequency monitoring.

2.3.1. Nitrogen

Nitrate, ammonium and nitrite were recorded periodically in the 5 sampling stations at 50 cm below water surface (Table 2). Concentrations were homogeneous in the whole pond, with the exceptions of periods characterised by low concentrations or bloom events. Nitrate concentrations decreased progressively between February and April to values below 0.1 mg L^{-1}. Between May and August, erratic variations were observed with maximal concentrations similar to those recorded in winter.

As a consequence, the development of algal species during spring was followed by punctual nitrification processes. For instance, a green algae rapid growth occurred at the end of May–beginning of June. The mineralisation of the phytoplankton biomass resulted in a release of ammonium rapidly oxidised in nitrate due to the presence of dissolved oxygen. This process resulted also in the production of nitrite with concentrations raising 0.2 mg L^{-1}. These high contents corresponded to an enrichment of approximately 10 times the background nitrite concentrations.

However, the increases of nitrite and ammonium concentrations were limited because of an efficient oxygenation of the pond during all seasons. Between June and September, nitrate concentrations remained low indicating that this nutritive salt could limit algal development (see thereafter). In pore-water sediments, nitrate was quantitatively consumed and concentrations dropped generally down to the LOD (LOD = 0.1 mg L^{-1}) within the first centimetre. However, some local maxima were observed probably due to bioturbation and local nitrification processes. Overall, the results clearly showed that the sediments did not act as a source of nitrate for the pond water column, but rather as a denitrification reactor producing gaseous di-nitrogen (N_2) by the bacterial activity as mentioned previously (Mortimer *et al.* 2004).

2.3.2. Phosphorus

In shallow water bodies, where the water depth does not exceed a few metres (including rivers, ponds and lakes), phosphorus cycle is dependent on a multitude of biogeochemical parameters both in the water column and in surface sediments.

In these environments, the recycling of DRP in the water column during winter is barely observed and more complex pathways are rather proposed to explain the behaviour of phosphorus and its link with primary production (House 2003). In the Pont-Rouge Pond, DRP concentrations were lower in winter than in summer (Table 2), suggesting an input of phosphate

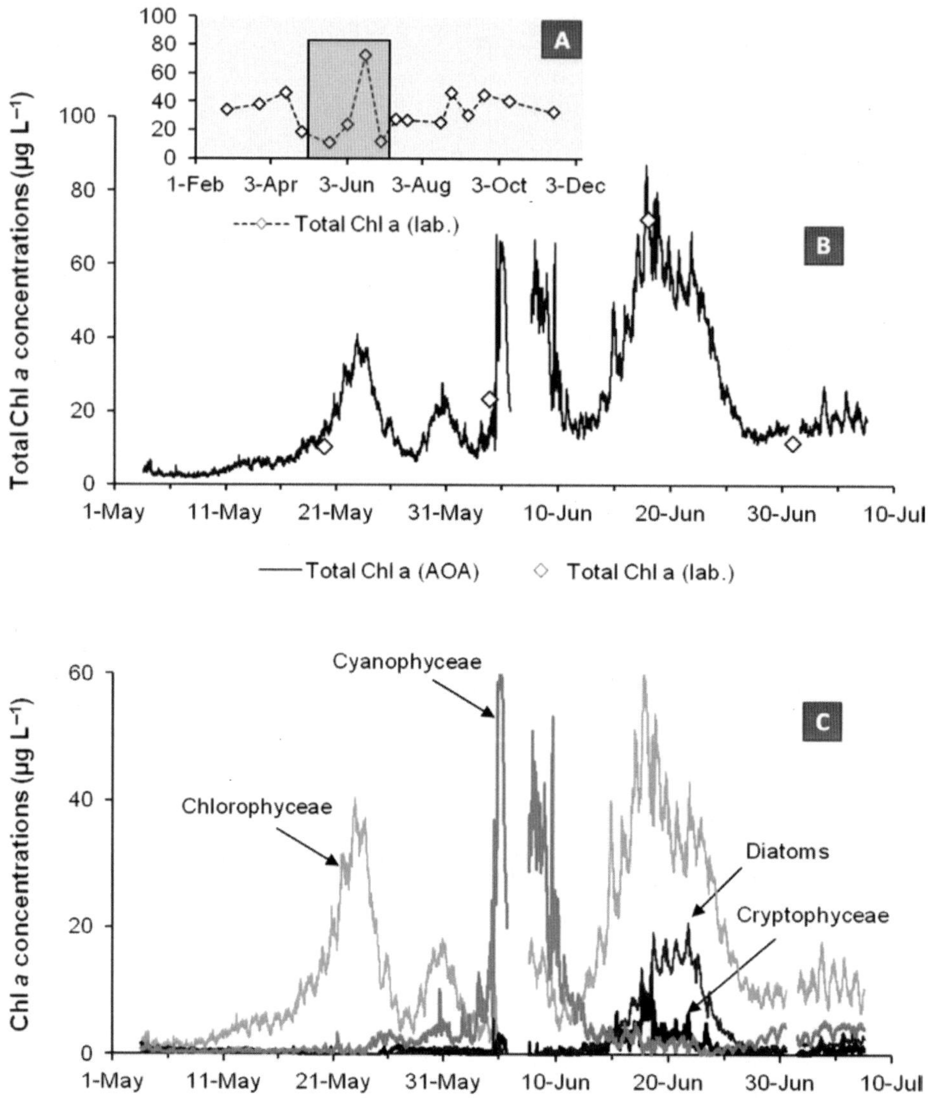

Fig. 5. Chlorophyll *a* concentrations (in µg L^{-1}) in the overlying water of the pond. A, total concentrations determined by the laboratory procedure (period February–November 2010, average for the 5 stations). B, total concentrations monitored by the Algae Online Analyser (AOA) (period May–July 2010). C, concentrations of phytoplankton species monitored by the AOA (period May–July 2010)

from sediment to the overlying water. As phosphorus cycle depends on numerous physical, chemical and biological parameters (House 2003), a complete understanding of this cycle was not achieved. However, several essential points, closely related to the main purposes of this article, need to be underlined in this relatively closed system: (1) during winter, DRP concentrations in overlying waters were quite high indicating that the pond can be considered as eutrophic to hypereutrophic according to the Carlsons's Trophic State Index; (2) in early spring, an increasing primary production corresponded to a decrease of DRP concentrations due to assimilation processes; and (3) from the beginning of May, DRP concentrations increased again to values up to 300–400 µg L^{-1}.

This relevant input of DRP in the pond is probably not bound to external inputs, since no important tributaries or agricultural activities charged the pond significantly. Most probably, surface sediments should play a key role in the cycle of phosphorus.

In sediment pore waters, DRP concentrations increased with depth (Fig. 3) because of (1) the mineralisation of organic matter by bacterial activity that produced inorganic phosphorus; (2) the reduction of iron (hydr)oxides on which exchangeable phosphate was initially bound to and; (3) a drop of pH values of approximately 0.5 unit compared to those measured in the overlying water that both contributed to desorb a fraction of phosphate and to partly dissolve carbonates in which phosphate was co-precipitated (House 2003). However, in the first centimetre of sediments, DRP concentrations remained low as a consequence of the presence of an iron oxide layer that additionally hampered, at least partially, the diffusion of phosphate

Table 2. Concentrations of nitrate, nitrite, ammonium, phosphate, iron and manganese (in $\mu g\ L^{-1}$) in the overlying water from February to November 2010: average values for the 5 stations and relative standard deviations for these 5 results (RSD, in%)

	Avge	*RSD*	Avge	*RSD*	Avge	*RSD*	Avge	*RSD*	Avge	*RSD*	Avge	*RSD*	Avge	*RSD*	Avge	*RSD*
	26-Feb		25-Mar		15-Apr		28-Apr		20-May		4-Jun		18-Jun		1-Jul	
NO_3^- ($\mu g\ L^{-1}$)	6099	10.3	3274	11.6	472	25.7	275	82.9	1070	6.5	2245	100.1	207	38.8	182	16.9
NH_4^+ ($\mu g\ L^{-1}$)	41	66.6	21	7.5	26	12.4	16	8.9	216	8.6	51	11.8	19	33.6	15	18.9
NO_2^- ($\mu g\ L^{-1}$)	10	29.5	20	9.2	3	31.3	3	38.3	47	13.3	33	7.0	<2		11	29.1
PO_4^{3-} ($\mu g\ L^{-1}$)	46	18.4	31	105.1	19	27.8	11	22.9	60	12.2	92	14.2	153	18.9	118	13.9
Fe ($\mu g\ L^{-1}$)	32	16.0	24	10.2	22	3.2	21	11.6	25	26.4	63	14.4	26	14.1	22	15.8
Mn ($\mu g\ L^{-1}$)	5	31.7	5	46.3	2	23.2	2	13.8	12	42.2	62	24.2	15	24.1	22	27.9
	13-Jul		22-Jul		18-Aug		26-Aug		9-Sep		22-Sep		11-Oct		15-Nov	
NO_3^- ($\mu g\ L^{-1}$)	117	24.5	144	21.4	378	6.6	99	0.0	101	0.0	188	42.6	100	0.0	3845	27.5
NH_4^+ ($\mu g\ L^{-1}$)	39	18.4	22	13.3	166	9.5	10	21.7	18	51.2	18	56.6	12	28.5	79	5.9
NO_2^- ($\mu g\ L^{-1}$)	21	13.3	14	19.6	55	22.9	35	31.0	8	7.8	6	24.9	4	60.9	37	17.1
PO_4^{3-} ($\mu g\ L^{-1}$)	382	4.3	280	4.4	122	10.7	94	9.5	178	2.5	100	23.5	100	0.0	262	16.5
Fe ($\mu g\ L^{-1}$)	19	12.0	17	5.7	9	26.6	21	42.8	11	26.0	7	38.0	15	7.5	98	21.1
Mn ($\mu g\ L^{-1}$)	31	44.5	77	43.8	88	6.6	64	41.1	<1		<1		<1		12	39.8

back to overlying waters. Analyses of total phosphorus (TP) concentrations in sedimentary particles revealed that concentrations were maximal close to the water interface at around 1.0–1.2 g kg^{-1}, and decreased with depth down to 0.6–0.8 mg kg^{-1}, depending on the sampling station (data not shown). These profiles indicated that the (hydr)oxides layer (and probably to a lesser extent carbonates' minerals) present in surface sediments accumulated efficiently a large amount of DRP, resulting in strong opposite gradient of dissolved DRP and TP concentrations in the sediment column (Anschutz *et al.* 2007; Kleeberg *et al.* 2010). The high amount of exchangeable phosphate at the water-sediment interface can be considered in this system as a temporal sink for phosphorus, but also as a potential source of DRP for the pond in case of anoxic processes. Anoxic sediment resuspension (for instance, by bioturbation) or degradation of detrital organic matter after a phytoplankton bloom could consume quantitatively the remaining oxygen present in sediment, and particularly that contained in iron (hydr)oxides, leading to a quick release of DRP in overlying water (Palmer-Felgate *et al.* 2011, and references therein).

Previous studies pointed out the scavenging of DRP excessive input by calcite co-precipitation (House 2003). According to Merz-Preiß and Riding (1999), this generally occurs when calcite saturation indexes ($SI_{calcite}$) are higher than 0.8. In this study, $SI_{calcite}$ punctually reached values higher than 1; CO_2 consumption by the photosynthetic activity resulted in a pH enhancement favourable for $CaCO_3$ precipitation. However, no significant correlation was found between $SI_{calcite}$ and DRP concentrations and a direct precipitation of DRP with Ca was also possible since hydroxyapatite SI was generally positive. Nevertheless, as the concentration of DRP remained much lower than 4.8 mg L^{-1}, the crystal nucleation of apatite phases did not occur easily (Plant, House 2002). These findings suggest that, in the overlying waters, the chemical control of DRP mobility by Ca minerals should probably not be the main process. Note further that, despite the high primary production occurring in the pond, dissolved phosphate concentrations were generally higher than 50–100 $\mu g\ L^{-1}$ and the $[NO_3^-]/[PO_4^{3-}]$ ratio values were higher than 16 until beginning of June, and then dropped down to 5. From this time, nitrate became the limiting factor and its low concentrations ($<100\ \mu g\ L^{-1}$) resulted in the relative limitation of algal development that gave heterocystous cyanobacteria the opportunity to grow because of their ability to assimilate gaseous di-nitrogen. However, DRP concentrations in the pond still increased during this period, suggesting a significant release of dissolved phosphate from sediments due to anoxic processes at the water-sediment interface (Lake *et al.* 2007; Xie *et al.* 2003). That occurs particularly after an algal bloom when high amounts of detrital organic matter settled onto surface sediments and are degraded by bacterial activity.

2.4. Impact of the barley straw treatment

In the Artois-Picardie watershed, cyanobacteria blooms were limited due to relatively bad weather conditions occurring in this period. For instance, the monthly average air temperatures were 19.3 °C in June, 20.2 °C in July, 17.0 °C in August and 14.5 °C in September. As for precipitation, rainy periods occurred mainly in August with 132 mm of water equivalent against 29, 67 and 75 mm in June, July and September, respectively. Additionally, it has been shown that, during the monitoring period, blooms were limited in time and a two-week sampling frequency was not sufficient to account for all of them in the Pont-Rouge Pond.

According to CEH (2004), the effects of barley straw strongly depend on water temperature. For an average temperature higher than 20 °C, the effect of straw degradation starts within 4–5 weeks after its watering and the renewing of the treatment is necessary after 4–6 months. For temperatures lower than 20 °C, the degradation of straw is slower and the effect on phytoplankton, probably linked to the production of polyphenols, starts 8–10 weeks after its watering. The renewing of straw is, in this latter case, recommended after 8–10 months. In our study, the temperature of the overlying water was on average equal to 18.5 °C in May–June 2010 that would promote an algistatic action of the straw from the beginning of July. This action would also end approximately on December 2010 because of an average temperature between July and October of 19 °C.

In the beginning of June, the bloom of *A. flos aquae* cannot be stopped, or at least hampered, by the degradation of straw. The development of this species and that of *Anabaena sp* as well is however known to be inhibited by barley straw decomposition (Brownlee *et al.* 2003; Rajabi *et al.* 2010) and in July, they completely disappeared. However, in October, *Anabaena sp* reappeared and dominated the algal population. Overall, the diversity of the cyanobacteria species was poor from April to June (1–2 species) and strongly increased in the beginning of July when six species were registered, although the decomposition of barley straw should have hampered the development of cyanobacteria. However, this richness disappeared quickly and only one species, *Aphanizomenon sp* dominated in summer. In the literature (Brownlee *et al.* 2003; Rajabi *et al.* 2010), this species should have been inhibited by barley straw decomposition that was not the case in this pond.

The values of Chl-*a* concentrations and number of cells per mL were also monitored every two weeks in the pond (Figs 4 and 5). A maximum of both parameters was observed in the middle of June (Chl-*a* = 72 μg L^{-1} and 66,800 cell mL^{-1}) and the values dropped down quickly to approximately 25 μg L^{-1} of Chl-*a* and 12,600 cell mL^{-1} from the middle of July to the end of August. Although the effect of barley straw decomposition cannot be totally excluded, the bad weather conditions in July and August may be the main factors limiting algal development. Indeed, the increase of irradiance at the end of August resulted in a quick increase of Chl-*a* concentrations by a factor 2.

The algal species present in 2010 have already been observed for several years (*Anabaena sp* in 1990 and 2003, *Aphanizomenon sp* in 2003, 2004 and 2005), but some others were totally absent this year (for instance, *Plankthotrix agardhii*). The interpretation of the data is however not so trivial: barley straw has indeed an algistatic effect on some species like *Microcystis, Anabaena* or *Aphanizomenon sp.* (Brownlee *et al.* 2003; Rajabi *et al.* 2010), while it may stimulate the development of some others, for instance, *Anabaena cylindrica* or *Oscillatoria* (Rajabi *et al.* 2010). Additionally, several studies counteracted these findings, especially concerning *Microcystis sp* (Ferrier *et al.* 2005; Ó hUallacháin, Fenton 2010). In parallel, the monitoring of phytoplankton carried out in the Artois-Picardie basin (data not shown) has clearly shown that in 2010, *A. flos-aqua* was never observed in other ponds, while *Aphanizomenon sp* was strongly present in summer, like in the Pont-Rouge Pond.

In a more technical and operational point of view, the introduction of barley straw treatment in water bodies of this size appears to be difficult. Indeed, the packaging and deployment of barley straw on the pond required a total workforce estimated to 250 hours per year. The choice of the location of the straw rolls was also not optimised because the pond is used for leisure activities and it was not possible to spread the rolls everywhere. The preservation of barley straw rolls at the water surface was operated by the use of 50 buoys and 5 anchorage floating structures that finally were not sufficient enough to ensure the complete flotation of straw. Furthermore, the degradation of barley straw was not as rapid as predicted in previous studies (CEH 2004). It was shown that the degradation occurred within three to four months, while in our case, a significant quantity of straw remained after seven months of deployment. This relatively low degradation rate may be due to the swelling of the straw and to the high suspended matter levels in the pond resulting in the rapid covering of straw by particles. Altogether, this could have limited the penetration of oxygen (average concentration measured inside the bales was approximately 6 mg L^{-1}) and UV-light and thus the efficiency of the treatment (Boylan, Morris 2003). Moreover, the treatment appeared to be more efficient when applied with finely chopped straw (2-mm size) (Iredale *et al.* 2012). In that case, a dosage of 5 g L^{-1} (dry weight) was sufficient to observe an inhibitory activity. However, it must be highlighted that this value largely exceeds our field dosage (approximately 12.5 g m^3). Another weak point was the monitoring of barley straw degradation due to the unknown operating conditions which lead to the release of algistatic compounds. For instance, polyphenolic compounds are not easy to measure at trace levels and the spectrophotometric method we tested (Grubešić *et al.* 2005) on several samples was not sensitive enough.

Conclusion

1. Small and shallow floodplain ponds are very reactive environments characterised by intense exchanges between sediments, interstitial water

and open water. This kind of pond is very common in Northern France and large and specific attention has been paid to this problem because of phytoplankton developments and particularly harmful cyanobacteria blooms, which have consequences both on aquatic environments and water usage.

2. Numerous algal treatments have been tested for more than 20 years in the Pont-Rouge Pond (Le Quesnoy city) but remained unsuccessful. In 2010–2011, the efficiency of the treatment based on barley straw presented in this paper, was not clearly evidenced and the question arose on the interest of barley straw to help fighting against algae and cyanobacteria for large ponds, especially because they require enormous investments in terms of means and time.
3. The persistence of an algal activity, including during the winter period, allowed a high oxygen content in the water column and thus the presence of Fe and Mn oxides at the water-sediment interface that prevent sediment phosphorus release.
4. The development of cyanobacteria blooms from early June did not result in anoxia due to the rapid and effective recycling of organic matter. However, the mineralisation capacities of the pond could be overpassed if algae blooms become larger and more frequent and, consequently, a phosphorus release from sediment could occur as its concentration is about 1 g kg^{-1} dry weight.
5. For small and shallow flood plain ponds, it is therefore essential to take into account both the phosphorus internal load and sediment-water phosphorus exchange dynamics when defining water phosphorus thresholds as required by the EU water Framework Directive.

Acknowledgements

This study was supported by the cooperation agreement between the Artois-Picardie French Water Agency and the University of Lille 1. The authors thank the town council of Le Quesnoy, Espace Naturel Lille Metropole (ENLM) and the local fishing association for their logistic help during the preparation and deployment of barley straw. The technical help of Michel Arold (French Water Agency), Olivier Boulinguez and Damien Comblez (ENLM), and Christine Grare (University of Lille 1) were also appreciated. We dedicate this work to our colleague, Guy Thoumelin, who contributed to the supervision of this project until his sudden death in April 2010.

References

Anschutz, P.; Chaillou, G.; Lecroart, P. 2007. Phosphorus diagenesis in sediment of the Thau Lagoon, *Estuarine, Coastal and Shelf Science* 72(3): 447–456. http://dx.doi.org/10.1016/j.ecss.2006.11.012

Bai, X. L.; Ding, S. M.; Fan, C. X.; Liu, T.; Shi, D.; Zhang, L. 2009. Organic phosphorus species in surface sediments of a large, shallow, eutrophic lake, Lake Taihu, China, *Environmental Pollution* 157: 2507–2513. http://dx.doi.org/10.1016/j.envpol.2009.03.018

Boylan, J. D.; Morris, J. E. 2003. Limited effects of barley straw on algae and zooplankton in a midwestern pond, *Lake and Reservoir Management* 19(3): 265–271. http://dx.doi.org/10.1080/07438140309354091

Brownlee, E. F.; Sellner, S. G.; Sellner, K. G. 2003. Effects of barley straw (Hordeum vulgare) on freshwater and brackish phytoplankton and cyanobacteria, *Journal of Applied Phycology* 15(6): 525–531. http://dx.doi.org/10.1023/b:japh.0000004353.15684.25

Cao, Z.; Zhang, X.; Ai, N. 2011. Effect of sediment on concentration of dissolved phosphorus in the Three Gorges Reservoir, *International Journal of Sediment Research* 26(1): 87–95. http://dx.doi.org/10.1016/S1001-6279(11)60078-4

CEH. 2004. *Information sheet 1: control of algae with barley straw*. Wallingford, UK: Centre for Ecology and Hydrology.

Everall, N. C.; Lees, D. R. 1997. The identification and significance of chemicals released from decomposing barley straw during reservoir algal control, *Water Research* 31(3): 614–620. http://dx.doi.org/10.1016/S0043-1354(96)00291-6

Ferrier, M. D.; Butler Sr., B. R.; Terlizzi, D. E.; Lacouture, R. V. 2005. The effects of barley straw (Hordeum vulgare) on the growth of freshwater algae, *Bioresource Technology* 96(16): 1788–1795. http://dx.doi.org/10.1016/j.biortech.2005.01.021

Grover, J. P.; Baker, J. W.; Ureña-Boeck, F.; Brooks, B. W.; Errera, R. M.; Roelke, D. L.; Kiesling, R. L. 2007. Laboratory tests of ammonium and barley straw extract as agents to suppress abundance of the harmful alga Prymnesium parvum and its toxicity to fish, *Water Research* 41(12): 2503–2512. http://dx.doi.org/10.1016/j.watres.2007.03.025

Grubešić, R. J.; Vuković, J.; Kremer, D.; Vladimir-Knežević, S. V. 2005. Spectrophotometric method for polyphenols analysis: prevalidation and application on Plantago L. species, *Journal of Pharmaceutical and Biomedical Analysis* 39(3–4): 837–842. http://dx.doi.org/10.1016/j.jpba.2005.05.004

Gustafsson, J. P. 2012. *Visual MINTEQ*, version 2.61 [online]. Available from Internet: http://www2.lwr.kth.se/English/OurSoftware/vminteq/index.html

Hagström, J. A.; Sengco, M. R.; Villareal, T. A. 2010. Potential methods for managing Prymnesium parvum blooms and toxicity, with emphasis on clay and barley straw: a review, *Journal of the American Water Resources Association* 46(1): 187–198. http://dx.doi.org/10.1111/j.1752-1688.2009.00402.x

House, W. A. 2003. Geochemical cycling of phosphorus in rivers, *Applied Geochemistry* 18(5): 739–748. http://dx.doi.org/10.1016/S0883-2927(02)00158-0

Iredale, R. S.; McDonald, A. T.; Adams, D. G. 2012. A series of experiments aimed at clarifying the mode of action of barley straw in cyanobacterial growth control, *Water Research* 46(18): 6095–6103.

Khan, F.; Ansari, A. 2005. Eutrophication: an ecological vision, *The Botanical Review* 71(4): 449–482. http://dx.doi.org/10.1663/0006-8101(2005)071%5B0449:EAEV%5D2.0.CO;2

Kisand, A. 2005. Distribution of sediment phosphorus fractions in hypertrophic strongly stratified Lake Verevi, *Hydrobiologia* 547: 33–39. http://dx.doi.org/10.1007/s10750-005-4141-2

Kleeberg, A.; Herzog, C.; Jordan, S.; Hupfer, M. 2010. What drives the evolution of the sedimentary phosphorus cycle?, *Limnologica – Ecology and Management of Inland Waters* 40(2): 102–113. http://dx.doi.org/10.1016/j.limno.2009.11.001

Lake, B. A.; Coolidge, K. M.; Norton, S. A.; Amirbahman, A. 2007. Factors contributing to the internal loading of phosphorus from anoxic sediments in six Maine, USA, lakes, *Science of the Total Environment* 373(2–3): 534–541. http://dx.doi.org/10.1016/j.scitotenv.2006.12.021

Le Jeune, A.-H.; Charpin, M.; Deluchat, V.; Briand, J.-F.; Lenain, J.-F.; Baudu, M.; Amblard, C. 2006. Effect of copper sulphate treatment on natural phytoplanktonic communities, *Aquatic Toxicology* 80(3): 267–280. http://dx.doi.org/10.1016/j.aquatox.2006.09.004

Merz-Preiß, M.; Riding, R. 1999. Cyanobacterial tufa calcification in two freshwater streams: ambient environment, chemical thresholds and biological processes, *Sedimentary Geology* 126(1–4): 103–124. http://dx.doi.org/10.1016/S0037-0738(99)00035-4

Mortimer, R. J. G.; Harris, S. D.; Krom, M. D.; Freitag, T. E.; Prosser, J. I.; Barnes, J.; Anschutz, P.; Hayes, P. J.; Davies, I. M. 2004. Anoxic nitrification in marine sediments, *Marine Ecology Progress Series* 276: 37–51. http://dx.doi.org/10.3354/meps276037

Newcombe, G.; Chorus, I.; Falconer, I.; Lin, T. F. 2012. Cyanobacteria: impacts of climate change on occurrence, toxicity and water quality management, *Water Research* 46(5): 1347–1348. http://dx.doi.org/10.1016/j.watres.2011.12.047

O'Neil, J. M.; Davis, T. W.; Burford, M. A.; Gobler, C. J. 2012. The rise of harmful cyanobacteria blooms: the potential roles of eutrophication and climate change, *Harmful Algae* 14: 313–334. http://dx.doi.org/10.1016/j.hal.2011.10.027

Ó hUallacháin, D.; Fenton, O. 2010. Barley (*Hordeum vulgare*)-induced growth inhibition of algae: a review, *Journal of Applied Phycology* 22(5): 651–658. http://dx.doi.org/10.1007/s10811-009-9492-z

Paerl, H. W.; Fulton, R. S.; Granéli, E.; Turner, J. T. 2006. Ecology of Harmful Cyanobacteria, in Granéli, E.; Turner, J. T. (Eds.). *Ecology of Harmful Algae.* Berlin: Springer Verlag, 95–109. http://dx.doi.org/10.1007/978-3-540-32210-8

Palmer-Felgate, E. J.; Mortimer, R. J. G.; Krom, M. D.; Jarvie, H. P.; Williams, R. J.; Spraggs, R. E.; Stratford, C. J. 2011. Internal loading of phosphorus in a sedimentation pond of a treatment wetland: effect of phytoplankton crash, *Science of the Total Environment* 409: 2222–2232. http://dx.doi.org/10.1016/j.scitotenv.2011.02.034

Pęczuła, W. 2013. Influence of barley straw (*Hordeum vulgare L.*) extract on phytoplankton dominated by Scenedesmus species in laboratory conditions: the importance of the extraction duration, *Journal of Applied Phycology* 25(2): 661–665. http://dx.doi.org/10.1007/s10811-012-9900-7

Plant, L. J.; House, W. A. 2002. Precipitation of calcite in the presence of inorganic phosphate, *Colloids and Surfaces A: Physicochemical and Engineering Aspects* 203(1–3): 143–153. http://dx.doi.org/10.1016/S0927-7757(01)01089-5

Purcell, D.; Parsons, S. A.; Jefferson, B.; Holden, S.; Campbell, A.; Wallen, A.; Chipps, M.; Holden, B.; Ellingham, A. 2013. Experiences of algal bloom control using green solutions barley straw and ultrasound, an industry perspective, *Water and Environment Journal* 27(2): 148–156. http://dx.doi.org/10.1111/j.1747-6593.2012.00338.x

Rajabi, H.; Filizadeh, Y.; Soltani, M.; Fotokian, M. H. 2010. The use of barley straw for controlling of cyanobacteria under field application, *Journal of Fisheries and Aquatic Science* 5(5): 394–401. http://dx.doi.org/10.3923/jfas.2010.394.401

Schwaderer, A. S.; Yoshiyama, K.; Pinto, P. de T.; Swenson, N. G.; Klausmeier, C. A.; Litchman, E. 2011. Eco-evolutionary differences in light utilization traits and distributions of freshwater phytoplankton, *Limnology and Oceanography* 56(2): 589–598. http://dx.doi.org/10.4319/lo.2011.56.2.0589

von Sperling, E.; da Silva Ferreira, A. C.; Nunes Ludolf Gomes, L. 2008. Comparative eutrophication development in two Brazilian water supply reservoirs with respect to nutrient concentrations and bacteria growth, *Desalination* 226(1–3): 169–174. http://dx.doi.org/10.1016/j.desal.2007.02.105

Waybright, T. J.; Terlizzi, D. E.; Ferrier, M. D. 2009. Chemical characterisation of the aqueous algistatic fraction of barley straw (*Hordeum vulgare*) inhibiting *Microcystis aeruginosa*, *Journal of Applied Phycology* 21: 333–340. http://dx.doi.org/10.1007/s10811-008-9373

Willame, R.; Hoffmann, L. 2005. *Les cyanobactéries (algues bleues) responsables d'efflorescences dans les eaux stagnantes. Aspects systématiques-écologiques-toxicologiques* [online]. Centre de Recherches Gabriel Lippmann, 125 p. Available from Internet: http://www.eau-artois-picardie.fr/IMG/BaseDoc/aegis/871/B%2010727.pdf. (in French).

Xiang, S. L; Zhou, W. B. 2011. Phosphorus forms and distribution in the sediments of Poyang Lake, China, *International Journal of Sediment Research* 26(2): 230–238. http://dx.doi.org/10.1016/S1001-6279(11)60089-9

Xie, L. Q.; Xie, P.; Tang, H. J. 2003. Enhancement of dissolved phosphorus release from sediment to lake water by Microcystis blooms: an enclosure experiment in a hyper-eutrophic, subtropical Chinese lake, *Environmental Pollution* 122(3): 391–399. http://dx.doi.org/10.1016/S0269-7491(02)00305-6

Emilie PRYGIEL. PhD. University Lille 1. Publications: One paper in an international journal and five communications in international and national conferences. Research fields: water pollution, phytoplankton evolution, sediment resuspensions impacts on water quality, relationships between contamination and toxicity on invertebrates.

Adeline CHARRIAU. Associate Professor. University of Limoges. Publications: author/co-author of four scientific papers, nine communications in international conferences. Research fields: environmental chemistry, sediment and water contamination, passive sampling techniques.

Romain DESCAMPS. Technician. University Lille 1. Publications: two national technical reports for the French Water Agency and acknowledged six times within international papers. Technical interests: atomic absorption and emission spectroscopies, ICP-MS, ionic chromatography and wide water and sediment sampling expertise.

Jean PRYGIEL. Regional Water Agency (northern France), Associate Professor (University Lille 1). Publications: 20 articles in international journals and more than 30 communications in international conferences. Research fields: status and functioning of aquatic environments, networks monitoring, Framework Water Directive, eutrophication, biological indices, diatoms, impact of water and sediment contamination on biological quality.

Baghdad OUDDANE. Professor. University Lille 1. Publications: 78 papers in international journals and more than 120 communications in national and international conferences. Research fields: a wide range of environmental areas including: trace element speciation in natural waters and sediments, metal-ligand complexes, biogeochemical processes in river and marine environments and analytical chemistry development as a tool for monitoring of environmental processes.

Gabriel BILLON. Professor. University Lille 1. Publications: 40 articles in international journals and more than 60 communications in international conferences. Research fields: sulphur cycle and metal pollutants in the estuarine and riverine sediments during early diagenetic processes; biogeochemical transformations at the water-sediment interface and the effect of sediment resuspension on metal and Phosphorus mobility.

AERATION AT OVERFLOW DAMS WITH CURVED SURFACES BY DIFFERENT FLASHBOARD SPILLWAYS

Kejian Chu[a, b, c], Zulin Hua[a, b, c], Lijun Ji[a, c]

[a]*Key Laboratory of Integrated Regulation and Resource Development on Shallow Lake of Ministry of Education, College of Environment, Hohai University, Nanjing City, P. R. China*

[b]*National Engineering Research Center of Water Resources Efficient Utilization and Engineering Safety, Nanjing City, P. R. China*

[c]*College of Environment, Hohai University, 1 Xikang Road, Nanjing City, 210098 Jiangsu Province, P. R. China*

Abstract. Aeration using overflow dams is an eco-friendly and economical method of improving dissolved oxygen levels in polluted urban streams. Laboratory investigations of aeration performance in smooth spillways, as well as in parallel flashboard and interlaced flashboard spillways with different flashboard intervals, for overflow dams with curved surfaces have been conducted separately. Aeration efficiencies of the different types of spillways, in particular the effects of varying discharge rate, total spillway height, and flashboard interval, have been discussed in detail. The test data illustrate that aeration efficiency in all spillways increases with spillway height and decreases with increasing discharge. Flashboard spillways appear to provide significantly higher aeration efficiency than smooth ones, and aeration efficiency increases with the number of flashboards, but with a continuously declining growth rate. By combining Fd_s with h_d^*, a new dimensionless parameter Ψ is created to characterize the comprehensive effects of hydraulic conditions on aeration. Empirical formulas for oxygen-transfer efficiency in smooth, parallel flashboard, and interlaced flashboard spillways have been developed with Ψ and the dimensionless number of flashboards per unit area of spillway N^* as independent variables, and the properties of these formulas are discussed.

Keywords: water cleaning technologies, aeration, flashboard spillways, overflow dam with curved surface, oxygen-transfer efficiency, spillway Froude number.

Introduction

Dissolved oxygen (DO) is the key important water-quality index for stream aquatic ecosystems. DO is usually lower in urban streams because they receive more pollutant inputs than rural streams. Increasing DO concentration in urban streams effectively promotes the activities of aerobic bacteria, improves the local living conditions of aquatic organisms, inactivates contaminants held in sediments, and even eliminates foul odors in water, thus contributing to improving stream ecosystems.

Artificial aeration has been widely used as an efficient oxygenation method to increase DO levels in urban streams. It can be technically divided into two categories: aeration through gas entrainment and aeration using hydraulic structures. The former approach has provided good aeration performance (Dalla Santa, Vinatea 2007; Ghaly, Thistle 2009; Dong *et al.* 2009; Moulick, Mal 2009; Kumar *et al.* 2010), but special devices are needed to entrain or blow air (pure oxygen) into the water body, which inevitably incur excessive energy consumption and relatively high operating cost. On the other hand, hydraulic structures are often constructed across urban streams to raise the water level. Low-overflow dams and weirs are among the main types of hydraulic structures used. These overflow dams and weirs increase available DO concentrations in streams without additional gas-blowing devices, even though the water is in contact with them only for a short time (Kramer, Hager 2005; Ghare *et al.* 2008; Bagatur 2009; Baylar *et al.* 2010; Crookston, Tullis 2013).

Spillway morphology is the key factor affecting the aeration efficiency of an overflow dam, and many studies on this topic have been conducted and reported. For example, Chanson and Toombes (2002) pointed out that a flat

Corresponding author: Hua Zulin
E-mail: zulinhua@hhu.edu.cn

stepped cascade for operation of nappe flows might be an effective means of water aeration by experimental investigation. Further laboratory studies have suggested that channel slope and step height are the important factors influencing the aeration performance of stepped chutes (Baylar *et al.* 2003, 2006; Takahashi, Ohtsu 2012), and mass-transfer equations were developed to predict aeration efficiency in terms of dissolved oxygen (Toombes, Chanson 2005; Baylar *et al.* 2007; Felder, Chanson 2009). Aeration on stepped chutes with end sills has also been explored, in particular the effects of the relative size of the end sill and the length/depth ratio of the chute (Emiroglu, Baylar 2003). Aras and Berkun (2010) determined empirical correlations between the length of the non-aerated flow region and aeration efficiency. Moulick *et al.* (2010) determined experimentally the optimum number of steps, the slope of the entire cascade, and hydraulic loading rate at which the maximum overall aeration efficiency occurs for the stepped cascade.

However, previous studies have focused mainly on oxygen transfer in stepped chutes and channels. Aeration performance of overflow dams with curved surfaces – typical low hydraulic structures widely used in rivers and lakes – has been insufficiently studied. Moreover, little attention has been paid to the spillway effect on oxygen transfer at overflow dams, although this problem should be investigated in depth for optimal design of dams in terms of aeration efficiency. In this research, experimental investigations of aeration on overflow dams with curved surfaces by spillways with different morphological characteristics, including smooth surfaces, parallel flashboards, and interlaced flashboards, have been carried out. The aeration performance of flashboard spillways was investigated by varying the flow discharge Q, spillway height h, and flashboard interval l, and results were compared with those for smooth spillways. The effects of spillway morphology on oxygen-transfer efficiency E_{20} have also been discussed. Moreover, relationships between overflow hydraulic conditions, number of flashboards, and E_{20} of the various types of spillways have been established.

1. Experimental apparatus

The flow system is presented schematically in Fig. 1. All experiments were performed in an open rectangular organic-glass flume with a cross section of 20 cm $W \times$ 30 cm H. The experimental flume consisted of an inlet section, a test dam section, and an outlet section. One 120-cm long straight section was used as an inlet, and the test dam was 80 cm high, with the shape of the curved surface designed according to the Japanese Technical Standards for inflatable rubber dams. Seven spillways were fitted onto the test dam for the experiments, including one smooth spillway, three parallel flashboard spillways (with 3 cm, 6 cm, and 9 cm streamwise flashboard intervals), and three interlaced flashboard spillways (also with 3 cm, 6 cm, and 9 cm streamwise intervals). The flashboards are thin square plates with a side length of 2 cm. Fig. 2 shows a schematic plot of the test dams with smooth spillway, parallel flashboard, and interlaced flashboard spillways separately.

2. Theoretical analysis

The overall oxygen-transfer efficiency E of a hydraulic structure at any water temperature can be defined as (Toombes, Chanson 2005):

$$E = (C_2 - C_1) / (C_s - C_1), \quad (1)$$

where: C_1 is the upstream DO concentration; C_2 is the downstream DO concentration; C_s is the saturation concentration.

To account for the temperature dependence of oxygen transfer and to provide a uniform basis for comparison, the overall oxygen-transfer efficiency is normalized to a standard temperature (20 °C) using Eqn (2) below (Gulliver *et al.* 1990):

$$\begin{cases} E_{20} = 1-(1-E)^{1/f} \\ f = 1.0+0.02103(T-20)+8.261\times10^{-5}(T-20)^2, \end{cases} \quad (2)$$

where: E_{20} is the oxygen-transfer efficiency at 20 °C; T is the water temperature in the test.

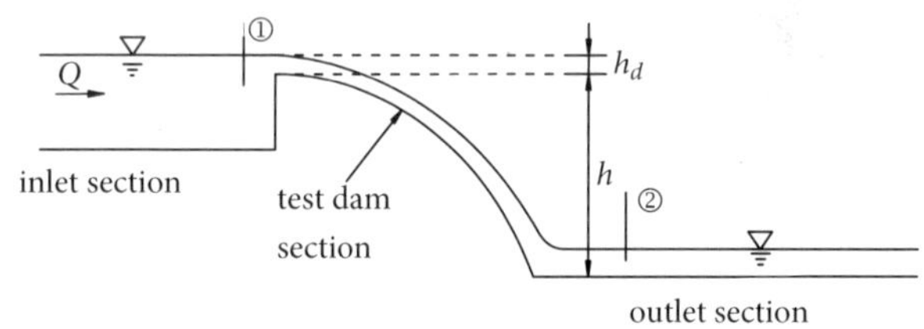

Fig. 1. Schematic diagram of the experimental apparatus

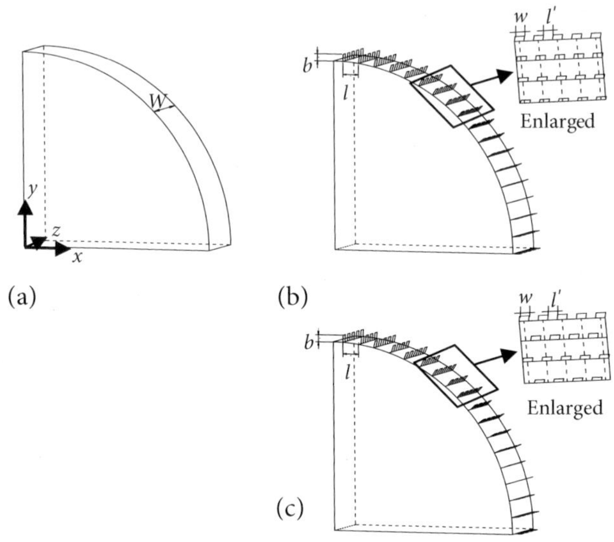

Fig. 2. Test dam with: (a) smooth spillway; (b) parallel flashboard spillway; (c) interlaced flashboard spillway

Aeration performance of the overflow dam depends on flow conditions, spillway morphology, and experimental medium properties. Therefore, the characteristic variables affecting aeration in the present experiments can be outlined as:

$$E_{20} = f(Q, h_d, h, L, W, b, w, l, l', \rho_w, \rho_a, \mu_w, \tau, n, g), \quad (3)$$

where: Q is the discharge rate; h_d is the water depth on the crest of the dam; h is the total height of the spillway; L is the curve length of the spillway; W is the width of the spillway; b is the height of the flashboards; w is the width of the flashboards; l is the streamwise interval of the flashboards; l' is the transverse interval of the flashboards; ρ_w is the density of water; ρ_a is the density of air; μ_w is the dynamic viscosity of water; τ is the surface tension of water; n is the absolute roughness of the spillway; g is the acceleration of gravity.

Assuming L, ρ_w and g to be fundamental dimensions and applying the Buckingham Pi-theorem, Eqn (3) can be rewritten as follows:

$$E_{20} = f\left(\frac{q}{\sqrt{gLL}}, h_d^*, h^*, \frac{L}{W}, \frac{b}{L}, \frac{w}{L}, \frac{L}{l}, \frac{l'}{L}, \frac{\rho_a}{\rho_w}, \frac{q\rho_w}{\mu_w}, \frac{\tau}{g\rho_w L^2}, \frac{n}{L}\right), \quad (4)$$

where: q is the discharge per unit width, $q = Q/W$; h_d^* is the relative depth on the crest, $h_d^* = h_d/L$; h^* is the relative total height of the spillway, $h^* = h/L$.

The experiments were conducted on the same fluids (air and tap water) and spillways of the same material at water temperatures varying from 11.8 °C to 14.8 °C. This means that there was little change in the variables characterizing the medium properties of ρ_w, ρ_a, μ_w, τ and n during the test runs. Therefore, the ratios ρ_a/ρ_w and n/L remain almost invariant and can be dropped from Eqn (4). The term $\tau/(g\rho_w L^2)$ can also be omitted because it can be replaced by the Morton number $Mo = g\tau^4/(\rho_w \tau^3)$ (Chanson 2009). The term $q{\cdot}\rho_w/\mu_w$ denotes the Reynolds number and can be ignored because the overflows are always in a state of turbulent motion (Moulick *et al.* 2010).

The term L/l characterizes the total number of flashboards in the stream direction along the spillway. The terms L/W, l'/L, and w/L can be combined as follows:

$$[(L/W){\cdot}(l'/L)+(L/W){\cdot}(w/L)]^{-1} = W/(l' + w), \quad (5)$$

where the term on the right-hand side of Eqn (5) can be used to characterize the total number of flashboards in the transverse direction across the spillway. The number of flashboards per unit area of spillway can thus be obtained as $N = (L/l) \cdot W / (l' + w) / (W \cdot L) = [l \cdot (l' + w)]^{-1}$. Using a dimensionless treatment with flashboard area, a dimensionless number N^* can be defined as:

$$N^* = N \cdot (b \cdot w) = b \cdot w / [l \cdot (l' + w)]. \quad (6)$$

Because the flashboard is square ($b = w$) in this study, Eqn (6) can be rewritten as:

$$N^* = b^2 / [l \cdot (l' + b)]. \quad (7)$$

The spillway Froude number, Fd_s, can be deduced as shown below by combination of the first and third terms in Eqn (4):

$$Fd_s = \frac{q}{\sqrt{gLL}} / h^* = \frac{q}{\sqrt{gLL}} / (\frac{h}{L}) = \frac{q}{\sqrt{gLL}} \cdot \frac{L}{h_d} \cdot \frac{h_d}{L} \cdot \frac{L}{h} = \frac{q}{\sqrt{gLh_d}} \frac{h_d}{h} = \frac{u}{\sqrt{gL}} \frac{h_d}{h}, \quad (8)$$

where: u is the average velocity at the crest. Hence, Eqn element (4) can be simplified as follows:

$$E_{20} = f(Fd_s, h_d^*, N^*). \quad (9)$$

The above equation indicates that the principal dimensionless variables determining aeration performance can be reduced to the spillway Froude number Fd_s, the relative depth on the crest h_d^*, and the dimensionless number of flashboards per unit area N^*.

3. Experiments and parameters

Tap water was used throughout these experiments to avoid disturbances to aeration by surfactants, organic substances, and suspended solids in a natural stream. First, the experimental water was poured into a storage tank, allowed to stand for several hours, and then deoxygenated by the sodium sulfite method to 2.0–3.0 mg/l of dissolved oxygen (DO). Afterward, the deoxygenated water was pumped to a stilling tank and then entered the test spillways through the inlet section. During the experiments, DO concentration measurements of the water were carried out upstream (point ①) and downstream (point ②) of the spillways simultaneously when the flow patterns being stable (Fig. 1), using Hach 'HQ' Series portable dissolved oxygen meters with a dissolved oxygen range of 0.00–20.00 mg/L (0–200% saturation); resolution: 0.01 mg/L; and measurement precision: ±0.1 mg/L (DO < 8 mg/L), ±0.2 mg/L (DO > 8 mg/L). When readings of the dissolved oxygen meters were stable, 10 measuring data of C_1 at point ① and C_2 at point ② respectively were recorded in each run. The discharge flowrate and temperature of the test water were measured by an LZB-80 glass rotor flowmeter installed in the supply line and by a digital thermometer respectively.

Throughout the experiments, the geometric size b (or w), the transverse interval l' of the flashboards, and the spillway width were kept constant. With variations in either Q or h, Fd_s and h_d^* were changed synchronously, and variations in l led to variations in N^*. In this way, 175 experimental runs were designed by varying Q, h at five different levels, and l at three different levels, to be

Table 1. Combinations of test conditions

Spillway type	Discharge rate Q (l/s)	Spillway height (length) h(L) (cm)	Stream-wise interval l (cm)	Width of spillway W (cm)	Length of flash-board b (cm)	Trans-verse interval l' (cm)
Smooth	0.4	60.0 (99.0)	–			
Parallel flashboard	0.6	65.0 (104.4)	3.0	20.0	2.0	2.0
	0.8	70.0 (109.7)				
Interlaced flashboard	1.0	75.0 (114.9)	6.0			
	1.2	80.0 (120.0)	9.0			

carried out on parallel flashboard, interlaced flashboard, and smooth spillways. The various combinations of test conditions are given in Table 1.

4. Results and discussion

Table 2. Expectation and relative standard deviation of E_{20}

Spillway type	Discharge rate Q (l/s)	Expectation	Relative standard deviation (%)
Smooth	0.4	0.125~0.171	0.39~2.80
	0.6	0.125~0.163	1.82~3.17
	0.8	0.120~0.159	0.87~4.86
	1.0	0.116~0.157	1.28~3.34
	1.2	0.116~0.154	1.42~4.72
Parallel flashboard	0.4	0.426~0.589	0.53~2.95
	0.6	0.421~0.584	0.50~3.62
	0.8	0.419~0.578	1.51~4.87
	1.0	0.411~0.576	0.65~4.02
	1.2	0.407~0.570	1.53~3.77
Interlaced flashboard	0.4	0.414~0.581	0.81~3.25
	0.6	0.410~0.578	0.81~3.85
	0.8	0.404~0.577	1.00~4.12
	1.0	0.401~0.575	2.15~3.95
	1.2	0.400~0.574	2.08~3.71

Table 2 describes the variation ranges of mathematical expectation and relative standard deviation of the set of E_{20} in each experiment, of which elements are calculated by Eqns (1) and (2) with the 10 pairs of DO measuring data respectively, at different Q in the smooth, parallel flashboard, and interlaced flashboard spillways.

Figure 3 illustrates the variation in the oxygen-transfer efficiency E_{20} with discharge rate Q and total height h (and h^*) of the smooth spillway. In these tests, the levels of E_{20} ranged from 0.116 to 0.171. In Figure 3(a), E_{20} decreases with increasing discharge rate Q at the same h, with an overall attenuation rate (calculated by $\Delta E_{20}/\Delta Q$) of 0.0190 from $Q = 0.4$ l/s to $Q = 1.2$ l/s. This can be explained by noting that increasing the discharge rate means increasing the flow velocity through the spillway, thus inevitably shortening the oxygen-transfer time at the air-water interface. On the other hand, E_{20} increases with h (or h^*) at the same Q (Fig. 3(b)), which can be attributed to extension of the oxygen-transfer length from air into water. The overall growth rate of E_{20} with h^* (calculated by $\Delta E_{20}/\Delta h^*$) is about 0.631 from $h^* = 0.606$ to $h^* = 0.667$.

By combining Fd_s with h_d^*, a new dimensionless parameter Ψ can be created to characterize the comprehensive effects of the hydraulic conditions of the overflow on aeration in the spillway, as outlined below:

$$\Psi = Fd_s^{1/3} \cdot h_d^* . \tag{10}$$

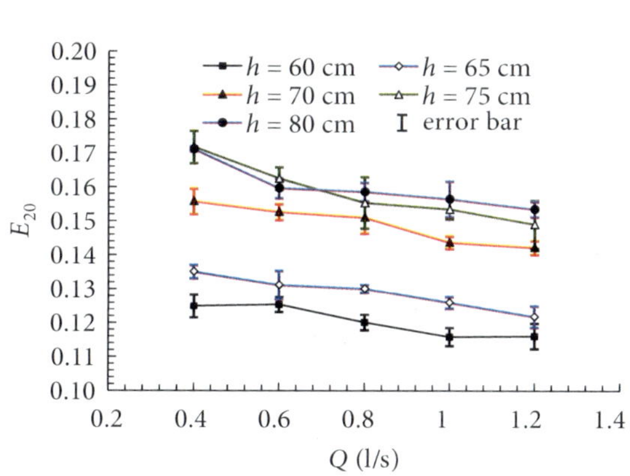

(a)

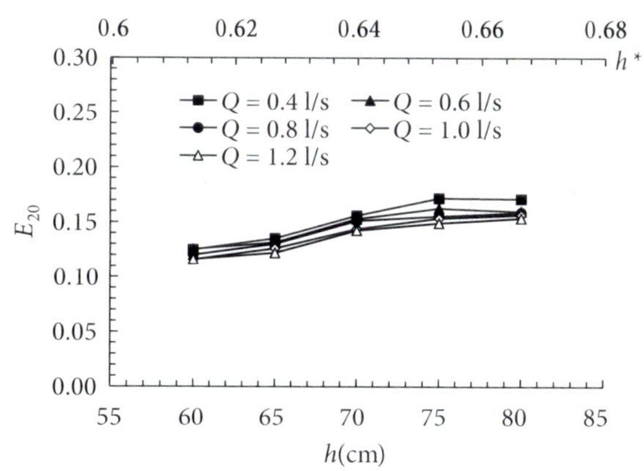

(b)

Fig. 3. Variation of oxygen-transfer efficiency E_{20} for a smooth spillway: (a) E_{20} vs. Q; (b) E_{20} vs. h. The error bars reflect the standard deviations of mean E_{20} derived from all measuring data in each experiment

The relationship between Ψ and E_{20} is displayed in Figure 4. On the whole, there appears to be an inverse relationship between the two, and the experimental correlation with E_{20} as the response and Ψ as the independent variable can be determined as follows:

$$E_{20} = 0.870\times10^{-3} \cdot \Psi^{-1};\ R^2 = 0.908. \qquad (11)$$

As Eqn (11) indicates, E_{20} decreases to a very small value with increasing Ψ. It can be determined that aeration performance is negatively correlated with discharge rate and with water depth on the crest, but positively correlated with the total height and length of the spillway. Because water flow is relatively slow and stable in urban streams, increasing physical size is almost the only method available to improve the aeration capabilities of curved overflow dams with smooth spillways, but the effectiveness of this approach is rather low, and it also entails a sharp rise in reconstruction investments.

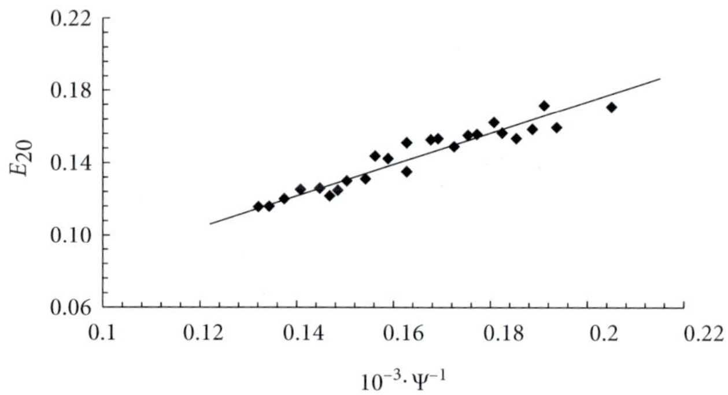

Fig. 4. Relationship between E_{20} and Ψ for a smooth spillway

Variations of E_{20} with discharge rate Q and height h of water through the parallel flashboard and interlaced flashboard spillways with different flashboard intervals are given in Figs 5 and 6 respectively. The trend in E_{20} is similar to that observed for smooth spillways: E_{20} also decreases with increasing Q at the same h, and increases with h (or h^*) at the same Q. However, the range of variation for E_{20} in the flashboard spillways was observed to be about 0.400 to 0.589, which is significantly higher than the levels in the smooth spillways. The overall attenuation rate with Q and growth rate with h^* of E_{20} were 0.0187 and 1.687 respectively in the parallel flashboard spillways and 0.0192 and 1.312 respectively in the interlaced flashboard spillways. On the basis of this, the rates of decrease of E_{20} with Q were almost the same in the smooth and flashboard spillways, but the growth rates of E_{20} with h^* actually differed from each other in the following descending order: parallel flashboard spillway > interlaced flashboard spillway > smooth spillway.

Comparisons of oxygen-transfer efficiency E_{20} in the smooth, parallel flashboard, and interlaced flashboard

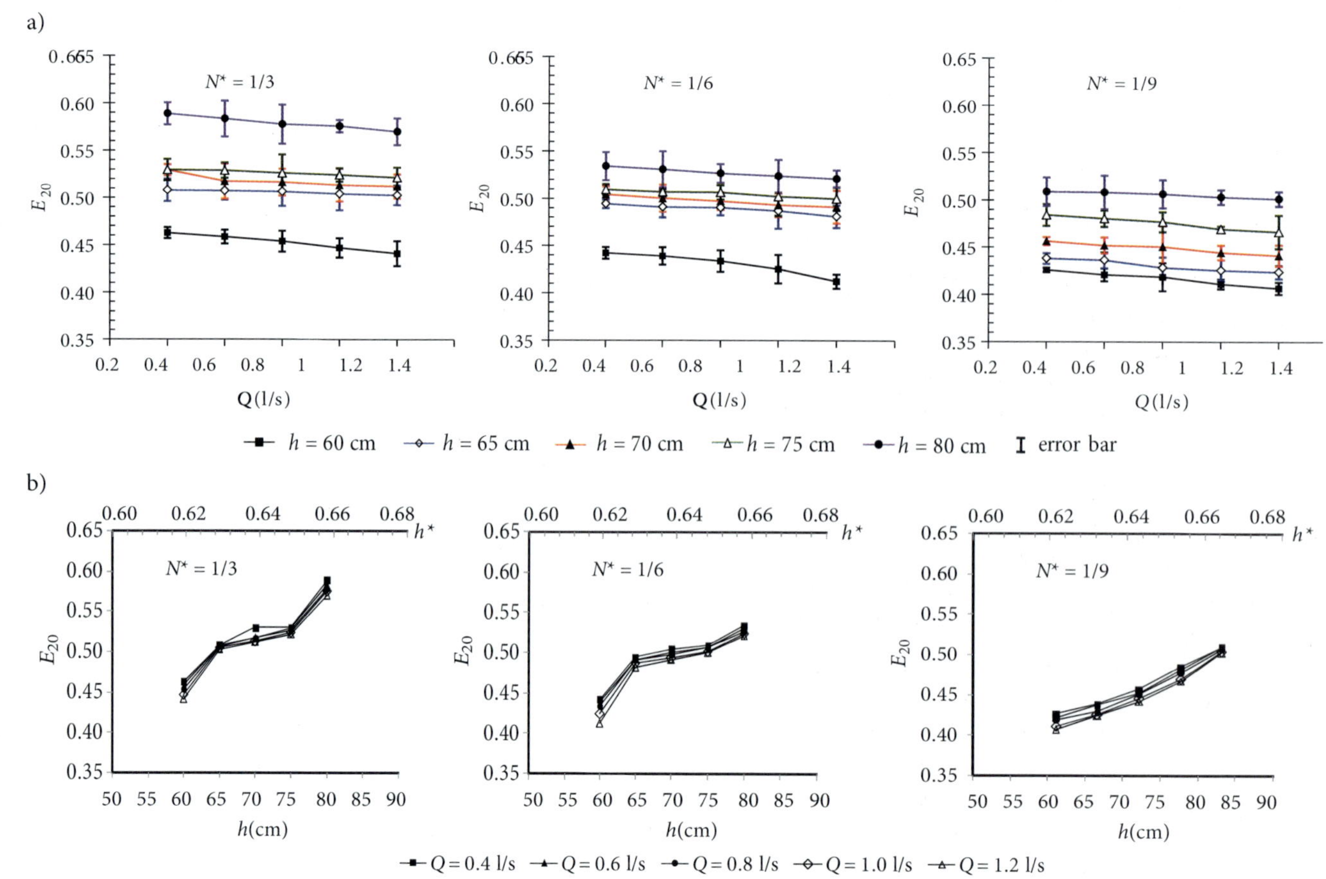

Fig. 5. Variation of oxygen-transfer efficiency E_{20} in parallel flashboard spillways: (a) E_{20} vs. Q; (b) E_{20} vs. h

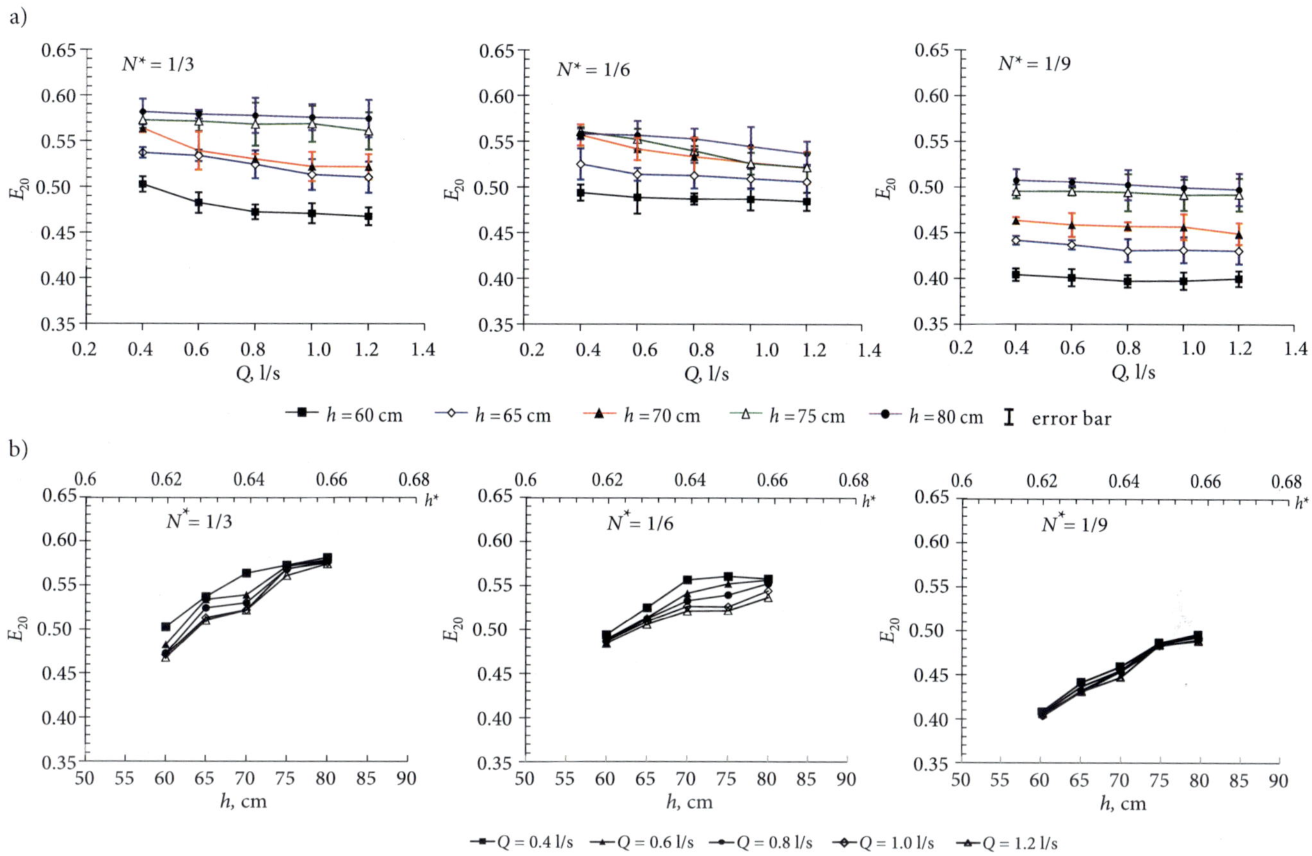

Fig. 6. Variation of oxygen-transfer efficiency E_{20} in interlaced flashboard spillways: (a) E_{20} vs. Q; (b) E_{20} vs. h

spillways under different flow conditions are displayed in Figure 7. There are 15 sub-graphs corresponding respectively to the 15 different combinations of Q (0.4 l/s, 0.6 l/s, 0.8 l/s, 1.0 l/s and 1.2 l/s) with N^* (1/3, 1/6, and 1/9), each one of which exhibits the values of E_{20} in 15 experimental runs of 5 different h^* (0.606, 0.623, 0.638, 0.653 and 0.667) by 3 spillways respectively, for a particular combination of Q and N^*. Flashboard spillways significantly outperformed smooth ones, with average increases in aeration performance (E_{20}) of 271.6–376.1% for parallel flashboard spillways and 286.4–402.3% for interlaced flashboard spillways. The flashboards strongly disturb and convert the skimming flow into a kind of milky and non-transparent gas-liquid mixture. They act like a series of air-entrainment points, greatly enhancing aeration efficiency in the spillway. It was thereby demonstrated that mounting flashboards on a spillway may indeed be an effective measure for promoting its aeration capability for water flowing through the overflow dam.

Calculated from Figure 7, the differences in E_{20} between the parallel and interlaced flashboard spillways range from 0.00044 to 0.0719, much less than those between the flashboard and smooth spillways which are 0.276–0.424. E_{20} levels in the parallel and interlaced flashboard spillways are almost the same at $N^* = 1/9$. With increasing N^*, the interlaced flashboard spillway appears to offer slightly higher aeration efficiency than the parallel one.

Figure 8 illustrates the variations in oxygen-transfer efficiency E_{20} for different N^* in the parallel and interlaced flashboard spillways separately under the test condition $Q = 0.4$ l/s (similar patterns have been found under other discharge rates), where $N^* = 0$ represents a smooth spillway. A significant rise in E_{20} is apparent between the smooth spillway and the $N^* = 1/9$ flashboard spillway because of the transition of overflow regime from skimming flow with a distinct free surface over the smooth spillway to a strongly discontinuous and turbulent nappe flow over the flashboard spillway. A relatively smooth improvement in aeration performance of the spillway then takes place, with a continuously declining growth rate as the number of flashboards increases.

Histograms of E_{20} growth rate with N^* under $Q =$ 0.4 l/s (similar patterns were observed under other discharge rates) in the parallel and interlaced flashboard spillways are shown in Figure 9. As shown, the growth rate in E_{20} decreases sharply with increasing N^*. As N^* increases from 0 to 1/9, the growth rate of E_{20} reaches almost 3.0, but it decreases to less than 0.1 when N^* increases from 1/6 to 1/3. It can be assumed that E_{20} would tend towards stability if N^*

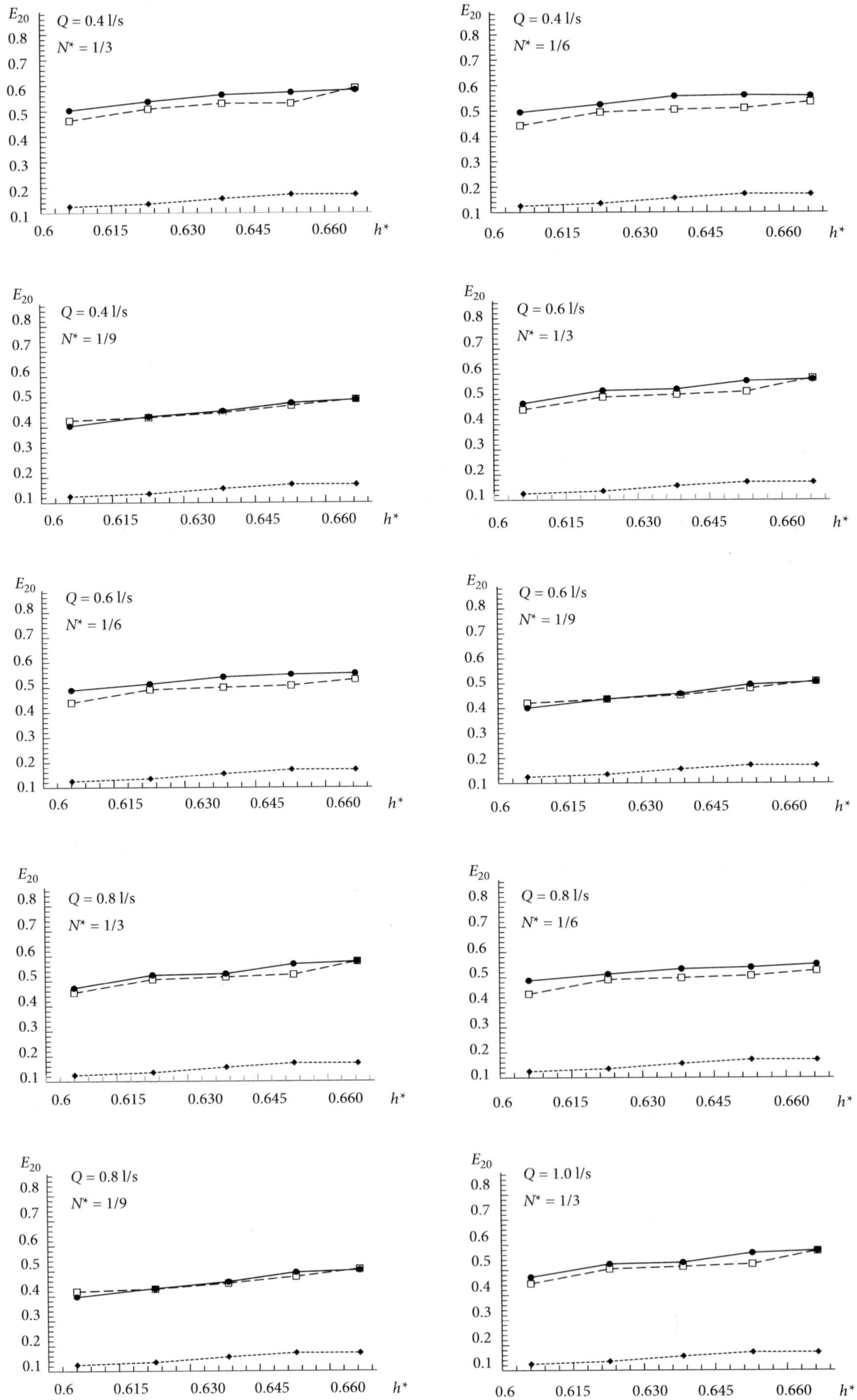
E_{20}
$Q = 0.4$ l/s
$N^* = 1/3$
$Q = 0.4$ l/s
$N^* = 1/6$
$Q = 0.4$ l/s
$N^* = 1/9$
$Q = 0.6$ l/s
$N^* = 1/3$
$Q = 0.6$ l/s
$N^* = 1/6$
$Q = 0.6$ l/s
$N^* = 1/9$
$Q = 0.8$ l/s
$N^* = 1/3$
$Q = 0.8$ l/s
$N^* = 1/6$
$Q = 0.8$ l/s
$N^* = 1/9$
$Q = 1.0$ l/s
$N^* = 1/3$
0.1 0.2 0.3 0.4 0.5 0.6 0.7 0.8
0.6 0.615 0.630 0.645 0.660
h^*

→

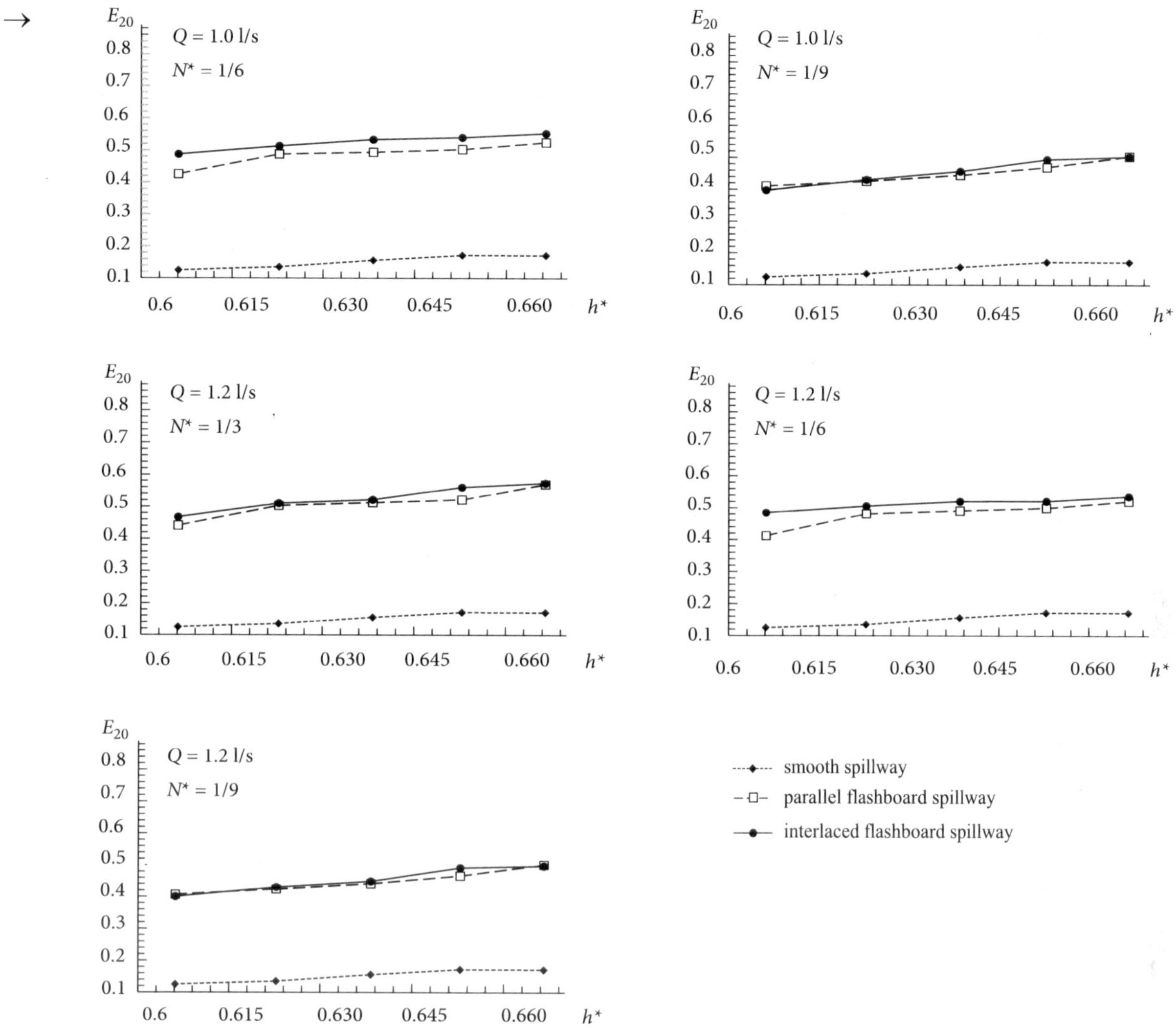

Fig. 7. Comparison between E_{20} in parallel flashboard and inter-laced flashboard spillways

exceeded 1/3 for the overflow dam with flashboard spillway.

Fig. 10 presents the change trend with Ψ of the oxygen-transfer efficiency E_{20} for flashboard spillways with different values of N^*. E_{20} also decreases with Ψ according to a power law. By multivariate regression analysis, the following empirical formulae for oxygen-transfer efficiency at the flashboard spillway were determined:

1) parallel flashboard spillway:

$$E_{20} = 0.0428 \cdot \Psi^{-0.5} \cdot 0.980^{(1/N^*)};\ R^2 = 0.924, \quad (12)$$

2) interlaced flashboard spillway:

$$E_{20} = 0.0458 \cdot \Psi^{-0.5} \cdot 0.975^{(1/N^*)};\ R^2 = 0.909. \quad (13)$$

Thus, by knowing the number of flashboards (N), the discharge per unit width (q), the water depth on the crest (h_d), and the total height of the spillway (h) the overall aeration efficiency of the spillway at 20 °C (E_{20}) can be determined.

It can be deduced from Eqns (12) and (13) that E_{20} decays gradually to stability with increasing Ψ with a fixed number of flashboards in the spillway, and also that E_{20} grows to another stable value with increasing N^* when Ψ is maintained constant. Figure 11 shows the oxygen-transfer efficiency E_{20} as a function of Ψ for a particular N^*, as well as E_{20} for a particular Ψ as a function of N^*. On the one hand, the delta E_{20} between the interlaced and parallel flashboard spillways is slightly affected by variation in Ψ when N^* remains constant (Figure 11(a)); on the other hand, for a particular Ψ, the delta E_{20} goes to zero for $N^* \leq 0.1$, increases smoothly as N^* increases from 0.1 to approximately 0.4, and gradually stabilizes after N^* exceeds 0.4 (Figure 11(b)).

Conclusions

Laboratory experiments were carried out to investigate the aeration performance of overflow dams with a curved surface and three types of spillways: smooth, parallel flashboard, and interlaced flashboard, over a range of flow from 0.4 to 1.2 l/s and with total spillway heights from 0.60 to

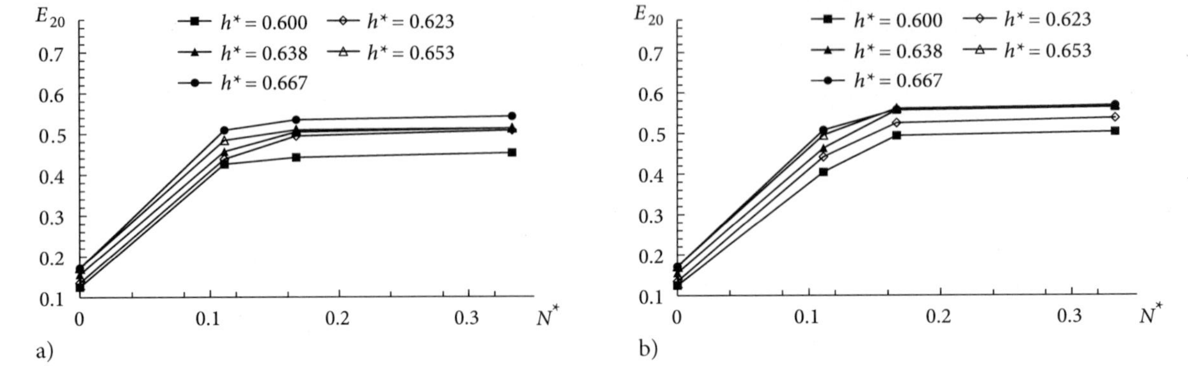

Fig. 8. Variation of E_{20} with N^* when Q = 0.4 l/s in: (a) parallel flashboard spillways; (b) interlaced flashboard spillways

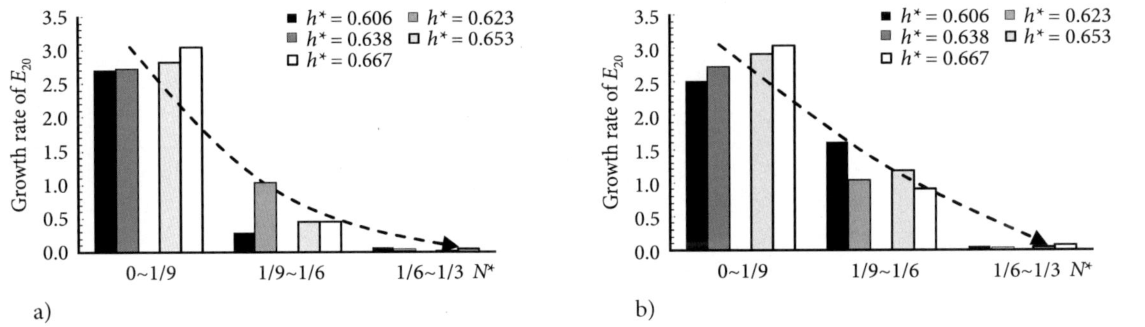

Fig. 9. Growth rate of E_{20} with N^* when Q = 0.4 l/s in: (a) parallel flashboard spillways; (b) interlaced flashboard spillways

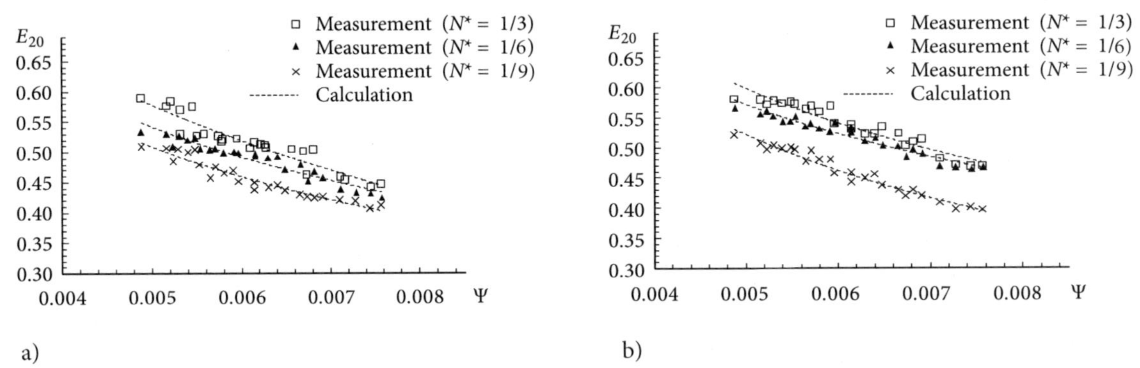

Fig. 10. Relationships between E_{20} and Ψ for: (a) parallel flashboard spillways; (b) interlaced flashboard spillways

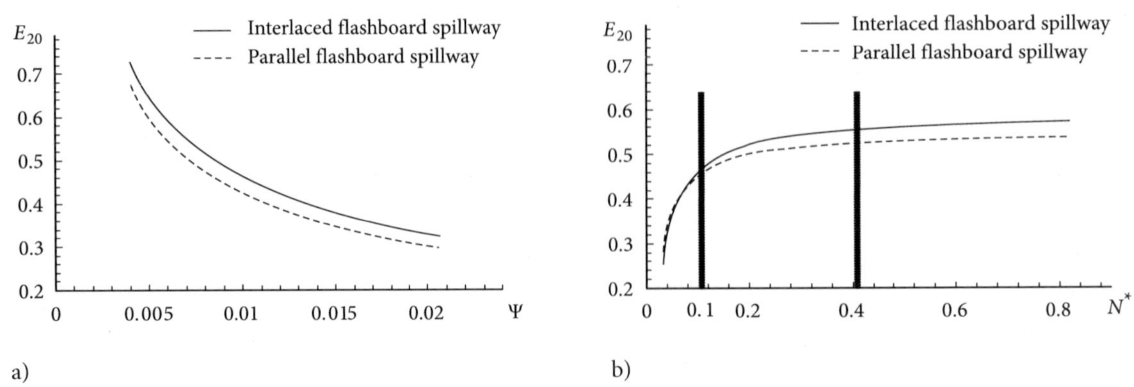

Fig. 11. E_{20} of flashboard spillways as a function of: (a) Ψ for a particular N^*; (b) N^* for a particular Ψ

0.80 m. A new dimensionless parameter Ψ was created to characterize the comprehensive effects on aeration of the hydraulic conditions of the overflow. Empirical correlations predicting the aeration efficiency of these differently shaped spillways on the overflow dam have been developed. The following conclusions can be drawn:

(1) Flashboard spillways significantly outperformed smooth ones in aeration performance (E_{20}) with average increases of 271.6–376.1% and 286.4–402.3% for parallel and interlaced flashboard spillways respectively. Note that the value E_{20} of the interlaced one is slightly higher than the parallel one. E_{20} increases as the dimensionless number of flashboards per unit area of spillway N^* increases at the growth rate declining sharply with N^*. Furthermore, E_{20} decreases with Ψ in the negative power-law relationship. Empirical formulas for predicting E_{20} in spillways as a function of Ψ and N^* were established separately, the properties of which indicate that E_{20} will achieve stability with increasing Ψ or N^* in isolation.

(2) Laboratory investigations demonstrated that installing flashboards in a spillway is an efficient method of promoting aeration performance at an overflow dam because of the strong turbulent mixing and air-entrainment points created in the overflow by the flashboards. Stream water flowing through a flashboard spillway is much more highly aerated than water flowing through a conventional smooth one. The work can be used for overflow dams with a curved surface optimally designed in terms of aeration efficiency.

Acknowledgments

This work was financially supported by the National Key Technology R&D Program (Grant No. 2012BAB03B04), Major Science and Technology Program for Water Pollution Control and Treatment (2012ZX07103-005), National Nature Science Foundation of China (Grant No. 51179052, 51379060 and 51009048), Natural Science Foundation of Jiangsu Province of China (BK20130828), Ministry of Water Resources' Special Funds for Scientific Research on Public Causes (201001028) and the Project of "Six Talent Peak" of Jiangsu Province (08-C).

References

Aras, E.; Berkun, M. 2010. Comparison of stepped and smooth spillway effects on stream reaeration, *Water SA* 36(3): 309–313.

Bagatur, T. 2009. Evaluation of preaeration with V-Notch Weir and cascade structures in clarifiers, *Journal of Environmental Engineering* ASCE 135(3): 176–184. http://dx.doi.org/10.1061/(ASCE)0733-9372(2009)135:3(176)

Baylar, A.; Bagatur, T.; Emiroglu, M. E. 2007. Prediction of oxygen content of nappe, transition, and skimming flow regimes in stepped-channel chutes, *Journal of Environmental Engineering and Science* 6(2): 201–208. http://dx.doi.org/10.1139/s06-048

Baylar, A.; Emiroglu, M. E. 2003. Study of aeration efficiency at stepped channels, *Proceedings of the Institution of Civil Engineers-Water Management* 156(3): 257–263. http://dx.doi.org/10.1680/wame.2003.156.3.257

Baylar, A.; Emiroglu, M. E.; Bagatur, T. 2006. An experimental investigation of aeration performance in stepped spillways, *Water and Environment Journal* 20(1): 35–42. http://dx.doi.org/10.1111/j.1747-6593.2005.00009.x

Baylar, A.; Unsal, M.; Ozkan, F. 2010. Hydraulic structures in water aeration processes, *Water Air and Soil Pollution* 210(1–4): 87–100. http://dx.doi.org/10.1007/s11270-009-0226-2

Chanson, H. 2009.Turbulent air-water flows in hydraulic structures: dynamic similarity and scale effects, *Environmental Fluid Mechanics* 9(2): 125–142. http://dx.doi.org/10.1007/s10652-008-9078-3

Chanson, H.; Toombes, L. 2002. Energy dissipation and air entrainment in stepped storm waterway: experimental study, *Journal of Irrigation and Drainage Engineering* 128(5): 305–315. http://dx.doi.org/10.1061/(ASCE)0733-9437(2002)128:5(305)

Crookston, B. M.; Tullis, B. P. 2013. Hydraulic design and analysis of labyrinth weirs. II: nappe aeration, instability, and vibration, *Journal of Irrigation and Drainage Engineering* ASCE 139(5): 371–377. http://dx.doi.org/10.1061/(ASCE)IR.1943-4774.0000553

Dalla Santa, K.; Vinatea, L. 2007. Evaluation of respiration rates and mechanical aeration requirements in semi-intensive shrimp Litopenaeus vannamei culture ponds, *Aquacultural Engineering* 36(1): 73–80. http://dx.doi.org/10.1016/j.aquaeng.2006.07.002

Dong, C.; Zhu, J.; Miller, C. F. 2009. Evaluation of six aerator modules built on venturi air injectors using clean water test, *Water Science & Technology* 60(5): 1353–1359. http://dx.doi.org/10.2166/wst.2009.420

Emiroglu, M. E.; Baylar, A. 2003. An investigation of effect of stepped chutes with end sill on aeration performance, *Water Quality Research Journal of Canada* 38(3): 527–539.

Felder, S.; Chanson, H. 2009. Energy dissipation, flow resistance and gas-liquid interfacial area in skimming flows on moderate-slope stepped spillways, *Environmental Fluid Mechanics* 9(4): 427–441. http://dx.doi.org/10.1007/s10652-009-9130-y

Ghaly, A.; Thistle, A. 2009. Performance evaluation of an oxidation ditch system with a disc aerator, *American Journal of Applied Sciences* 8(7): 662–674. http://dx.doi.org/10.3844/ajassp.2011.662.674

Ghare, A. D.; Wadhai, P. J.; Mistry, N. J.; Porey, P. D. 2008. Hydraulic and environmental aspects of long crested weirs, *Global Journal of Environmental Research* 2(3): 122–125.

Gulliver, J. S.; Thene, J. R.; Rindels, A. J. 1990. Indexing gas transfer in self-aerated flows, *Journal of Environmental Engineering* 116(3): 503–523. http://dx.doi.org/10.1061/(ASCE)0733-9372(1990)116:3(503)

Kramer, K.; Hager, W. H. 2005. Air transport in chute flows, *International Journal of Multiphase Flow* 31(10–11): 1181–1197. http://dx.doi.org/10.1016/j.ijmultiphaseflow.2005.06.006

Kumar, A; Moulick, S; Mal, B. C. 2010. Performance evaluation of propeller-aspirator-pump aerator, *Aquacultural Engineering* 42(2): 70–74. http://dx.doi.org/10.1016/j.aquaeng.2009.12.001

Moulick, S.; Mal, B. C. 2009. Performance evaluation of double-hub paddle wheel aerator, *Journal of Environmental Engineering* ASCE 135(7): 562–566. http://dx.doi.org/10.1061/(ASCE)0733-9372(2009)135:7(562)

Moulick, S.; Tambada, N. V.; Singh, B. K.; Mal, B. C. 2010. Aeration characteristics of a rectangular stepped cascade system, *Water Science & Technology* 61(2): 415–420. http://dx.doi.org/10.2166/wst.2010.828

Takahashi, M.; Ohtsu, I. 2012. Aerated flow characteristics of skimming flow over stepped chutes, *Journal of Hydraulic Research* 50(4): 427–434. http://dx.doi.org/10.1080/00221686.2012.702859

Toombes, L.; Chanson, H. 2005. Air-water mass transfer on a stepped waterway, *Journal of Environmental Engineering* ASCE 131(10): 1377–1386. http://dx.doi.org/10.1061/(ASCE)0733-9372(2005)131:10(1377)

Kejian CHU. PhD, College of Environment, Hohai University. He has published about 30 academic papers, 3 monographs, and attended 6 international conferences in recent years. Research interests include water cleaning technologies, contaminant transport modelling and simulation, and environmental impact assessment

Zulin HUA. PhD, College of Environment, Hohai University. He has published more than 60 academic papers, 3 monographs, and attended 10 international conferences in recent years. Research interests include water environment treatment and water environment simulation.

Lijun JI. Master, College of Environment, Hohai University. Her research interests include water cleaning technologies and cm environmental processes modelling.

18

A NOVEL PROCESS FOR NUTRIENTS REMOVAL AND PHOSPHORUS RECOVERY FROM DOMESTIC WASTEWATER BY COMBINING BNR WITH INDUCED CRYSTALLIZATION

Haiming ZOU[a, b], Xiwu LU[a], Ting LI[a]

[a]*School of Energy and Environment, Southeast University, Nanjing 210096, PR China*

[b]*Department of Resource and Environment, Anhui Science and Technology University, Fengyang 233100, PR China*

Abstract. An excessive discharge of phosphorus from wastewater to water bodies may potentially contribute to eutrophication. On the other hand, mineral phosphorus resources will be depleted in the near future, because of difficulty to automatically recycle from water to land, unlike nitrogen. A new process for nutrients removal coupled with phosphorus recovery was proposed in this study by combining biological nutrients removal (BNR) with induced crystallization (IC), BNR-IC for short later, differently from conventional phosphorus recovery process. Our results showed that the BNR-IC system can maintain not only high and stable carbon, nitrogen and phosphorus removal efficiencies, all presenting above 90%, but also good phosphorus recovery performance from synthetic domestic wastewater, displaying about 70.2% of phosphorus recovery rate. When the COD, TN, NH_4–N and P concentrations of 250 mg L^{-1}, 42 mg L^{-1}, 40 mg L^{-1}, and 10 mg L^{-1}, respectively were given in the influent, a stable removal efficiencies of 92.5% COD, 78.6% TN, 85.9% NH_4–N and 95.2% P were obtained for the BNR-IC process and correspondingly the COD, TN, NH_4–N and P concentrations of 18.75 mg L^{-1}, 8.99 mg L^{-1}, 5.64 mg L^{-1}, 0.42 mg L^{-1} were monitored in the effluent, meeting the Chinese National Class I (grade A) Sewage Discharge Standard. Analyses of SEM and EDS, moreover, also demonstrated that the surface of seed crystal (calcite used here) was completely covered by hydroxyl calcium phosphate (HAP) produced during the induced crystallization process to recover phosphorus. Although our study involved only a small-scale trial, the proposed BNR-IC process here may be a promising technology, and can potentially aid in improvement of the effluent quality from WWTP and in recycle of mineral phosphorus resources when applied to practice.

Keywords: wastewater treatment, nutrients removal, phosphorus recovery, denitrifying phosphorus removal, induced crystallization, hydroxyl calcium phosphate.

Introduction

Phosphorus (P) is an essential element for the development of biomass constituting one of the key nutrients necessary for growth. However, an excessive discharge of P from wastewater to water bodies may adversely affect water resources in some ways, especially potential contributions to eutrophication, which usually occurs in Taihu Lake, Chaohu Lake and Dianchi Lake, China. Removal of P from domestic sewage is one of the major strategies for preventing eutrophication. Therefore, legislation on P disposal into the water environment is becoming stricter worldwide. In Europe and North America, TP concentration of 0.1 mg L^{-1} or lower in the effluent from wastewater treatment plants (WWTP) is being required to achieve (Boltz *et al.* 2012). Similarly, 0.5 mg L^{-1} or lower P of effluent from WWTP is also imposed by Chinese government, compulsively meeting the National Class I (grade A) Sewage Discharge Standard (GB 18918-2002 published by State Environmental Protection Administration of China).

Moreover, the quantities of mineral phosphorus resources are rapidly decreasing in the world, presenting a global concern that P resources will be depleted in the near future, probably next 50–100 years (Chen *et al.* 2009; Takeda *et al.* 2010). Unlike nitrogen, P is not widely circulated on the globe due to the fact that flux of P from land to water is substantial but the reverse flux is extremely

Corresponding author: Xiwu Lu
E-mail: xiwulu@seu.edu.cn

limited, only via fishery or collection of guano (Childers *et al.* 2011). On the other hand, there has been an increase in the global demand for P with the world population growing, approximately 80% of which is mostly used to produce fertilizer for agricultural activities (De-Bashan, Bashan 2004; Kröger *et al.* 2013). Consequently, the recovery of P from wastewater is highly essential for enhancing the quality of water environment and for recycling of mineral phosphorus resources.

In recent years, studies focusing increasingly on combined processes, removing P from domestic wastewater and recovering it simultaneously in the form of a valuable product, such as struvite (MAP) or hydroxyl calcium phosphate (HAP), have been performed in lab-scale or full-scale reactors (Antonini *et al.* 2012; Cao *et al.* 2007; Cusick, Logan 2012; Daniels *et al.* 2013; Hug, Udert 2012). Most of studies have mainly focused on the combination of biological nutrients removal (BNR) with chemical precipitation process. In these processes, phosphorus in the P-rich solution generated by anaerobic sludge digestion was recovered via precipitation of MAP or HAP (Mañas *et al.* 2011). The disadvantages of this approach are obvious by the following two points: (a) operation of anaerobic sludge digestion performed out of biological nutrient removal process can suffer from high costs; (b) the precipitation product containing P with high moisture content is not easy to use directly in agricultural or industrial activities.

For this, a new process for P removal and recovery was proposed in this study by combining BNR with induced crystallization, BNR-IC for short here, expecting an achievement of stable and efficient carbon, nitrogen and phosphorus removal and maximum P recovery simultaneously. Compared with the conventional process linking BNR with chemical precipitation, the advantages of the new process may display: (a) a P-rich supernatant generated by denitrifying polyphosphate accumulating organisms (DPAO) in the anaerobic tank, included the BNR-IC system, omitting the procedure of anaerobic sludge digestion; (b) the crystallization product can be dewatered easily and could be potentially used directly as phosphorus fertilizer or industrial raw materials; (c) moreover, the induced HAP crystallization section can greatly reduce the P loading in the anoxic environment (presenting the excessive P uptake by DPAO), thus resulting in a minimize sludge production. Several studies have demonstrated that P can be removed from solution by crystallization of HAP in the presence of appropriate concentration of Ca^{2+} (Jang, Kang 2002; Kim *et al.* 2006).

Therefore, it was a hypothesized that the new process, proposed in this study, for nutrients removal and phosphorus recovery from domestic sewage by combining BNR with IC may be an effective strategy and could achieve the stable removal efficiencies of carbon, nitrogen and phosphorus, and successfully recover P from domestic wastewater. The aim of this study was to demonstrate the effectiveness of the novel process by monitoring carbon, nitrogen and phosphorus removal performances using chemical analysis and investigating the crystallization product using scanning electron micrograph (SEM) and energy dispersive spectrometry (EDS). The results obtained from here may serve as a new suggestion for linking the nutrients removal with P recovery from domestic wastewater.

1. Materials and methods

1.1. Experimental set-up and operation

The BNR-IC (Fig. 1) consisted of biological nutrients removal system adopting a double-sludge process running in anaerobic-anoxic/aerobic described by (Peng *et al.* 2004) and induced crystallization column, as shown in Figure 2, was designed by us, which has been applied for a parent (No. 201110431802.4) in China. The IC column with the total working volume of 2.5 L consisted of reaction zone for the induced crystallization, buffer zone and settling zone for turning of seed crystal (calcite used in this study) back to reaction zone again. In addition, a calcium chloride solution was given to provide Ca^{2+} for the induced HAP crystallization process and air was pumped for fluidization of seed crystal in reactor zone and stripping of carbon dioxide from P-rich solution, which can increase the pH value, for the better of induced HAP crystallization.

In the BNR-IC system, domestic wastewater is introduced into anaerobic tank, with a working volume of 3.3 L, where P is released from cells of DPAO responsible for phosphorus removal and most of readily

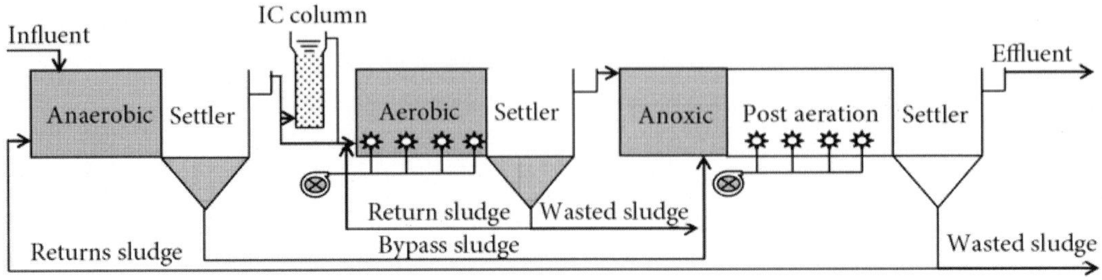

Fig. 1. Schematic diagram of combining BNR with IC process

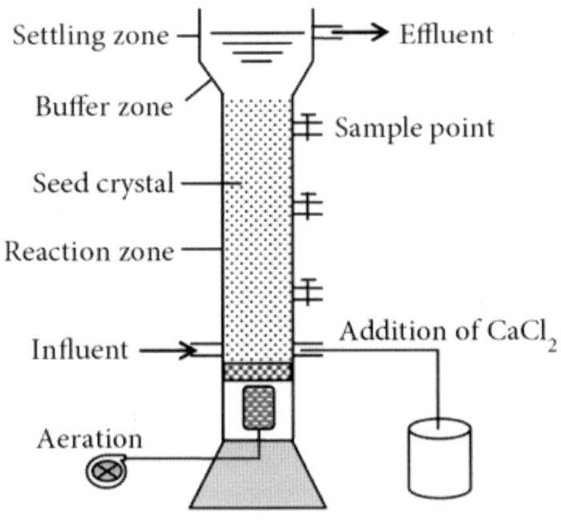

Fig. 2. Configuration of IC column

biodegradable COD is taken up and stored as poly-β-hydroxybutyrate (PHB). After settling, the P-rich supernatant is partly led into the IC column to recover P and the rest of that with effluent of the IC column altogether flow into aerobic tank (3.3 L) for nitrification. Sequentially, the clarified supernatant from nitrification tank is introduced into anoxic tank (3.3 L) for denitrifying nitrogen and phosphorus simultaneously, completed by DPAO, here presenting the excessive P uptake using the nitrate as electron acceptor. Lastly, a post aeration tank (3.3 L) is adopted for nitrogen gas stripping and the residual nutrients removal in order to ensure or enhance the quality of effluent.

The whole experiment was divided into three main phases according to the operational conditions, as described below:

Phase 1 (0~55 days): Seed sludge (collected from an aerobic basin of the Chengdong Municipal Wastewater Treatment Plant, Nanjing, China with AA–aerobic (A^2O) configuration) were introduced into aerobic tank and anoxic tank respectively for enrichments of nitrification bacteria and DPAO in their respective tanks. These were preformed according to the method (Peng *et al.* 2004) with slight modification, in which no waste sludge was discharged from tanks in this phase to maintain a lager amount of biomass. During the enrichment of DPAO, influent concentrations of COD and P were 250 mg L^{-1}, and 10 mg L^{-1}, respectively in anaerobic and addition of 45 NO_3-N mg L^{-1} was conducted in anoxic, and for nitrification bacteria cultivation, that of COD, NH_4-N and alkalinity ($CaCO_3$) were 80 mg L^{-1}, 40 mg L^{-1} and 120 mg L^{-1}, respectively.

Phase 2 (56~85 days): The BNR-IC system proposed in this study was connected and operated continuously until the combined process reach steady nutrients removal efficiencies and P recovery rate, where the influent concentrations of COD, TN, NH_4-N and P were 250 mg L^{-1}, 42 mg L^{-1}, 40 mg L^{-1}, and 10 mg L^{-1}, respectively. In this phase, the flow rates of influent, bypass sludge, return sludge, controlled by variable speed peristaltic pumps, were maintained at 18 L d^{-1}, 7.2 L d^{-1} and 7.2 L d^{-1}, respectively. In addition, the lateral flow ratio was set at 35.0%, representing that 35.0% of supernatant from anaerobic tank flowed into IC column and 65.0% of that directly into aerobic tank. Air was supplied at a flow rate of 1.5 L min^{-1} to meet the dissolved oxygen (DO) concentration higher than 2 mg L^{-1} at the aerobic and post-aeration tank. The mixed liquid suspended solid (MLSS) in the anaerobic tank, anoxic tank and post aeration tank were maintained at 3.4~3.6 g L^{-1} while that in aerobic tank was controlled at 3.0~3.2 g L^{-1}, resulting in the SRT of nitrifying bacteria and DPAO were approximately 16 days and 12 day, respectively. Moreover, the amount of Ca^{2+} (calcium chloride used here) was provided into IC column following the molar ratio 2.5 of Ca to P to enable HAP crystallization.

Phase 3 (86~115 days): After the BNR-IC process reached the stable steady, the concentrations of carbon, nitrogen and phosphorus were investigated along the flow and induced crystallization product was recovered and analyzed by using SEM coupled with EDS, where the process parameters were given similar to the phase 2.

1.2. Assessment of phosphorus release and uptake capacities by DPAO

After completing the cultivation of microorganisms, the capacities of anaerobic phosphorus release and anoxic phosphorus uptake by DPAO were assessed through a batch test under anaerobic/anoxic environments, which was conducted according to the following procedures: a) 0.5 L of activated sludge was taken from anoxic tank, immediately washed twice with the nutrients solution (Table 1) not containing the basic medium, and then, filled into one 1 L test device (transformed triangular flask); b) 0.5 L of synthetic wastewater was added to achieve an initial concentration of COD 250 mg L^{-1}, P 10 mg L^{-1} for anaerobic phosphorus release; c) after a 2 h anaerobic period, nitrogen gas was supplied in the anoxic condition for 4 h, with the addition of 45 NO_3-N mg L^{-1} as well for providing the electron acceptor to denitrify nitrogen and phosphorus removal.

1.3. Synthetic domestic sewage

The synthetic domestic wastewater, composition detailed in Table 1, was fed to the BNR-IC system in this study. In phase 1, the influent concentrations of COD, P and NO_3-N were 250 mg L^{-1}, 10 mg L^{-1} and 45 mg L^{-1}, respectively and in the phase 2 and 3, the influent concentrations of COD, PO_4-P, NH_4-N, TN were 250 mg L^{-1}, 10 mg L^{-1}, 40 mg L^{-1} and 42 mg L^{-1}, respectively, and correspondingly the ratios of COD/P, COD/TN were 25:1, 6:1, respectively, belonging to the lower C/N ratio wastewater similar to real domestic sewage.

Table 1. Composition of wastewater used for experiment

Composition of feeds	Content g/L	Composition of nutrient solution	Content g/L
CH_3COONa	0.322	$FeCl_3{\cdot}6H_2O$	1.50
KH_2PO_4	0.044	H_3BO_3	0.15
$(NH_4)_2SO_4$	0.047	$CuSO_4{\cdot}5H_2O$	0.03
$CaCl_2$	0.005	KI	0.18
$MgSO_4{\cdot}7H_2O$	0.05	$MnCl_2{\cdot}4H_2O$	0.12
Nutrient solution	0.30 mL/L	$Na_2MoO_4{\cdot}2H_2O$	0.06
		$ZnSO_4{\cdot}7H_2O$	0.12
		$CoCl_2{\cdot}6H_2O$	0.15
		EDTA	10.00

1.4. Analytical methods

Samples of activated sludge and effluent from different tanks were taken regularly from the BNR-IC system. COD and MLSS were determined according to the standard methods for the examination of water and wastewater (APHA 2005). PO_4-P, NH_4-N and TN were analyzed by segmented flow analysis (AutoAnalyzer3, SEAL, UK). DO, pH and Oxidation-reduction potential (ORP) were measured by a DO meter analyzer (YSI DO200, USA) and a pH and ORP meter analyzer (YSI pH100, USA).

For analysis of crystallization product, used calcite, probably covered by HAP crystal onto the surfaces of those, was collected at the end of running 30 days in the phase 3 (described above) and dried at 60 °C. The treated samples were observed after spray-gold treatment, by using SEM (JSM-6360LV, Japan) coupled with EDS (GENESIS 2000 XMS, USA). For comparison purposes, raw seed crystal was also observed according to the same methods.

2. Results and discussion

2.1. The performance of BNR-IC process

Nitrifying bacteria and DPAO were successfully enriched in their respective tanks during the 55 days cultivation process. At the steady state, in the BNR-IC system, when the COD, TN, NH_4-N and P concentrations of 250 mg L^{-1}, 42 mg L^{-1}, 40 mg L^{-1}, and 10 mg $L{-}^{1}$, respectively were given in the influent, a stable removal efficiencies of 92.5% COD, 78.6% TN, 85.9% NH_4-N and 95.2% P were obtained for the BNR process and correspondingly the COD, TN, NH_4-N and P concentrations of 18.75 mg L^{-1}, 8.99 mg L^{-1}, 5.64 mg L^{-1}, 0.42 mg L^{-1} were monitored in the effluent, meeting the Chinese National Class I (grade A) Sewage Discharge Standard.

A typical key phosphorus biochemical transformation, namely anaerobic P release and anoxic P uptake, was observed through a cycle batch test (2 h of anaerobic and 4 h anoxic) performed at the end of 55 days, as shown in Figure 3. The variations of carbon, nitrogen and phosphorus concentrations during a whole anaerobic-anoxic cycle were similar to what is typically observed in EBPR systems (Podedworna, Żubrowska-Sudoł 2012), strongly suggesting that DPAO was predominant in the BNR-IC process. In this process, the anaerobic P release rate and anoxic P uptake rate were 8.47 mg P (g MLSS)$^{-1}$ and 11.13 mg P (g MLSS)$^{-1}$, presenting a good P biochemical transfer performance indicated by the ratio of anoxic P uptake to anaerobic P release (1.3), thereby creating a preferable condition that could support the growth of DPAO in the BNR-IC system (Guerrero *et al.* 2012a). Figure 3 shows that throughout the 6h-long batch test, there were a gradual decrease in COD concentration during the anaerobic phase with the increase of P concentration and a rapidly decline in nitrate concentration under the anoxic condition with the P removal simultaneously. Moreover, the significantly positive correlations between COD consumption and P release under anaerobic condition with 0.9972 of R^2, and between nitrate removal and P uptake in the anoxic mode (R^2 of 0.9874) also demonstrated that DPAO was dominant in the tanks included the BNR-IC system. These results are consistent with previous reports (Wang *et al.* 2010; Guerrero *et al.* 2012b; Lanham *et al.* 2013) which suggested the existence of significantly positive correlations between anaerobic COD consumption and P release, anoxic nitrate removal and P uptake.

Compared with DPAO, nitrification bacteria can easily be cultivated in the aerobic tank due to the fact that seed sludge collected from an aeration pond of A^2O process had a high capacity for nitrification, displaying the 78.5% of NH_4-N removal efficiency. At the 25th days from

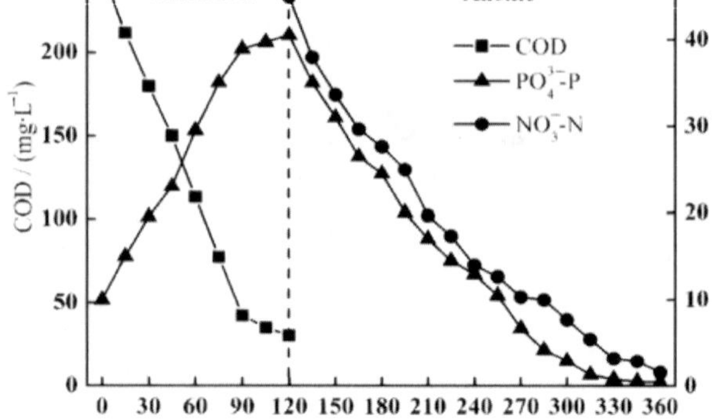

Fig. 3. Variations of carbon, nitrogen and phosphorus concentrations during a whole anaerobic-anoxic cycle

the beginning of cultivation, the NH_4-N removal efficiency reached above 95%, suggesting that enrichment of aerobic microorganisms may be easier than that of anaerobic microbe in WWTP.

2.2. Performance of BNR-IC process in each stage

In the phase 3, as described section 1.1, 30 days continuous operation was performed for investigating nutrients removal efficiencies and P recovery rate in each tank along the flow. A stable and good performance of nutrients removal and P recovery was obtained in this phase, and the variations of COD, TN, NH_4-N, NO_3-N and P concentrations in the effluent of different tanks along the flow in the BNR-IC process were shown Figure 4 and Table 2, respectively, and carbon, nitrogen and phosphorus removal efficiencies in each stage contributions to that in the total BNR-IC system were illustrated in Figure 5, where the negative values indicated that nutrients in this stage were not removed but produced by microbes.

As shown in Figure 5, the COD removal efficiency in anaerobic tank was significantly higher than that in other tanks, accounting for 85.9% of total COD removal rate (92.5%) in the BNR-IC system, which is agreement well with the characteristic of DPAO (Acevedo *et al.* 2012). In anaerobic condition, carbon source, especially VFA, was taken up by DPAO and stored them as PHB, resulting in the P release from the cells of DPAO. So there was a linear relationship between COD consumption and P release in the anaerobic condition, demonstrated by the results (described in Section 2.1) obtained in this study and (Chuang *et al.* 2011).

It should be noted that the COD concentration in the effluent from anoxic tank was more than that from aerobic tank, presenting 41.00 mg L^{-1} and 14.25 mg L^{-1} (Fig. 4 and Table 2), respectively. Consistently, the removal efficiencies of COD in anoxic tank and aerobic tank were – 11.6% and 16.0%, respectively (Fig. 5). It may be explained that a portion of COD in anaerobic tank was introduced into anoxic tank via the bypass sludge return from anaerobic tank to anoxic tank, thus leading to the increase of COD concentration in anoxic tank. There is a growth competition between DPAO and denitrifying bacteria when carbon sources provided in anoxic condition (Guerrero *et al.* 2011). Thereby creating a lower ratio of bypass sludge may support the growth of DPAO in anoxic tank. Although the COD concentration in the effluent from anoxic tank in the BNR-IC process proposed in this study increased slightly, the residual COD can mostly be biodegraded in the post-aeration tank (contributing to 9.6% of total COD removal efficiency found here). Generally, the post-aeration tank was often used as the key equipment for further nutrients removal and discharge to meet standard (Kapagiannidis *et al.* 2011; Smidt, Parravicini 2009; Zafiriadis *et al.* 2012), thus resulting in enhancement of water quality of effluent before discharge to water bodies.

In this BNR-IC system proposed here, NH_4-N removal mainly occurred in the aerobic tank, accounting for 95.4% of total NH_4-N removal efficiency, while TN and NO_3-N were mostly removed in the anoxic phase, as shown in Table 2, Figure 4 and Figure 5, respectively, indicating that good nitrification and denitrifying nitrogen and phosphorus removal simultaneously performances were present in their respective tanks.

Similar to COD concentration variation in the each stage of the BNR-IC process, there was a significant increase in NH_4-N concentration in the anoxic tank compared to that in the aerobic tank, 10.03 mg L^{-1} and 0.92 mg L^{-1}, respectively, probably also resulting from the sludge bypass return. The residual NH_4-N can be mostly removed in the post-aeration tank, thus ensuring that the NH_4-N concentration in the final effluent was about or less than 5 mg L^{-1} throughout the running period. The design of a post-aeration tank placed after a wastewater treatment process can effectively remove residue organic substances, nitrogen and phosphorus compound, thereby may further enhancing effluent quality (Zhang *et al.* 2013).

Considerable NO_3-N or TN was removed in anoxic tank, as given in Table 2 and Figure 5, respectively. At the same time, the previously stored PHB was used as the carbon resources, namely electron donor, utilized by DPAO for denitrifying nitrogen and phosphorus removal simultaneously (Coats *et al.* 2011). Also, post-aeration contributed to low concentrations of NO_3-N and TN in the effluent.

Table 2. Concentrations of carbon, nitrogen and phosphorus along the flow in BNR-IC

	Influent	Anaerobic		Aerobic		Anoxic	Post-aeration
COD / mg L^{-1}	250.00	51.25		14.25		41.00	18.75
TN / mg L^{-1}	42.00	32.17		28.77		9.95	8.99
NH_4-N / mg L^{-1}	40.00	31.70		0.92		10.03	5.64
NO_3-N / mg L^{-1}	0.58	0.51		22.70		1.95	4.82
	Influent	Anaerobic	IC	Aerobic (in)	Aerobic (out)	Anoxic	Post-aeration
P / mg L^{-1}	10.00	28.31	8.25	21.29	20.37	0.84	0.42

Figure 4 shows the variation of P concentration in each stage of BNR-IC system. In anaerobic tank, a good P release performance conducted by DPAO was found, in which the P concentration in the mixed liquor rapidly increased. However, there was a significant difference in amount of anaerobic P release between BNR-IC and BNR, ranging from 10.00 mg L^{-1} to 28.31 mg L^{-1} in BNR-IC lower than that in BNR (from 10.00 mg L^{-1} to 40.95 mg L^{-1}, as illustrated in Fig. 3). This was due to the fact that Poly-p stored in the DPAO decreased when IC column was coupled with BNR process. A part of P released by DPAO in anaerobic condition was recovered via induced HAP crystallization in IC column, thus leading to the reduction of P uptake (stored as Poly-p) in anoxic environment and subsequently P release (hydrolysis of Poly-p) in anaerobic condition. After 30 days running, a new balance

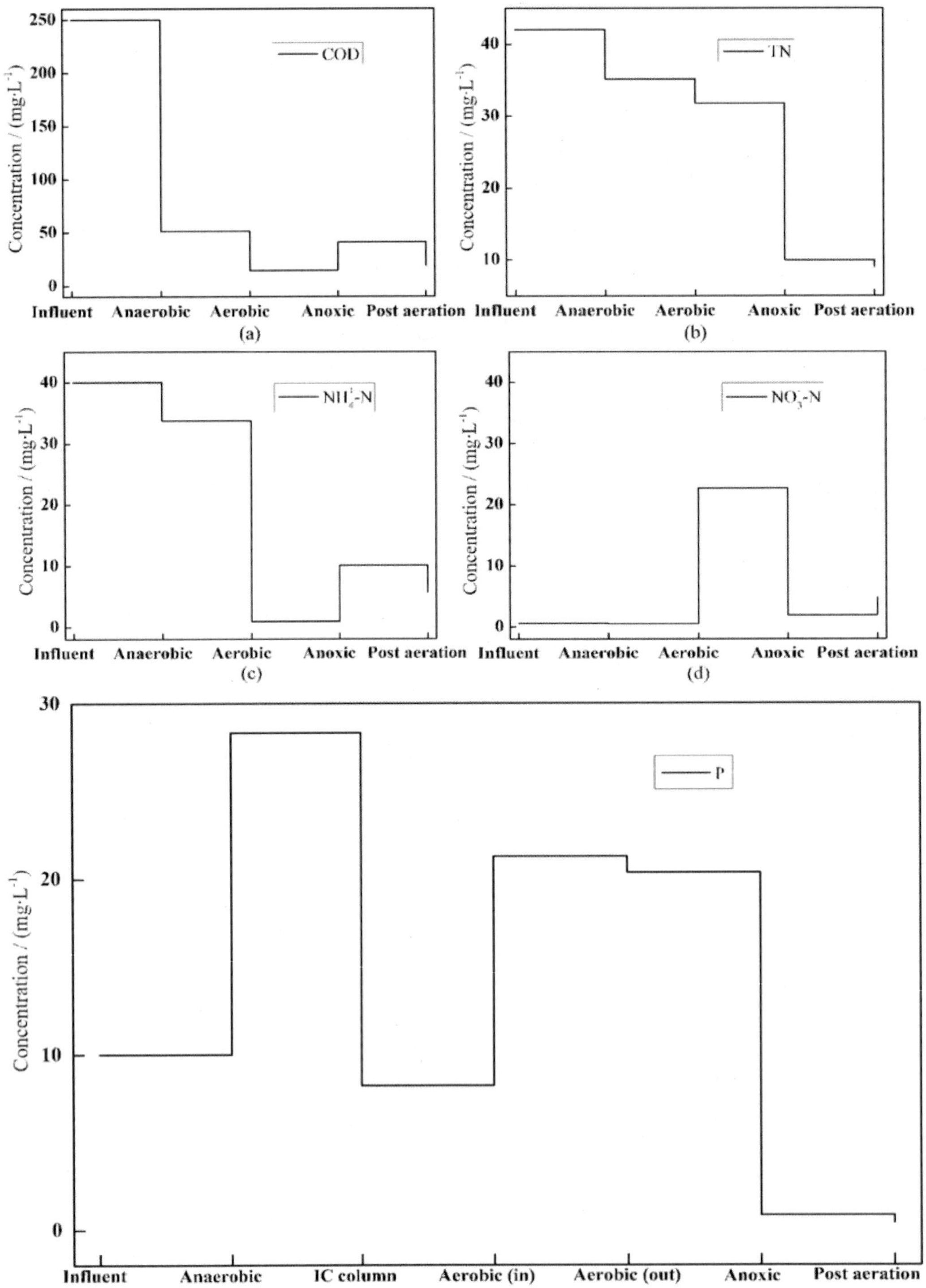

Fig. 4. Variation of carbon, nitrogen and phosphorus concentrations in each phase of BNR-IC process

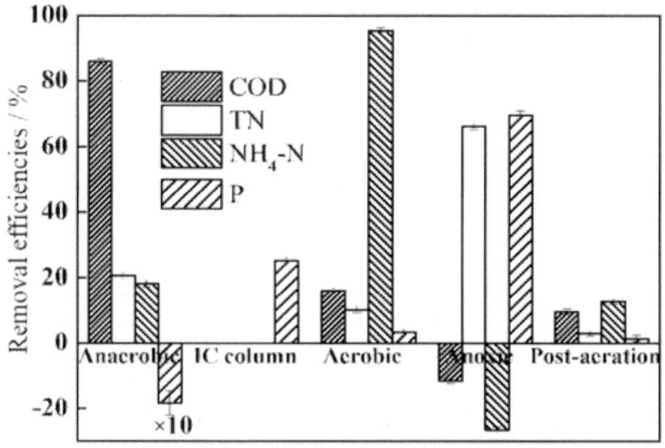

Fig. 5. Contributions to removal of carbon, nitrogen and phosphorus in each phase of BNR-IC process

among anaerobic P release, P recovery and anoxic P uptake was found in the new BNR-IC system proposed in this study. Although the IC column was introduced into the BNR leaded to the reduction of Poly-p stored in cells, the BNR-IC system still displayed a good characteristic of denitrifying nitrogen and phosphorus simultaneously when the lateral flow ratio was 35.0%. It was noteworthy that determination of the maximum lateral flow ratio should follow the principle that normal biological performance of denitrifying nitrogen and phosphorus simultaneously in BNR-IC system cannot be damaged when maximizing the P recovery efficiency.

In the IC column, 70.9% of P recovery rate was calculated according to the P concentration of in the influent (28.31 mg L^{-1}) and effluent (8.25 mg L^{-1}) of the reactor when the lateral flow ratio was 35.0%, namely that 35.0% of supernatant from anaerobic tank flowed into IC column and the rest 65.0% of that directly into aerobic tank, thus leading to 21.29 mg L^{-1} of P concentration in the influent of aerobic tank.

Based on the content of P in mixed liquor after anaerobic P release, about 69.7% of total P removal in BNR-IC system occurred in anoxic tank with 66.1% TN consumption (Table 2 and Fig. 5), obviously completed by DPAO performing denitrifying nitrogen and phosphorus removal simultaneously, which is agreement with the appearance in various EBPR systems (Fang *et al.* 2012; Marcelino *et al.* 2009; Slater *et al.* 2010), indicating that a good P uptake performance was also existent in the BNR-IC system. This result further demonstrated that combing BNR with IC proposed in this study was feasible for nutrients removal and P recovery simultaneously from domestic wastewater.

According to the P removal performance in the IC column described above, about 70.2% of P in the domestic wastewater (10 mg L^{-1} given in this study) was recovered depending on the induced HAP crystallization in the BNR-IC system, namely about 7 g P per ton wastewater recovered. Compared with well known P recovery process of PhoStrip (Bertanza *et al.* 2013), where the P-rich solution is generated by anaerobic treatment of only a part of return activated sludge through the addition of external carbon source such as acetic acid and then P is recovered with chemical precipitation via the adding of lime, the BNR-IC process was designed to recovery P from the supernatants (including influent and total return sludge anaerobic treatment) by induced crystallization, where directly used the carbon sources in the influent as the electron donor for DPAO anaerobic P release, thereby exhibiting a better significant economic and environmental benefits.

2.3. Crystallization product investigation

To investigate the physicochemical changes of surfaces of seed crystal (calcite used in this study) before and after induced HAP crystallization, microscopic analyses were conducted by using SEM and EDS, as illustrated in Fig. 6. There was a significant difference in the surface of calcite between before (Fig. 6a) and after (Fig. 6c) induced HAP crystallization, where the surface of used calcite was coarse greater than that in raw calcite. The SEM micrograph in Figure 6c revealed that the surface of calcite was covered with a finely distributed crystallization product layer, which was regarded as HAP according to the EDS analysis. The EDS analyses demonstrated that the crystallization layer consisted predominantly of O, P and Ca, as depicted in Figure 6d, with a Ca/P molar ratio of 1.79 slightly higher than that of 1.67 in theoretical value for HAP, which might be affected by the $CaCO_3$ formation. Figure 6d shows that carbon content obviously increased in the crystallization layer compared with the raw seed crystal, suggesting the formation of HAP crystallization with the $CaCO_3$ production, which are in accordance with the reported studies (Barca *et al.* 2012; Jang, Kang 2002; Johansson, Gustafsson 2000). The formation of $CaCO_3$ involved the consumption of Ca^{2+}, thus leading to molar ratios of Ca reduction to P reduction higher than 1.67 (Barca *et al.* 2012).

To investigate the growth of HAP in the surface calcite during the induced crystallization process, the amount of calcite before and after this process in IC column was measured in a long-term test of 120 days. The initial calcite at a dosage of 60 mg L^{-1} occupied about 6.5% of the IC volume. At the end of 120 days running, the volume of calcite increased significantly, accounting for 12% of the total IC volume due to the HAP production during the induced crystallization process.

Conclusions

1. This study showed that the new BNR-IC system combining biological nutrients removal with induced crystallization proposed here was an effective strategy for carbon, nitrogen and phosphorus removal and P recovery

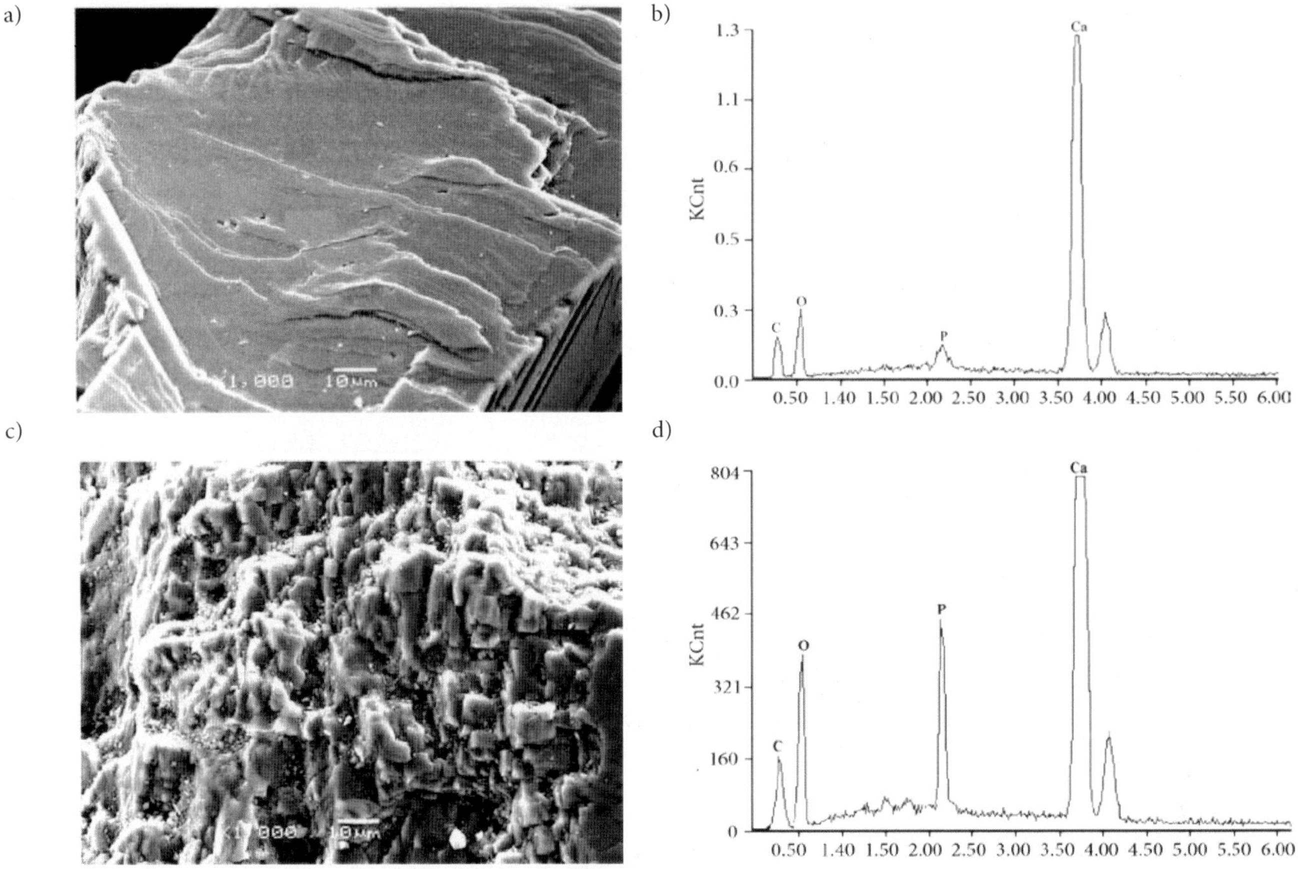

Fig. 6. SEM observation and EDS analyses of seed crystal before and after induced P crystallization: a) surface of seed crystal before P recovery; b) chemical analysis of seed crystal before P recovery; c) surface of seed crystal after P recovery at the 30th days; d) chemical analysis of seed crystal after P recovery at the 30th days

from domestic wastewater, exhibiting a good performance of denitrifying nitrogen and phosphorus removal simultaneously coupled with HAP crystallization.

2. In the BNR-IC system, the removal efficiencies of COD, TN, NH_4-N and P were maintained at about 90%, 78%, 85% and 95%, respectively and P recovery rate achieved 70.2% via induced HAP crystallization.
3. From these results, the proposed BNR-IC system, probably being a promising technology, can potentially aid in improvement of the effluent quality from WWTP and in recycle of mineral phosphorus resources. However, future research is necessary to focus on the optimization of operation parameters and practical applications of crystallization product.

Acknowledgments

We wish to thank M.M.SU for assistance with activated sludge treatment and Dr. Li for enlightening discussion. This research is supported by grant 2012ZX07101-005 from National Key Technology in Water Pollution Control and Treatment in 12th Five-year Plan of China and grant 51078074 from National Natural Science Foundation of China.

References

Acevedo, B.; Oehmen, A.; Carvalho, G.; Seco, A.; Borras, L.; Barat, R. 2012. Metabolic shift of polyphosphate-accumulating organisms with different levels of polyphosphate storage, *Water Research* 46(6): 1889–1900. http://dx.doi.org/10.1016/j.watres.2012.01.003

APHA. 2005. *Standard methods for the examination of water and wastewater*. 20th ed. American Public Health Association, American Water Works Association and Water Environment Federation, Washington, DC.

Antonini, S.; Arias, M. A.; Eichert, T.; Clemens, J. 2012. Greenhouse evaluation and environmental impact assessment of different urine-derived struvite fertilizers as phosphorus sources for plants, *Chemosphere* 89(10): 202–1210. http://dx.doi.org/10.1016/j.chemosphere.2012.07.026

Barca, C.; Gérente, C.; Meyer, D.; Chazarenc, F.; Andrès, Y. 2012. Phosphate removal from synthetic and real wastewater using steel slags produced in Europe, *Water Research* 46(7): 2376–2384. http://dx.doi.org/10.1016/j.watres.2012.02.012

Bertanza, G.; Pedrazzani, R.; Manili, L.; Menoni, L. 2013. Bio-P release in the final clarifiers of a large WWTP with coprecipitation: key factors and troubleshooting, *Chemical Engineering Journal* 230(15): 195–201. http://dx.doi.org/10.1016/j.cej.2013.06.067

Boltz, J. P.; Morgenroth, E.; Daigger, G. T.; DeBarbadillo, C.; Murthy, S.; Sørensen, K. H.; Stinson, B. 2012. Method to identify potential phosphorus rate-limiting conditions in post-denitrification biofilm reactors within systems designed for simultaneous low-level effluent nitrogen and phosphorus concentrations, *Water Research* 46(19): 6228–6238. http://dx.doi.org/10.1016/j.watres.2012.08.020

Cao, X.; Harris, W. G.; Josan, M. S.; Nair, V. D. 2007. Inhibition of calcium phosphate precipitation under environmentally-relevant conditions, *Science of The Total Environment* 383(1–3): 205–215. http://dx.doi.org/10.1016/j.scitotenv.2007.05.012

Chen, X.; Kong, H.; Wu, D.; Wang, X.; Lin, Y. 2009. Phosphate removal and recovery through crystallization of hydroxyapatite using xonotlite as seed crystal, *Journal of Environmental Sciences* 21(5): 575–580. http://dx.doi.org/10.1016/S1001-0742(08)62310-4

Childers, D. L.; Corman, J.; Edwards, M.; Elser, J. J. 2011. Sustainability challenges of phosphorus and food: solutions from closing the human phosphorus cycle, *Bioscience* 61(2): 117–124. http://dx.doi.org/10.1525/bio.2011.61.2.6

Chuang, S.; Chang, W.; Huang, Y.; Tseng, C.; Tai, C. 2011. Effects of different carbon supplements on phosphorus removal in low C/P ratio industrial wastewater, *Bioresource Technology* 102(9): 5461–5465. http://dx.doi.org/10.1016/j.biortech.2010.11.118

Coats, E. R.; Watkins, D. L.; Brinkman, C. K.; Loge, F. J. 2011. Effect of anaerobic HRT on biological phosphorus removal and the enrichment of phosphorus accumulating organisms, *Water Environment Research* 83(5): 461–469. http://dx.doi.org/10.2175/106143010X12851009156402

Cusick, R. D.; Logan, B. E. 2012. Phosphate recovery as struvite within a single chamber microbial electrolysis cell, *Bioresource Technology* 107: 110–115. http://dx.doi.org/10.1016/j.biortech.2011.12.038

Daniels, Y.; Zhu, X.; Alexandratos, S. D. 2013. Distinguishing between organic and inorganic phosphorus in hydroxyapatite by elemental analysis, *Microchemical Journal* 110: 263–265. http://dx.doi.org/10.1016/j.microc.2013.04.005

De-Bashan, L. E.; Bashan, Y. 2004. Recent advances in removing phosphorus from wastewater and its future use as fertilizer (1997–2003), *Water Research* 38(19): 4222–4246. http://dx.doi.org/10.1016/j.watres.2004.07.014

Fang, J.; Sun, P.; Xu, S.; Luo, T.; Lou, J.; Han, J.; Song, Y. 2012. Impact of Cr (VI) on P removal performance in enhanced biological phosphorus removal (EBPR) system based on the anaerobic and aerobic metabolism, *Bioresource Technology* 121: 379–385. http://dx.doi.org/10.1016/j.biortech.2012.07.001

Guerrero, J.; Guisasola, A.; Baeza, J. A. 2011. The nature of the carbon source rules the competition between PAO and denitrifiers in systems for simultaneous biological nitrogen and phosphorus removal, *Water Research* 45(16): 4793–4802. http://dx.doi.org/10.1016/j.watres.2011.06.019

Guerrero, J.; Tayà, C.; Guisasola, A.; Baeza, J. A. 2012a. Understanding the detrimental effect of nitrate presence on EBPR systems: effect of the plant configuration, *Journal of Chemical Technology and Biotechnology* 87(10): 1508–1511. http://10.1002/jctb.3812

Guerrero, J.; Taya, C.; Guisasola, A.; Baeza, J. A. 2012b. Glycerol as a sole carbon source for enhanced biological phosphorus removal, *Water Research* 46(9): 2983–2991. http://dx.doi.org/10.1016/j.watres.2012.02.043

Hug, A.; Udert, K. M. 2012. Struvite precipitation from urine with electrochemical magnesium dosage, *Water Research* 47(1): 289–299. http://dx.doi.org/10.1016/j.watres.2012.09.036

Jang, H.; Kang, S. 2002. Phosphorus removal using cow bone in hydroxyapatite crystallization, *Water Research* 36(5): 1324–1330. http://dx.doi.org/10.1016/S0043-1354(01)00329-3

Johansson, L.; Gustafsson, J. P. 2000. Phosphate removal using blast furnace slags and opoka-mechanisms, *Water Research* 34(1): 259–265. http://dx.doi.org/10.1016/S0043-1354(99)00135-9

Kapagiannidis, A. G.; Zafiriadis, I.; Aivasidis, A. 2011. Upgrading the efficiency of an external nitrification BNR system – the modified Dephanox process, *Chemical Engineering Journal* 175: 124–135. http://dx.doi.org/10.1016/j.cej.2011.09.080

Kim, E.; Yim, S.; Jung, H.; Lee, E. 2006. Hydroxyapatite crystallization from a highly concentrated phosphate solution using powdered converter slag as a seed material, *Journal of Hazardous Materials* 136(3): 690–697. http://dx.doi.org/10.1016/j.jhazmat.2005.12.051

Kröger, R.; Dunne, E. J.; Novak, J.; King, K. W.; McLellan, E.; Smith, D. R.; Strock, J.; Boomer, K.; Tomer, M.; Noe, G. B. 2013. Downstream approaches to phosphorus management in agricultural landscapes: regional applicability and use, *Science of The Total Environment* 442: 263–274. http://dx.doi.org/10.1016/j.scitotenv.2012.10.038

Lanham, A. B.; Oehmen, A.; Saunders, A. M.; Carvalho, G.; Nielsen, P. H.; Reis, M. A. 2013. Metabolic versatility in full-scale wastewater treatment plants performing enhanced biological phosphorus removal, *Water Research* 47(19): 7032–7041. http://dx.doi.org/10.1016/j.watres.2013.08.042

Mañas, A.; Biscans, B.; Spérandio, M. 2011. Biologically induced phosphorus precipitation in aerobic granular sludge process, *Water Research* 45(12): 3776–3786. http://dx.doi.org/10.1016/j.watres.2011.04.031

Marcelino, M.; Guisasola, A.; Baeza, J. A. 2009. Experimental assessment and modelling of the proton production linked to phosphorus release and uptake in EBPR systems, *Water Research* 43(9): 2431–2440. http://dx.doi.org/10.1016/j.watres.2009.03.003

Peng, Y.; Wang, Y.; Ozaki, M.; Pan, M. 2004. Denitrifying phosphorus removal in a continuously-flow A^2N two-sludge process, *Journal of Environmental Science and Health, Part A* 39(3): 703–715. http://dx.doi.org/10.1081/ESE-120027736

Podedworna, J.; Żubrowska-Sudoł, M. 2012. Nitrogen and phosphorus removal in a denitrifying phosphorus removal process in a sequencing batch reactor with a forced anoxic phase, *Environmental Technology* 33(2): 237–245. http://dx.doi.org/10.1080/09593330.2011.563428

Slater, F. R.; Johnson, C. R.; Blackall, L. L.; Beiko, R. G.; Bond, P. L. 2010. Monitoring associations between clade-level variation, overall community structure and ecosystem function in enhanced biological phosphorus removal (EBPR) systems using terminal-restriction fragment length polymorphism (T-RFLP), *Water Research* 44(17): 4908–4923. http://dx.doi.org/10.1016/j.watres.2010.07.028

Smidt, E.; Parravicini, V. 2009. Effect of sewage sludge treatment and additional aerobic post-stabilization revealed by infrared spectroscopy and multivariate data analysis, *Bioresource Technology* 100(5): 1775–1780. http://dx.doi.org/10.1016/j.biortech.2008.10.003

Takeda, I.; Somura, H.; Mori, Y. 2010. Recovery of phosphorus from natural water bodies using iron-oxidizing bacteria and woody biomass, *Ecological Engineering* 36(8): 1064–1069. http://dx.doi.org/10.1016/j.ecoleng.2010.04.019

Wang, Y.; Jiang, F.; Zhang, Z.; Xing, M.; Lu, Z.; Wu, M.; Yang, J.; Peng, Y. 2010. The long-term effect of carbon source on the competition between polyphosphorus accumulating organisms and glycogen accumulating organism in a continuous plug-flow anaerobic/aerobic (A/O) process, *Bioresource Technology* 101(1): 98–104. http://dx.doi.org/10.1016/j.biortech.2009.07.085

Zafiriadis, I.; Ntougias, S.; Mirelis, P.; Kapagiannidis, A. G.; Aivasidis, A. 2012. Molecular characterization of denitrifying bacteria isolated from the anoxic reactor of a modified Dephanox plant performing enhanced biological phosphorus removal, *Water Environment Research* 84(6): 475–484. http://dx.doi.org/10.2175/106143012X13280358613543

Zhang, W.; Peng, Y.; Ren, N.; Liu, Q.; Chen, Y. 2013. Improvement of nutrient removal by optimizing the volume ratio of anoxic to aerobic zone in AAO-BAF system, *Chemosphere* 93(11): 2859–2863. http://dx.doi.org/10.1016/j.chemosphere.2013.08.047

Haiming ZOU. Dr, School of Energy and Environment, Southeast University; Department of Resource and Environment, Anhui Science and Technology University. He is the author of 29 publications and has participated in 12 conferences. His research interests include nitrogen and phosphorus removal from wastewater.

Xiwu LU. Dr, Professor, School of Energy and Environment, Southeast University. He is the author of 195 publications and has participated in 89 conferences. His research interests include nitrogen and phosphorus removal from wastewater; lake pollution control.

Ting LI. Master, School of Energy and Environment, Southeast University. He is the author of 3 publications and has participated in 2 conferences. His research interests include nitrogen and phosphorus removal from wastewater.

GEOFILTRATION STUDIES OF CLAY AT THE FUTURE RADIOACTIVE WASTE REPOSITORY FOR IGNALINA NUCLEAR POWER PLANT

Petras Klizas

Department of Hydrogeology and Engineering Geology, Faculty of Natural Sciences, Vilnius University, M. K. Čiurlionio g. 21/27, 03101 Vilnius, Lithuania

Abstract. Clay geofiltration studies became important in Lithuania when the clays began to be used as geofiltration barriers for waste disposal sites/repositories. The clay is planned to be used also in construction of the future radioactive waste repository for the Ignalina Nuclear Power Plant. The results of investigation on Stabatiškė subformation till clay occurring at the base of the repository are given in the paper. The investigation has been performed with the purpose to determine potential dependence of clay filtration coefficient values on the number of freeze-thaw cycles and duration of filtration. The filtration coefficient for the repository till clay has been found to be greatly varying at the start of the filtration and during the experiment that lasted several days. This confirms that, during the filtration, structural changeover of clay-forming aggregates takes place with pore space volumes changing and affecting the filtration coefficient values. The filtration parameters are even more affected by freeze-thaw cycles and their number. When the clay thaw is finished the colmatation of fissure system is very slow and incomplete.

Keywords: clay geofiltration, hydraulic conductivity, soil pollution, water contamination, freeze-thaw cycles.

Introduction

Lately the demand in geofiltration investigations on clay deposits have significantly increased in Lithuania. Being impermeable to water clay is used for constructing various barriers against percolation of hazardous waste, including radioactive ones, in repositories and landfills of various type and size. The EU laws contain strict geofiltration requirements to be applied for the clay used for such barriers. The hydroisolation capability of clay layers depends on many interrelated factors and parameters: permeability, porosity, microstructure peculiarities of pore space, moisture, thickness of natural or artificial hydroisolation, clay mineral composition, chemical composition of future leachate, hydrodynamic load, potential impact of freeze etc. However, the main indices affecting the efficiency of the future geofiltration barrier are hydraulic conductivity and clay layer thickness, which are regulated quantitatively. In France, for instance, there is a complex concept called 'passive safety barrier' (Cassan 2005). Depending on a degree of waste hazardousness, it defines the ratio between the barrier hydroisolation thickness and the hydraulic conductivity. Therefore, it is very important to have accurate values of hydraulic conductivity determined, because formation of necessary thickness of permeable layer is easier to be done. To select the type of clay satisfying percolation requirements is more difficult, because these characteristics are changing in time due to the above mentioned factors.

The percolation characteristics of Lithuanian clays are not widely studied, since there was almost no such demand. Another reason might be the fact that due to the very low percolation, experimental laboratory investigations would be long-lasting, hence, time- and funds-consuming. Our investigations have been performed on order by a private company, therefore, the information about sampling sites, depth etc., which is not important methodologically for the present study, are not provided in this paper. We shall analyse the dependence of clay geofiltration parameters on filtration duration, its stopping and renewal, as well as number of freeze-thaw cycles.

Corresponding author: Petras Klizas
E-mail: petras.klizas@gf.vu.lt

The abundance in scientific publications of late years shows lively interest in clay geofiltration studies in different geological regions. For one of such investigations a French expansive clay known as Fo-Ca, which is a natural Ca-smectite coming from the Paris basin of Ypersian age, was used. Permeability experiments were performed with distilled water and several copper concentrations (10^{-3} to 10^{-1} mol/l). The permeability of clay was found to increase with copper concentration. Using the syringe oedometer, permeability was 1.1E-12 m/s with distilled water, and 2.4E-12 m/s with the 0.1 mol/l copper solution (Jullien *et al.* 2002). The Mont Terri rock laboratory in Switzerland studied compacted Opalinus Clay core samples from 200 m depth. Permeability and porosity was determined to be a function of time at constant pressure (Jobmann, Wilsnack 2010). In a few studies it was shown that composition and compression of clay are very important for its hydraulic conductivity, permeability and filtration properties (Shafiee 2008; Ebina *et al.* 2004; Hamdi, Srasra 2008).

In this paper we show similar functional relationships for Grūda subformation till clay from Stabatiškė site in Lithuania. In another study (Heister *et al.* 2005), the material taken from the clay between the motorway and railroad on the northeast of the harbour in Rodbyhavn, owned by Dansk Bentonit, was examined and a new laboratory set-up for measurements of electrical, hydraulic and osmotic fluxes in clays and clay-rich materials was established. Researches performed in Italy on water vapour permeability of clay bricks showed relationships between a wide range of chemical and mineralogical compositions and particle size distributions. The correlations of vapour permeability, open and closed porosity, bulk density, mean pore diameter, pore size selection and specific surface were determined (Dondi *et al.* 2003). Soil air permeability is one of the most important parameters which govern the aeration in agricultural soils. The relationship between air permeability, soil capacity parameters and vertical stress was analysed by performing confined uniaxial compression tests and air permeability measurements on three different French soils (Tang *et al.* 2011).

Permeability studies for frozen rocks are also often performed in the regions where mean winter air temperatures are below zero (Al Houri *et al.* 2009; McCauley *et al.* 2002). In France, the permeability of non-cohesive rock mixtures of different grain-size has been studied. The laboratory tests were performed by a permeameter at varying negative temperatures and using different permeants: water, different NaCl solutions, bentonite and trapped decane mixture (Enssle *et al.* 2011). In another research performed in Canada an infiltration of snowmelt water under small depressions has been examined. These depressions stored most of the snowmelt runoff generated within their respective watersheds (Hayashi *et al.* 2003). In a current work we analyse rock permeability variations under the freeze and thaw impact for the first time in Lithuania. The oedometric laboratory and field infiltration studies on suitability of clay deposits from Šaltiškė, Pašaminė and Stabatiškė sites for the Ignalina nuclear power plant radioactive waste repository hydroisolation barrier were performed by using a double ring infiltrometer and injection into a borehole method after V. Nasberg (Gadeikis *et al.* 2012). Possibilities of transport and accumulation in soil of radionuclides and heavy metals were determined by properties of clay (Jakimavičiūtė–Maselienė *et al.* 2012; Pundytė *et al.* 2011).

The main objective of our study was to determine filtrational parameters of the clay from the future radioactive waste repository at Stabatiškė site. We aimed to determine how the filtration coefficient depends on the number of freeze-thaw cycles, the colmatation after freezing and the duration of filtration. Filtration coefficient is the main parameter, which determines the spread of radioactive nuclides in rocks of aeration zone and protects the below laying groundwater from possible pollution.

1. Composition and structure of clayey rocks

Structural characteristics of clayey rocks are determined by dispersed clay particles called micelles. The micelles in clayey rocks are notable for an electric double layer formed at the boundary between the solid particle and the liquid (water in clay). Structural diversity of clay minerals and crystal-chemistry specificity of particle surfaces at the microstructural level change the understanding of clay pore space. These features of clay cause different percolation course, if compared to other dispersed systems composed of similar particles with no interaction, such as sand, for instance. Clay beds at the microstructural level are notable for a higher degree of diversity and depend on ratio of three basic clay minerals – kaolinite, muscovite mica and sodic montmorillonite. The macrostructural diversity of beds is caused by clastic inclusions, macropores, lamination and fracturedness in highly lithified beds. The macropores are typical of the clay beds in the aeration zone, including those artificially formed as hydroisolation barriers at the land surface. Pores and cracks are main factors determining permeability anisotropy of clay beds. A characteristic feature of such macropores is their potential clogging (colmatation) during long period, therefore, hydraulic conductivity of clay sample gradually decreases during a long-term experiment. This process was observed by us during long-term permeability testing. During percolation through clay pores ranging from 3 to 20 Å, the following processes also take place: cation exchange, hydratation-dehydratation at the surface and inside of clay minerals and sticking of particles into larger aggregates. At the micro-aggregate level, when pores are within the range of 1–10 μm, changes in pore and structure space are observed in relatively homogenous samples and beds, and this causes anisotropicity of water percolation in clay (Oradovskaja 1983). During percolation, the structural changeover of particles takes place, and this process is more effective the more diverse is the initial structure. At larger pressure gradients, flat particles of clay minerals reorientate parallel to the water flow lines, and at lower pressure gradients they do not move back to their initial position. This effect is typical of high moisture and porosity clay and loam.

The abundance of scientific publications on various clay studies, especially on geofiltration, shows the significance of these rocks and their practical usefulness in solving environmental problems.

2. Methods and materials studied

The goal of the present study is to assess the variations of clay permeability properties: (1) during a long-term percolation; (2) repetition of percolation stopping and renewal cycles; (3) during sample freeze and thaw. The Grūda subformation till clay (gIIIgr) at the projected Stabatiškė waste repository site has been chosen for detailed investigations. Macroelement oxides and mineralogical composition of examined clay were determined in the laboratory of Center for Physical Sciences and Technology. Average percentage of macroelement oxides in clay fraction is as follows: Na_2O – 0.7%; MgO – 2.9%; Al_2O_3 – 10.8%; SiO_2 – 59.9%; K_2O – 3.7%; CaO – 8.1%; TiO_2 – 0.5% and Fe_2O_3 – 10.6%. The average percentage of clay minerals in clay fraction and deposits (in parenthesis) is as follows: illite – 83.3 (15.3)%, kaolinite – 15.8 (2.8)% and chlorite – 0.9 (0.2)%. The following physical properties have been determined as well: natural moisture content – 12%; natural density – 2.25 Mg/m^3; hydraulic conductivity obtained by oedometric measurements at 0.1 and 0.2 MPa load is, respectively, 2.3E-11 m/s and 7.1E-11 m/s. The clay analysed in this study was taken from the same place as one used in publication of Gadeikis *et al.* (2012). Laboratory permeability testing has been performed with non-compacted clay of natural structure. A scheme of the permeameter used is given in Figure 1.

This type infiltrometer is applied for filtering performed according to a non-stationary scheme. Maximum piezometric head is up to 35 cm, sample height is 4 cm, cross-section area is 40 cm^2; and initial of maximum hydraulic gradient is 8,75.

Hydraulic conductivity is calculated by the formula:

$$K = \frac{2.3AL}{ST\log\frac{H_1}{H_2}},$$

where: A is piezometer cross-section area; L is sample height; S is filtration ring cross-section area; T is filtering time, and H_1 and H_2 are hydraulic heads.

Starting the geofiltration investigations, first of all, we have assessed the variations in hydraulic conductivity depending on duration of filtering. Then an attempt was made to determine the filtering process during its stopping and renewal. The second cycle of experiments dealt with the impact of freezing on clay hydraulic conductivity, as well as the assessment of the role of the freeze and thaw cycles. All the filterings have been performed according to a non-stationary scheme.

Of late years, the alternation of periods with negative and positive air temperatures in Lithuania is frequently observed, therefore, it is important to assess the effect of these phenomena on clay permeability properties. The clay hydroisolation barriers formed for the near-surface waste repositories of various types are not thick. Their upper part, as a rule, freezes in winter and thaws in spring. During all the laboratory experiments, measurements of water evaporation, air temperature and humidity have been done, with the purpose to eliminate the influence of these factors on hydraulic conductivity values determined. The PV infiltrometer has an open orifice for filtrate outflow as the piezometric tubes used with open-ended

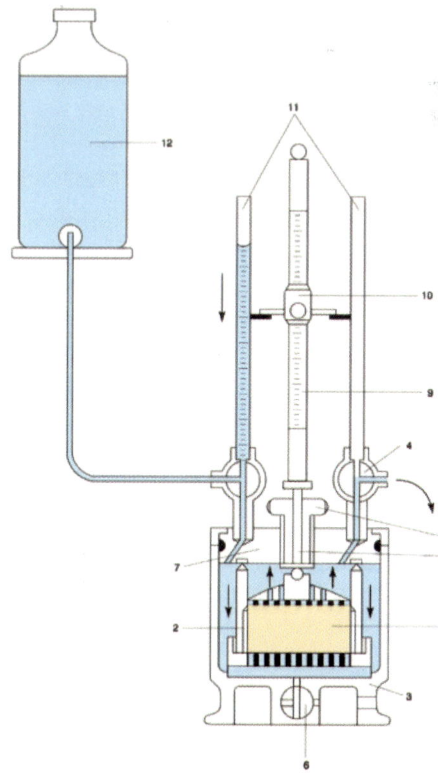

Fig. 1. A scheme of PV infiltrometer (permeameter): 1 – clay sample; 2– filtering chamber ring; 3 – a case; 4 – three-way faucet (tap); 5 – valve; 6 – two-way faucet (tap) for filling the infiltrometer with water; 7 – a case cover; 8 – thermometer; 9 – water level scale in piezometer; 10 – sliding marker; 11 – piezometers; 12 – a container for water; 13 – filtrate outflow orifice

top. Therefore it is very important to have data on evaporation intensity, especially for long-term measurements and at very low filtrate debits at low hydrodynamic head gradients. The longest filtering lasted 430 hours, and minimum hydrodynamic head gradients reached just some tenths of a digit one. The clay sample was frozen together with a filtration chamber ring in a freezer at a stable –7 °C temperature for 5 days. According to the data from Utena Meteorological Station, which belongs to the network of Lithuanian Hydrometeorological Service, the average minimal temperature of January and February, in the time period from 1971 until 2000, was –7.1 and –7.3 °C respectively (www.meteo.lt). There is no meteorological station in Visaginas, therefore Utenos Meteorological Station, which is the closest one to Visaginas, was chosen. The freezer, which was used for experiments, could be regulated with the accuracy of one degree of Celsius, thus –7 °C was used. By observing the freezing of the clay sample it was determined that after 5 days a stable system of cracks forms Figure 4 B, which does not change while freezing longer.

3. Results and analyses

The clay sample before the first filtering was saturated with distilled water in the filtering chamber. The PV infiltrometer allows filtering by forming percolation flows from top to the bottom and vice versa (Fig. 1). For saturation, as well as all filterings, the upward flow of water has been applied. This enabled to saturate the clay sample of the initial 15% natural humidity by creating a unit hydraulic head. Water level in piezometer coincided with the top of the

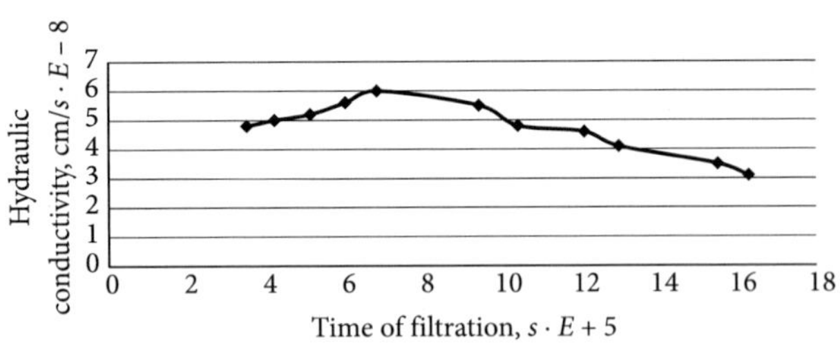

Fig. 2. Hydraulic conductivity of clay versus the filtering duration during the first filtering term

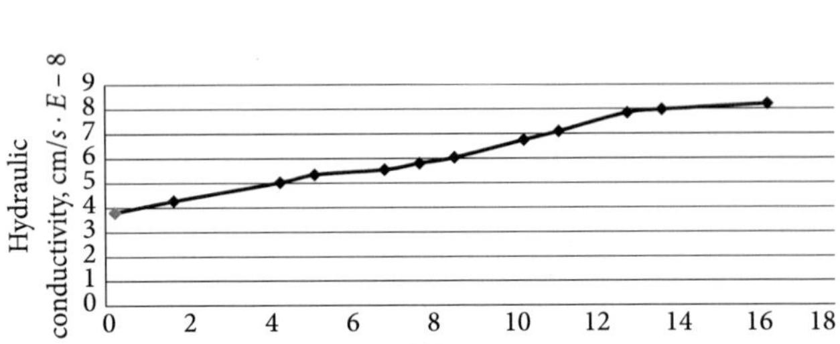

Fig. 3. Hydraulic conductivity of the same clay sample versus the filtering time at the renewal of the experiment after two weeks

sample. The course of the first filtering showed the increase in hydraulic conductivity during 194 hours of filtering and the decrease in head gradient from 8.75 to 3.5 (Fig. 2). Further, the hydraulic conductivity values decreased nearly linearly as far as the stopping of filtering after 430 hours with a decrease in head gradient to 1.2 (Fig. 2).

Such a course of filtering can be explained by incomplete saturation of a sample, although the process of saturation lasted 48 hours. This is confirmed by the increase in mass of the sample from 335 g to 338 g measured after the saturation and filtering. The second part of the filtering course, i.e. the decrease in hydraulic conductivity, can be explained by changes in clay pore spaces caused by structural changeovers in clay minerals and their aggregates. The second reason of this phenomenon can be explained so: due to a decrease in head gradient with time, a lower content of bound water in clay moves together with percolating flow and turns into an immobile phase. Therefore, as the head values decrease, the filtering debit declines unproportionally, as well as the hydraulic conductivity.

After the first filtering was over, the clay sample remained in the gauge in distilled water. The second filtering was renewed after two weeks. The results obtained are presented in Figure 3. The initial stage of filtering again shows increase in conductivity values, as in the first filtering, followed by stabilisation. The difference from the first case is that there was no increase in sample mass, and a change in hydraulic conductivity appeared earlier than in the first case. This can be explained by further structural changes in clay pore spaces during the filtering. The structural changes in the clay sample being in water for two weeks confirmed the fact that after the first filtering the hydraulic conductivity value was 3.1E-8 cm/s, and it grew to 3.9E-8 cm/s at its renewal after two weeks.

Such a change of hydraulic conductivity values with time can be explained by the changes in pore spaces taking place not only during filtering, but also when the sample is swelling in water, i.e. the aggregates of clay particles changeover structurally. This conclusion is confirmed by other authors, which obtained similar results. In France, the Callovo–Oxfordian clay formations consisting of the muscovite mica (Comptoir de Mineraux et Matieres premieres) are essentially composed of SiO_2 and sodium montmorillonite in bentonite (Oene, France). The analysis of this by Scattering Electron Microscopy showed that the structure of pore space changed if compared low humidity samples to those after swelling 24 h in water (Mammar *et al.* 2001). After the second filtering was stopped, the clay sample was taken out from the infiltrometer together with the filtering chamber ring and placed into a freezer with constant temperature 7 °C below freezing point. After 5 days, the sample was taken out. The picture of the sample after the freezing is given in Figure 4, where freeze cracks are well seen.

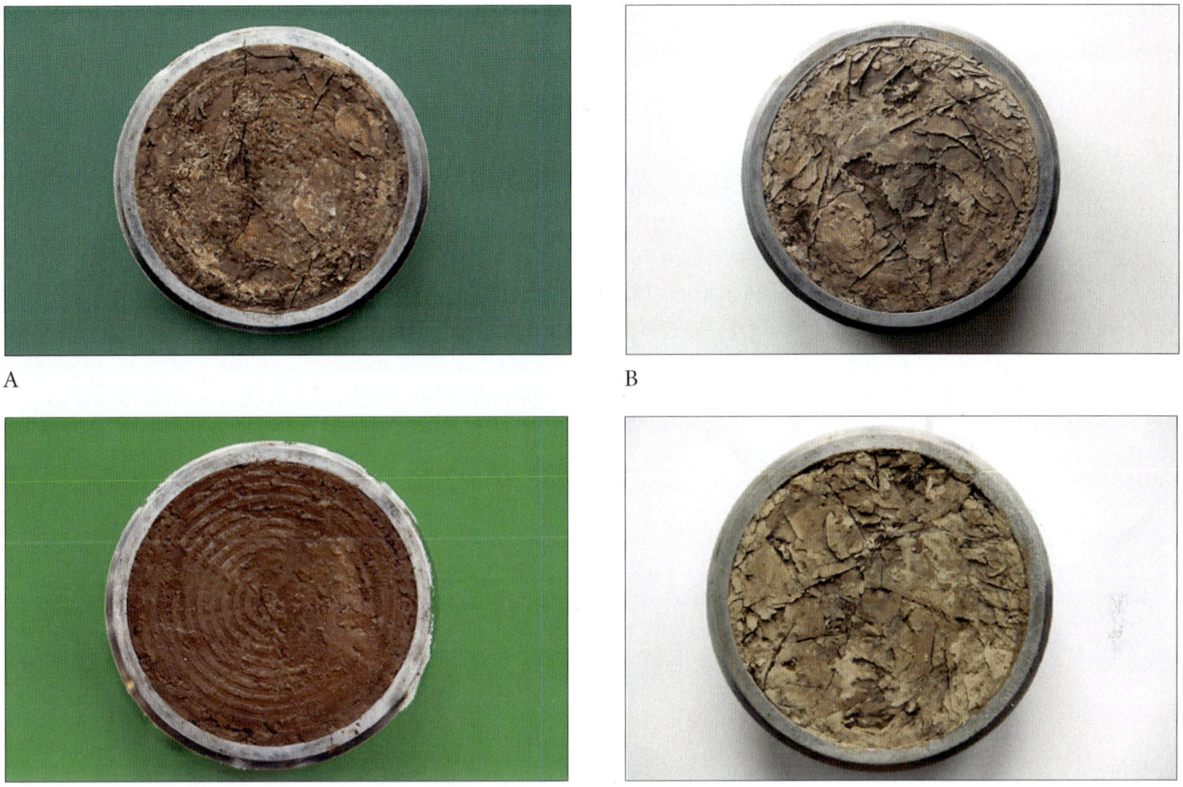

A B C D

Fig. 4. Picture of the sample with the filtering chamber ring: A – after first freezing; B – after second freezing; C – after second filtration; D – after third freezing

Weighing the sample after the freeze showed a decrease in its mass by 2 g. This confirms that, during freezing, water partly migrates from the sample towards the freeze zone at the walls of the freezer. The frozen sample was put outright into the infiltrometer to begin filtering. The initial results of this filtering are given in Figure 5. The data obtained show the course of thawing at the start of the filtering, i.e. cyclic acceleration and deceleration. The hydraulic conductivity values obtained ranged from 5E-6 cm/s to 3.5E-5 sm/s, thus, being higher by two orders of magnitude than before the freezing.

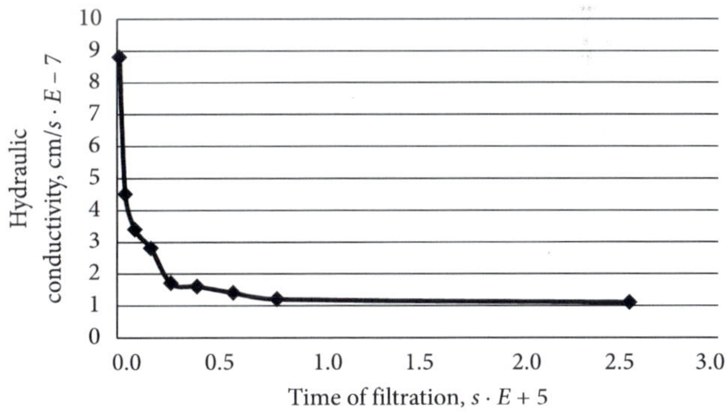

Fig. 5. Variations of clay hydraulic conductivity in time when the filtering and the thawing takes place at the same time in filtering chamber

After the renewal of filtering of completely thawed sample, as fixed according to water temperature course in filtering chamber, the changes in hydraulic conductivity are presented in Fig. 6. The filtering course shows a rather intense deceleration of percolation and stabilisation of hydraulic conductivity. This reveals a strong effect of cracks formed during freezing and rather rapid colmatation of the freeze crack system.

As the filtering after the first freezing is over, the sample was frozen once more under the identical temperature regime. A picture of the sample after the second freezing shows even more obvious freeze cracks, although the crack system looks alike as after the first freezing (Fig. 7). This indicates that the cracks clogged during filtering opened again and got wider and sharper. The filtering course is very similar, but the stabilisation goes much slower and

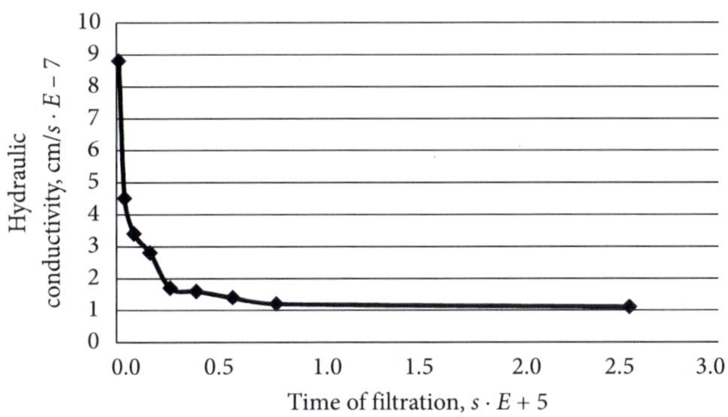

Fig. 6. Variations of clay hydraulic conductivity during the initial filtering performed after thawing of the sample

longer, thus confirming the conclusion that the cracks are wider, therefore, hydraulic conductivity values are higher.

The course of filtering after the stopping and renewal is identical to that before the freezing as the results given in Figure 8 indicate.

As the second filtering was completed, the sample was frozen for the third time. After five days of freeze the vertical cracks were wider but their scheme remained the same. Additional system of horizontal cracks was formed. Therefore, the sample began crumbling and falling from the filtering ring. Clay lamina began to peel off. Vertical

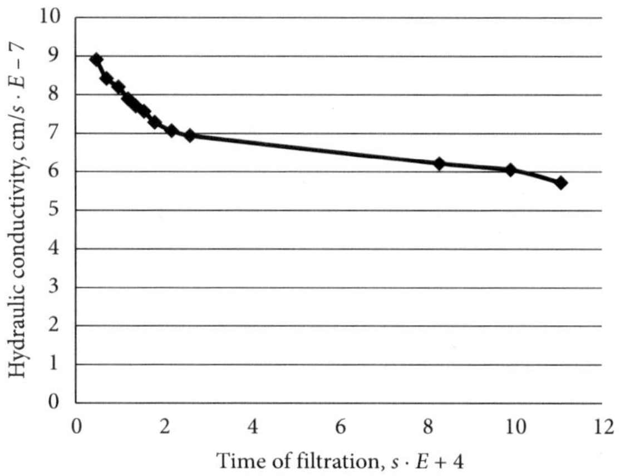

Fig. 7. Variations of hydraulic conductivity during the second filtering stage after the second freeze-thaw cycle

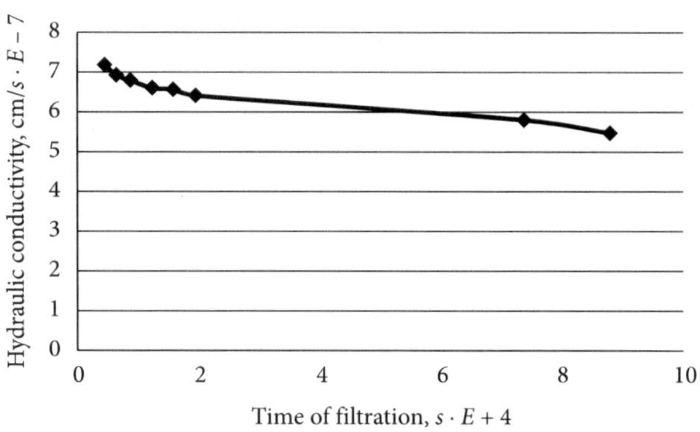

Fig. 8. Variations of hydraulic conductivity in time during the third stage of filtering after the second freeze-thaw cycle

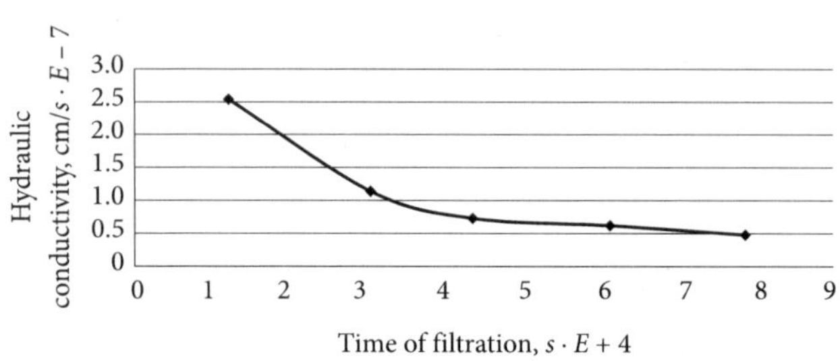

Fig. 9. Time variations in intensity of evaporation directly from the water surface in open-ended hydraulic head piezometer, as recalculated into the hydraulic conductivity corresponding to the evaporation intensity

and horizontal cracks joined in the sample over all its height, and the sample crumbled.

Performing the filtration at the laboratory, especially in winter, when the air humidity reached 35–45%, it was necessary to assess the effect of evaporation on the hydraulic head and filtrate debit. Therefore, these characteristics were measured in parallel. The water evaporation volumes and their variations with time, when the water level in piezometer was descending, enabled to assess how this could affect the determined hydraulic conductivity values. The time variations in the hydraulic head and debit values were recalculated into the hydraulic conductivity, if solely the evaporation took place. The results obtained are given in Fig. 9. It is obvious that it would be important, if the conductivity values during the experiments were at 3–4 orders of magnitude lower than the minimum values determined by us.

Conclusions

1. For the first time in Lithuania experimental research analysing the effect of the number and duration of freeze-thaw cycles on filtrational parameters of clay was performed. Grūda subformation till clay (gIIIgr), taken from the projected Stabatiškė waste repository site, was used in these experiments.

2. The generalisation of all the investigation results obtained for this type of clay allows us to formulate several methodological conclusions. The variation in hydraulic conductivity for clay of this type can range from 9.0E-7 to 3.0E-8 cm/s. Its higher values are observed at the start of the filtering. Later, during the experiment, they are decreasing. The repeated experiment with the same sample after some time shows some rise in conductivity values at its start with further possible stabilisation. Therefore, the hydraulic conductivity of clay determined by short-term infiltration studies will be imprecise and unsuitable for long-term prognostication.

3. The effect of clay freeze-thaw cycles on the hydraulic conductivity is not very high. During thawing of the test sample, the hydraulic conductivity ranges from 3.5E-6 to 5.0E-7 cm/s. The fully frozen clay sample is found to contain a newly formed vertical and horizontal crack system that raise the hydraulic conductivity in several times. The colmatation of crack system also takes place during the filtering. The hydraulic conductivity values decreasing with time confirm this, although they remain significantly higher than they were before the freezing. The second freeze-thaw cycle forms very similar system of cracks, but the same cracks are wider than before. The hydraulic conductivity grows further followed by even more slower colmatation causing a rise in clay filtering capacity (permeability).

4. The third freeze-thaw cycle destroyed our sample. Therefore, we could not perform filtering because the crack system formed had broken the structural monolithicness of

the sample and probably even more improved the filtering capacity (permeability) of clay. As the sample was not fully thawed, it was visually observed how water began percolate via the well seen frozen clay crack system. This filtering cycle was very unstable in time, and only after the sample thawed completely a slow colmatation of cracks began, but the hydraulic conductivity values remained higher than they were before the freezing of the sample.

References

Al Houri, Z. M.; Barber, M. E.; Yonge, D. R.; Ullman, J. L.; Beutel, M. W. 2009. Impacts of frozen soils on the performance of infiltration treatment facilities, *Cold Regions Science and Technology* 59(1): 51–57. http://dx.doi.org/10.1016/j.coldregions.2009.06.002

Cassan, M. 2005. *Les essais de permeabilite sur site dans la reconnaissance des sols.* Paris: Presses des Ponts et Chaussees. 568 p.

Dondi, M.; Principi, P.; Raimondo, M.; Zanarini, G. 2003. Water vapour permeability of clay bricks, *Construction and Building Materials* 17(4): 253–258. http://dx.doi.org/10.1016/S0950-0618(02)00117-4

Ebina, T.; Minja, R. J. A.; Nagase, T.; Onodera, Y.; Chatterjee, A. 2004. Correlation of hydraulic conductivity of clay-sand compacted specimens with clay property, *Applied Clay Science* 26(1–4): 3–12. http://dx.doi.org/10.1016/j.clay.2003.09.010

Enssle, C. Ph.; Cruchaudet, M.; Croise, J.; Brommundt, J. 2011. Determination of the permeability of the Callovo-Oxfordian clay at the metre to decametre scale, *Physics and Chemistry of the Earth* 36(17–18): 1669–1678. http://dx.doi.org/10.1016/j.pce.2011.07.031

Gadeikis, S.; Dundulis, K.; Žaržojus, G.; Gadeikytė, S.; Klizas, P.; Urbaitis, D.; Gribulis, D. 2012. Inžinerinių barjerų izoliacinių molio gruntų sluoksnių geotechniniai tyrimai, *Mokslas Gamtos Mokslų Fakultete* (7): 117–128.

Hamdi, N.; Srasra, E. 2008. Filtration properties of two Tunisian clays suspensions: effect of the nature of clay, *Desalination* 220(1–3): 194–199. http://dx.doi.org/10.1016/j.desal.2007.02.037

Hayashi, M.; Kamp, G., Schmidt, R. 2003. Focused infiltration of snowmelt water in partially frozen soil under small depressions, *Journal of Hydrology* 270(3–4): 214–229. http://dx.doi.org/10.1016/S0022-1694(02)00287-1

Heister, K.; Kleingeld, P. J.; Keijzer, T. J. S.; Loch, J. P. G. 2005. A new laboratory set-up for measurements of electrical, hydraulic, and osmotic fluxes in clays, *Engineering Geology* 77(3–4): 295–303. http://dx.doi.org/10.1016/j.enggeo.2004.07.020

Jakimavičiūtė–Maselienė, V.; Mažeika, J.; Motiejūnas S. 2012. Radionuclide and heat transport from hypothetical SNF canister in crystalline basement, case of South–Eastern Lithuania, *Journal of Environmental Engineering and Landscape Management* 20(2): 121–128. http://dx.doi.org/10.3846/16486897.2012.688363

Jobmann, M.; Wilsnack, Th.; Voigt, H.-D. 2010. Investigation of damage-induced permeability of Opalinus clay, *International journal of Rock Mechanics and Mining Sciences* 47(2): 279–285. http://dx.doi.org/10.1016/j.ijrmms.2009.11.009

Jullien, A.; Proust, Ch.; Le Forestier, L.; Baillif, P. 2002. Hydro-chemio-mechanical coupling effects on permeability and swelling behaviour of a Ca smectite soaked by Cu solutions, *Applied Clay Science* 21(3–4): 143–153. http://dx.doi.org/10.1016/S0169-1317(01)00084-9

Mammar, N.; Rosanne, M.; Prunet-Foch, B.; Thovert, J.-F.; Tevissen, E.; Adler, P. M. 2001. Transport properties of compact clays, *Journal of Colloid and Interface Science* 240(2): 498–508. http://dx.doi.org/10.1006/jcis.2001.7697

McCauley, C. A.; White, D. M.; Lilly, M. R.; Nyman, D. M. 2002. A comparison of hydraulic conductivities, permeabilities and infiltration rates in unfrozen soils, *Cold Regions Science and Technology* 34(2): 117–125. http://dx.doi.org/10.1016/S0165-232X(01)00064-7

Oradovskaja, A. E. 1983. Hydrogeological importance of water filtration in clays, *VSEGINGEO* 152(1): 14–19.

Pundytė, N.; Baltrėnaitė, E.; Pereira, P. 2011. Anthropogenic effects on heavy metals and macronutrients accumulation in soil and wood of Pinus sylvestris L, *Journal of Environmental Engineering and Landscape Management* 19(1): 34–43. http://dx.doi.org/10.3846/16486897.2011.557473

Shafiee, A. 2008. Permeability of compacted granule-clay mixtures, *Engineering Geology* 97(3–4): 199–208. http://dx.doi.org/10.1016/j.enggeo.2008.01.002

Tang, A. M.; Cui, Y.-J.; Richard, G.; Defossez, P. 2011. A study on the air permeability as affected by compression of three French soils, *Geoderma* 162(1–2): 171–181. http://dx.doi.org/10.1016/j.geoderma.2011.01.019

Petras KLIZAS. Dr, Assoc. Prof. Department of Hydrogeology and Engineering Geology, Vilnius University (VU). Doctor of Science (Natural Science) Vilnius University, Faculty of Natural Sciences, 1994. Master of Science, Neuchatel University (Switzerland), 1992, First degree in Hydrogeology and Geology Moscow State University (Russia) 1978. Publications: author of 4 textbooks, over 45 scientific publications. Research interests: geofiltration, aeration zone, humidity migration, hydraulic conductivity of soils.

Permissions

 Every chapter published in this book has been scrutinized by our experts. Their significance has been extensively debated. The topics covered herein carry significant findings which will fuel the growth of the discipline. They may even be implemented as practical applications or may be referred to as a beginning point for another development.

The contributors of this book come from diverse backgrounds, making this book a truly international effort. This book will bring forth new frontiers with its revolutionizing research information and detailed analysis of the nascent developments around the world.

We would like to thank all the contributing authors for lending their expertise to make the book truly unique. They have played a crucial role in the development of this book. Without their invaluable contributions this book wouldn't have been possible. They have made vital efforts to compile up to date information on the varied aspects of this subject to make this book a valuable addition to the collection of many professionals and students.

This book was conceptualized with the vision of imparting up-to-date information and advanced data in this field. To ensure the same, a matchless editorial board was set up. Every individual on the board went through rigorous rounds of assessment to prove their worth. After which they invested a large part of their time researching and compiling the most relevant data for our readers.

The editorial board has been involved in producing this book since its inception. They have spent rigorous hours researching and exploring the diverse topics which have resulted in the successful publishing of this book. They have passed on their knowledge of decades through this book. To expedite this challenging task, the publisher supported the team at every step. A small team of assistant editors was also appointed to further simplify the editing procedure and attain best results for the readers.

Apart from the editorial board, the designing team has also invested a significant amount of their time in understanding the subject and creating the most relevant covers. They scrutinized every image to scout for the most suitable representation of the subject and create an appropriate cover for the book.

The publishing team has been an ardent support to the editorial, designing and production team. Their endless efforts to recruit the best for this project, has resulted in the accomplishment of this book. They are a veteran in the field of academics and their pool of knowledge is as vast as their experience in printing. Their expertise and guidance has proved useful at every step. Their uncompromising quality standards have made this book an exceptional effort. Their encouragement from time to time has been an inspiration for everyone.

The publisher and the editorial board hope that this book will prove to be a valuable piece of knowledge for researchers, students, practitioners and scholars across the globe.

List of Contributors

Enrico Mancinelli
Faculty of Civil Engineering, Università Politecnica delle Marche, via Brecce Bianche 12, 60131 Ancona, Italy

Edita Baltrėnaitė, Pranas Baltrėnas and Dainius Paliulis
Department of Environmental Protection, Vilnius Gediminas Technical University, Saulėtekio al. 11, 10223 Vilnius, Lithuania

Giorgio Passerini
Department of Industrial Engineering and Mathematics Sciences, Università Politecnica delle Marche, via Brecce Bianche 12, 60131 Ancona, Italy

Åsgeir R. Almås
Department of Plant and Environmental Sciences, Norwegian University of Life Science, N–1432 Ås, Norway

Aušra Mažeikienė and Marina Valentukevičienė
Department of Water Engineering, Vilnius Gediminas Technical University, Saulėtekio al. 11, 10223 Vilnius, Lithuania

Arndt Wiessner, Jochen A. Müller, Peter Kuschk, Uwe Kappelmeyer, Matthias Kästner and Ulrich Stottmeister
UFZ-Helmholtz Centre for Environmental Research, Permoserstrasse. 15, D-04318 Leipzig, Germany

Yong-Jun Liu
Key Laboratory of Northwest Water Resource, Ecology and Environment, Ministry of Education, Xi'an University of Architecture & Technology, Xi'an, China

Mohamad Reza Khodadadi
Department of Nanotechnology, Semnan University, Semnan, Iran

Sarfaraz Hashemkhani Zolfani
Department of Management, Science and Technology, Technology Foresight Group, Amirkabir University of Technology (Tehran Polytechnic), Tehran, Iran

Morteza Yazdani
Department of Business Management, Faculty of Social Sciences, Universidad Europea de Madrid, 28670 Madrid, Spain

Edmundas Kazimieras Zavadskas
Research Institute of Smart Building Technologies, Vilnius Gediminas Technical University, Saulėtekio al. 11, LT-10223 Vilnius, Lithuania

Haiming ZOU
School of Resource and Environment, Anhui Science and Technology University, Donghua Road, Fengyang 233100, PR China
School of Energy and Environment, Southeast University, Sipailou Road, Nanjing 210096, PR China

Xiwu LU
School of Energy and Environment, Southeast University, Sipailou Road, Nanjing 210096, PR China

Vaidotas Valskys, Roberta Valskienė and Gytautas Ignatavičius
Ecology and Environmental Center, Vilnius University, Vilnius, Lithuania

Ömer Atabeyoğlu
Department of Landscape Architecture, Faculty of Agrigulture, Ordu University in Ordu, Turkey

Adam Piasecki and Rajmund Skowron
Department of Hydrology and Water Management, Nicolaus Copernicus University, Lwowska 1, 87-100 Toruń, Poland

Jakub Jurasz
Department of Engineering Management, Faculty of Management, AGH University of Science and Technology, ul. Gramatyka 10, 30-067, Kraków, Poland

Judita Koreivienė and Jūratė Karosienė
Nature Research Centre, Institute of Botany, Žaliųjų Ežerų g. 49, 08604 Vilnius, Lithuania

Robertas Valčiukas and Pranas Baltrėnas
Department of Environmental Protection, Faculty of Environmental Engineering, Vilnius Gediminas Technical University, Saulėtekio al. 11, 10223 Vilnius, Lithuania

Tamer Bagatur and Nizamettin Hamidi
Civil Engineering Department, Engineering Faculty, Dicle University, 21280 Diyarbakir, Turkey

Diana Sarauskiene and Jurate Kriauciuniene
Laboratory of Hydrology, Lithuanian Energy Institute, Breslaujos g. 3, 44403 Kaunas, Lithuania

Alvina Reihan
Institute of Environmental Engineering, Tallinn University of Technology, Ehitajate tee 5, 19086 Tallinn, Estonia

Maris Klavins
Faculty of Geography and Earth Sciences, University of Latvia, Raina Blvd. 19, 1586, Riga, Latvia

Fevzi Onen
Civil Engineering Department, Engineering Faculty, Dicle University, 21280 Diyarbakir, Turkey

Atieh Ebrahimi
Department of Civil Engineering, Institute of Pardisan Fereidonkenar Iran

Ghasem D. Najafpour
Biotechnology Research Lab., Faculty of Chemical Engineering, Noshirvani University of Technology, Babol, Iran

Manouchehr Nikzad
Faculty of civil Engineering, Noshirvani University of Technology, Babol, Iran

Roya Pishgar, Bahram Navayi Neya, Nafise Mousavi and Zeinab Bakhshi
School of Civil Engineering, Babol Noshirvani University of Technology, Babol, Mazandaran, Iran

Ghasem D. Najafpour
Biotechnology Research Center, Faculty of Chemical Engineering, Babol Noshirvani University of Technology, Babol, Babol, Iran

Anolda Četkauskaitė
Department of Biochemistry and Molecular Biology, Faculty of Natural Sciences, Vilnius University, Čiurlioniog. 21/27, LT-03101 Vilnius, Lithuania

Milda Zita Vosylienė and Nijolė Kazlauskienė
Nature Research Centre, Akademijos g. 2, LT-08412 Vilnius, Lithuania

Virginija Kalcienė
Environmental Studies Centre, Faculty of Natural Sciences, Vilnius University, Čiurlionio g., 21/27, LT-03101 Vilnius, Lithuania

Emilie Prygiel, Romain Descamps, Baghdad Ouddane and Gabriel Billon
aGéosystèmes, UMR 8217 CNRS University Lille 1, Villeneuve d'Ascq, France

Adeline Charriau
Géosystèmes, UMR 8217 CNRS University Lille 1, Villeneuve d'Ascq, France
Research Group on Water, Soil and the Environment EA 4330, Faculty of Sciences and Technology, University of Limoges, France

Jean Prygiel
Géosystèmes, UMR 8217 CNRS University Lille 1, Villeneuve d'Ascq, France
Artois-Picardie Water Agency, Douai, France

Kejian Chu and Zulin Hua
Key Laboratory of Integrated Regulation and Resource Development on Shallow Lake of Ministry of Education, College of Environment, Hohai University, Nanjing City, P. R. China
National Engineering Research Center of Water Resources Efficient Utilization and Engineering Safety, Nanjing City, P. R. China
College of Environment, Hohai University, 1 Xikang Road, Nanjing City, 210098 Jiangsu Province, P. R. China

Lijun Ji
Key Laboratory of Integrated Regulation and Resource Development on Shallow Lake of Ministry of Education, College of Environment, Hohai University, Nanjing City, P. R. China
College of Environment, Hohai University, 1 Xikang Road, Nanjing City, 210098 Jiangsu Province, P. R. China

Xiwu Lu and Ting Li
School of Energy and Environment, Southeast University, Nanjing 210096, PR China

Haiming Zou
School of Energy and Environment, Southeast University, Nanjing 210096, PR China
Department of Resource and Environment, Anhui Science and Technology University, Fengyang 233100, PR China

Petras Klizas
Department of Hydrogeology and Engineering Geology, Faculty of Natural Sciences, Vilnius University, M. K. Čiurlioniog. 21/27, 03101 Vilnius, Lithuania

Index